OCCUPATIONAL AND ENVIRONMENTAL CHEMICAL HAZARDS

Cellular and Biochemical Indices for Monitoring Toxicity

Titles also published by Ellis Horwood Limited
for the International Union of Pure and Applied Chemistry

Trinajstic:
Mathematical and Computational Concepts in Chemistry
Sakurai:
Organosilicon and Bioorganosilicon Chemistry
Brown and Kodama:
Toxicology and Clinical Chemistry of Metals

Published for the
INTERNATIONAL UNION
OF PURE AND APPLIED CHEMISTRY

Institute of Occupational Health
«Clinica del Lavoro L. Devoto»
University of Milan, Italy

Center for Occupational and
Environmental Health and
Division of Occupational Medicine
The Johns Hopkins Medical Institutions
Baltimore, Maryland, USA

Carlo Erba Foundation
Occupational and Environmental
Health Section
Milan, Italy

OCCUPATIONAL AND ENVIRONMENTAL CHEMICAL HAZARDS

Cellular and Biochemical Indices for Monitoring Toxicity

V. FOÁ, M.D., Associate Professor of Occupational Medicine
Head of Industrial Hygiene and Toxicology Unit,
Institute of Occupational Health "Clinica del Lavoro L. Devoto"
University of Milan, Italy.

E.A. EMMETT, M.D., Professor of Occupational Medicine,
Head of Center for Occupational and Environmental Health and Division
of Occupational Medicine, The Johns Hopkins Medical Institutions,
Baltimore, Maryland, USA.

M. MARONI, M.D., Associate Professor of Occupational Medicine
Institute of Occupational Health "Clinica del Lavoro L. Devoto"
University of Milan, Italy.

A. COLOMBI, Sc. D., Ph. D., Associate Professor of Industrial Hygiene
Institute of Occupational Health "Clinica del Lavoro L. Devoto"
University of Milan, Italy.

ELLIS HORWOOD LIMITED
Publishers.Chichester

Halsted Press: a divison of
JOHN WILEY & SONS
New York.Chichester.Brisbane.Toronto

First published in 1987 by
ELLIS HORWOOD LIMITED
Market Cross House, Cooper Street,
Chichester, West Sussex, PO19 1EB, England

The publisher's colophon is reproduced from James Gillison's drawing of the ancient Market Cross, Chichester.

Distributors:
Australia and New Zealand:
JACARANDA WILEY LIMITED
GPO Box 859, Brisbane, Queensland 4001, Australia
Canada:
JOHN WILEY & SONS CANADA LIMITED
22 Worcester Road, Rexdale, Ontario, Canada
Europe and Africa:
JOHN WILEY & SONS LIMITED
Baffins Lane, Chichester, West Sussex, England
North and South America and the rest of the world:
Halsted Press: a division of
JOHN WILEY & SONS
605 Third Avenue, New York, NY 10158, USA

British Library Cataloguing in Publication Data
OCCUPATIONAL AND ENVIRONMENTAL CHEMICAL HAZARDS:
Cellular and Biochemical Indices for Monitoring Toxicity. –
(Ellis Horwood series in chemical science)
1. Toxicity testing
I. Foá, Vito
II. International Union of Pure and Applied Chemistry
615.9'07 RA1199

ISBN 0–7458–0088–2 (Ellis Horwood Limited)
ISBN 0–470–20802–3 (Halsted Press)

LIBRARY OF CONGRESS Card No. 86–27801

Printed in Great Britain by Unwin Bros. of Woking

Contents

Foreword

The widespread welfare and the increased life expectancy occurring in the industrialized world are to a great extent due to the progress of science and production technologies and, in particular, of chemistry. This evolution however has also been responsible for a number of serious problems not only among the workers directly engaged into the productive activities but also in the population at large.

Cancer increase is one of the most striking and tragic phenomena and can in part be ascribed to altered environmental conditions.

Society and particularly scientific community has never accepted that such problems should be considered an unavoidable toll to pay to the progress. It is relevant to remember that scientific research on the toxicological risk has become well established just to give answer to such a problem. To day the complex methodologies developed in the field can be considered themselves a separate and specialized discipline. While the techniques for controlling pollution and emissions from industrial plants into the environment have been greatly developed, many questions still remain in the evaluation of the health impact of chemicals on the human organism. This gap of knowledge led the National Research

Council of Italy to promote the special research programme "Toxicological Risk" within the frame of the national research project on Preventive Medicine and Rehabilitation in 1981. The special programme has funded the work of research groups throughout Italy. Part of the results obtained by this programme have been reported to the International Symposium on Biochemical and Cellular Indices of Human Toxicity in Occupational and Environmental Medicine that was held in Milano from 19 to 22 May 1986. At this Symposium about one hundred papers including several original papers from leading international research groups were presented to an audience of more than 240 scientists including toxicologists, occupational doctors and hygienists.

The contributions excerpted in this book well illustrate the state of the art on this subject and review the most recent progresses in the field of surveillance and control of toxic effects induced in man by exogenous chemicals.

Luigi Rossi Bernardi
President of the
National Research Council of Italy

Preface

This volume contains the proceedings of the International Symposium on "Biochemical and Cellular Indices of Human Toxicity in Occupational and Environmental Medicine" held in Milan, Italy, June 19–22, 1986. There were 240 participants from 31 countries.

The Symposium was organized by the Institute of Occupational Health "Clinica del Lavoro L. Devoto" of the University of Milan, the Center for Occupational Environmental Health and Division of Occupational Medicine of the Johns Hopkins University, Baltimore, U.S.A. and the Carlo Erba Foundation, Occupational and Environmental Health Section, Milan. The meeting was a product of a collaborative agreement in Occupational Health between the University of Milan and the Johns Hopkins University. Financial assistance was given by the World Health Organization, of the Commission of the European Communities, the U.S. National Institute of Environmental Health Sciences, the Research Project on Toxicological Risk of the National Research Council of Italy and the International Agency for Research on Cancer. The meeting was held under the auspices of numerous major national and international societies including the International Commission on Occupational Health and the International Union of Pure and Applied Chemistry.

Outstanding research scientists and occupational and environmental health physicians were brought together to explore common ground to exchange information, and discuss issues raised by both the clinical and experimental sciences. There have been a large number of meetings devoted to the toxicity of single substances, groups of chemically related compounds or to the effects on single organs. However, this meeting and these proceedings are unique in that they deal with the tools applicable to the investigation of chemical toxicity in a variety of organs in the human host.

The proceedings are divided into an introductory section dealing with general concepts, emphasizing validation and user interpretation of tests. This is followed by discussions of indices in current or proposed use for toxic effects on hepatic, renal and nervous system functions in porphyrin metabolism.

The various methods for the evaluation of human exposure to genotoxic chemicals, including those recently proposed, are presented.

The state of the art in the light of the invited and contributed papers in each section have been summarised by a senior scientist in each field who has acted as rapporteur.

We trust that this volume will be of use to physicians, scientists and administrators involved this field.

Vito Foà
Edward A. Emmett
Marco Maroni
Antonio Colombi

Introduction

Coordination:
P. Bourdeau (CEC–Bruxelles)

Present and future of Occupational Medicine

A. Grieco – Director of the Institute of Occupational Health,
University of Milan, Italy

Wide and significant changes in techniques and also in organization have been achieved in the last ten years and further changes are still taking place in many areas of production.

The major changes in terms of their effects on the relationships with the human operator can be considered as those achieved through the introduction of new, computer-based machinery, since they have in many cases led to the operator no longer working directly at the site of production processes and therefore no longer in direct contact with the sources of chemical, physical and particulate pollutants.

Clear examples of these changes can already be seen in both the processing and manufacturing industries, especially in the iron and steel, basic chemical, automobile, pharmaceutical and printing industries.

Even more important changes have been achieved and others are under way in the services, transport and telecommunications sectors, where one of the most common and also most disputed phenomena are all those technical and organizational changes generally defined as "office automation".

Robots have long since passed the experimental stage and, although still subject to further perfection, they are widely used in Japan, the USA and Europe in the automobile industry and other production areas where repetitive tasks, movement of

materials of handling of dangerous substances are required.
All this will inevitably further develop over the few years between now and the beginning of the next century, with perhaps greater acceleration coefficients than those already seen and even more significant changes in the social structure of the labour force and in employment levels.
For proof of these transformations, it will suffice to quote just two phenomena, the consequences of which in social and health terms are already well known, although here, too, further research is needed.
The first is the fact that in the USA, Japan and Canada and in the majority of West European countries, more than half the working population is no longer employed in agriculture or industry but in other sectors generally defined as "services", with peaks already exceeding 70% in the USA, Belgium, Luxembourg and the Netherlands (Istituto Centrale di Statistica, 1983). The second phenomenon, which is different in nature and certainly a cause of concern, is the general fall in employment levels, and is reflected in the recent initiatives taken by the WHO and the ILO aimed at stimulating research on "unemployment disease".
The quality and quantity of the new types of skills demanded of the human operator already reveal clear trends: a progressive reduction in muscular activity, in handling of materials, in contact with intermediate products, along with a parallel increase in fixed and constrained postures, sensory load and mental effort due to an increase in perceptive, cognitive, memorization, decisional and control activities. In other words, what was previously done with the hands on physical objects, where we could use touch and our proven abilities, to-day and even more so tomorrow, will either be abandoned or will be done with machines by reading and checking numbers and symbols transmitted by a video display terminal.
Thus it is not surprising that human health risks will change, not only within but also without the workplace. However, the general situation is not simplified but, on the contrary, becomes more complicated, if only due to the fact that in many countries there is a tendency towards a transition phase, where the old and the new appear to be destined to coexist for no short time, as we have already seen in some cases in the developing countries. Furthermore, it seems that some of the "old" often persists in the "new", errors of long ago which have survived in the new interfaces with the human operator: for example, the first fatal accidents reported by the NIOSH

in the USA in 1985 that occurred where robots were used to move materials in areas that were not sufficiently protected (NIOSH, 1985).

Macropollution gives way to micropollution, so that situations involving large-scale release of pollutants become less frequent, while there are increasing probabilities of subtle, more long-lasting exposures, where the borders between living and working environments are often not so clear, the latency period longer, the preclinical evolution and interpretation of laboratory tests more complicated, and the differential diagnosis between occupational and common diseases more difficult. Serious intoxications are already less frequent, although they appear to be spreading at present in the developing countries; at the same time, the risk of allergies and cancer is increasing in association with the marked increase in the use of chemical substance and new formulations.

New techniques reduce the probability of accidents, especially those involving the limbs, but they can also lead to an increase in the destructive powers of any incompatibility within the man-machine system and of the deficiencies in quality and quantity of monitoring, to the point of catastrophies, such as recently occurred at Chernobyl and about which we still know very little.

However, although it is difficult to deny the significance of the points I have mentioned, however briefly, the most important phenomenon from our observation point is what appears as a set of complaints, disorders, or real diseases known as "aspecific diseases", better defined by the international scientific community as "work-related diseases", and to which to WHO assigned a new programme of research and measures in 1983 (El Batawi, 1984).

In 1982 and 1983, two WHO groups of experts issued the following statement on the subject: "Work-related diseases may be an appropriate term to describe disorders other than and in addition to recognized occupational diseases that occur among working people when the work environment and performance contribute significantly, but in varying magnitude, to disease condition". In 1984 El Batawi added: ".... work-related diseases, as distinguished from specific occupational diseases, usually affect the general population, including workers; they are caused by, or are associated with, risk factors which may at times be encountered in the work environment".

The diffusion and social significance of this phenomenon are now unanimously acknowledged, since hundreds of millions of people of working age are involved and the diseases included are among the most frequent causes of absenteeism in all countries, contrary to the traditional occupational diseases; at least this is what emerges from the available statistics.

As far as health effects are concerned, although there are still many areas of uncertainty, the international literature tends to identify 4 apparatuses as the sites of the most frequent work-related diseases: a) the psycho-emotional sphere (neuroses, psychoses, etc.); b) the cardiovascular apparatus (ischemic heart disease, hypertension, etc.); c) the digestive system (gastroduodenal diseases with or without ulcer, colitis, etc.); d) the locomotor apparatus (degenerative alterations of the spine, tendinitis, epicondylitis, etc.).

In my opinion, we should also include the respiratory apparatus in order to emphasize the importance of further research on the etiopathogenesis of certain bronchopulmonary diseases, where degenerative alterations and fibrotic evolution create difficulties in differential diagnosis with pneumoconiosis.

It is evident that these diseases are progressively increasing; the etiology is certainly multifactorial and sometimes also influenced by lifestyle, even though for some of the complaints mentioned the role of occupational factors has become increasingly clear in the last few years.

In this respect; we only have to consider, very briefly, the example of disorders of the locomotor apparatus, which include complaints where occupational factors were so clear that a group of Japanese authors recently suggested the definition "occupational cervico-brachial diseases" (Maeda et al., 1982). We came to similar conclusions in the case of spondiloarthropaty of the cervical and lumbar segments, where the relative risk (RR) among group of video display terminal (VDT) operators was just over double that of the reference population (Grieco, 1986).

But there is another aspect of human disease, undoubtedly connected with a large part of the disorders so far mentioned, that occupational medicine has never seriously considered, neither in the past nor at present, execpt for some rare and laborious approaches; and this is the set of genetically determinant neurohormonal and biochemical reactions which Hans Seyle, with his genius and intuition, already contemplated in 1936 and which were later described and demonstrated, under his guidance, as being always present along with the specific responses that living organisms put into action whenever they undergo various types of stimulation, be they physical, chemical or psychic in nature (Seyle, 1936). This set of reactions, that have been given the name of Stress-Selye in the literature, consists of aspecific responses that are objectively demonstrable and measurable, independent of the nature of the stimulus and correlated only with its intensity, and also of internal and external influencing factors which account for the variety of the clinical situations that often ensue.

In my opinion, although it is not possible here to make a more detailed analysis of the problem, a large part of the work-related diseases could be considered as the final state of a series of alterations where we have, in various combinations: a) alterations following the specific action of each of the external agents which come into contact with the humans organism and with the respective target organs in particular (specific effects); b) the set of signs and symptoms following repeated phases of exhaustion of Selye's General Adaptation Syndrome (aspecific effects); c) the results of the action of internal and external influencing factors.

If this is the background of the changes now taking place in what we usually define as health risks and consequent injury, whatever are our assumptions in attempting to identify their origins and relationships, there are without doubt two problems which modern occupational medicine must face now and even more so in the immediate future.

The first concerns the need to re-examine the theoretical models from which are derived the conceptual categories and the methods we use to analyse and describe work in real conditions. In fact, the current conception whereby work consists of a sequence of elementary operations of pre-established content and duration has not only never succeeded in explaining the presence of pollutants but, by excluding consideration of other variables, has never permitted evaluation of the real magnitude of exposure of operators to the individual toxic agents. Environmental and biological monitoring, although both still deficient compared with the complexity to which each is presumed to refer, have thus become two separate approaches to the problem that cannot be compared or controlled one in respect with the other; with the result that in cases where biological monitoring reveals an abnormal absorption of toxic agents, we still have no knowledge of the place and manner of the cause of the absorption; therefore any etiological diagnosis and precise preventive measured become impossible. If we add to this the fact that the biological effects of the individual toxic substances are looked for according to theoretical models which underestimate the synergic actions which are instead very frequent on account of the numerous physical and chemical variables present at the workplace in real situations, then it is clear that the levels of semplification that occupational medicine and in particular industrial toxicology have reached are unacceptable, especially with a view to using such models also for the study of the effects of microdoses.

The second problem, which I wish to mention only briefly, in order to offer an occasion for reflexion to the many distinguished experts participating in this Symposium, is even more closely concerned with the technical transformations taking place in nearly all sectors of production and with the changes in performance demanded of the human operator. As has already been said on many other occasions, technical innovation and the expansion of the services sector require new types of performance from the operators, involving increased strain on the organs of sense and the central nervous system, as well as new problems in the assignment of roles and in social relationships. Thus, more and more significance will be assigned to those psychosocial variables that occupational medicine has so far not adequately dealt with, in spite of the objectively demonstrated relationships between these variables and the physiopathological reactions occuring in the neurovegetative, central nervous, cardiovascular, digestive and immune systems. One explanation of this phenomenon can, to my mind, be found in the neopositivistic conception used in occupational medicine in nearly all countries in the interpretation of work, disease and man in general, in spite of the fact that this philosophical conception is now considered a closed chapter in nearly all other disciplines, including physics.

I believe that in the next few years we will not be able to avoid the necessity of finding a satisfactory solution to these problems, in respect of which we are all to a certain extent behind time. Occupational medicine must to this and seek above all a detailed comparison with the psychosocial disciplines, such as psychophysiology, perceptive and cognitive psychology, etc., so as to renew and further develop its own theoretical models, conceptual categories, analytical indicators, without losing its identity acquired and developed by means of laborious scientific research going back centuries.

I am sure that in some countries the cultural conditions required to face this new project already exist, and these are the same countries where in the last 20 years such a comparison with other basic disciplines, like physiology, biochemistry, pharmacology, etc. has already been achieved; with their help, industrial toxicology of to-day has developed and has already been able to make many invaluable contributions to the progress of our scientific knowledge and to safeguarding the health of the working man.

REFERENCES

EL BATAWI M.A. (1984): Work-related diseases, Scandinavian Journal of Work Environment and Health, 10, 341-346.
GRIECO A. (1986): Sitting posture: an old problem and new one, Ergonimics, vol. 29, n. 3, 345-362.
ISTITUTO CENTRALE DI STATISTICA (1983): Annuario statistico, Roma.
MAEDA K, HORIGUCHI S. and HOSOKAWA M. (1982): History of the studies on occupational cervicobrachial disorder in Japan and remaining problems, Journal of Human Ergology, 11, 17-29.
NIOSH (1984): DHHS Publication no. 85-103: Request for assistance in preventing the injury of workers by robots"".
SELYE H. (1936): A syndrome produced by diverse nocuous agents, Nature, 138, 32.

The Challenge of the twenty-first century for environmental health

Gareth M. Green – Professor and Chairman, Department of Environmental Health Sciences, The Johns Hopkins School of Hygiene and Public Health, Baltimore, Maryland, U.S.A.

The challenges of the twenty-first century to environmental health lie in three major areas: (1) preventing the spread of diseases of industrialization around the world; (2) addressing the unsolved problems of environmental health from the twentieth century; and (3) learning to anticipate the health impacts of new technologies that are already unleashed in the post-industrialized societies and are on the drawing boards for the twenty-first century. The field of environmental health is increasingly a concern of the effects of technology on health. The objectives of professionals in this field must be to understand, predict, and prevent the untoward biological effects of industrialization and technologic development, and to discover or design biologic, pharmacologic, behavioral, and engineering methods for treatment and control of such effects. The future of this field and its challenges in the twenty-first century, therefore, will be governed by the direction of scientific and technological development in the western post-industrialized societies and in the industrializing countries around the world. Projections for the twenty-first century can be made with the certainty that existing technology will spread around the world, and with the uncertainties of how future technologic developments will change post-industrialized societies.

The popular predictors for global development in the balance of this century and on into the twenty-first are that major demographic and technological trends of the past one hundred years will continue at an increasingly rapid pace as the spread of these technologies and demographic trends reaches into most societies of the world with adequate populations to provide skilled labor. Population growth, urbanization and super-urbanization, reduction in agricultural land through human settlement and factory construction, natural resource depletion, aging populations, and transporation practices are likely to follow current trends. The impact of these twentieth-century events on vast populations around the world and in many societies that are struggling through a thousand years of socioeconomic development in mere decades, is having staggering sociopsychological and behavioral effects in these societies, stimulating revolution, reactionary antisocietal behavior manifested in activities such as terrorism, and related pervasive manifestations of social discontent. Indeed, these problems may be greater than the ability to provide adequate nutrition to rapidly growing numbers of people, or to control human, agricultural, and industrial wastes and to avoid local and global biological and chemical poisoning of air and water supplies. These changes will not only place enormous burdens on economies struggling to provide employment and material betterment for increasing numbers of their citizens, but will pose threats to the environment, and thus to the health, in societies with little societal infrastructure to implement the necessary control measures to prevent effects on health not only in their own societies but outside cultures and nations. Current experiences both with terrorism and accidents in nuclear power-generating facilities are but two very dramatic examples of the regional or worldwide effects of technologic accidents and antisocial behaviors that are, at least in part, the products of industrialization and technologic development. Thus, the challenges to environmental health in the industrializing world will be close to overwhelming faced with the explosion of technology and development in those societies and the inevitability that such effects impact the more highly developed, post-industrialized nations of the world.

Challenges of the twenty-first century to environmental health in already industrialized countries are likely to reflect five unresolved issues of the twentieth century. The first of these is housing. The primary purpose of housing is to provide adequate protection from the adverse forces and conditions of the natural environment and for nurture of the family social unit. While construction and technology have essentially solved all the major problems of protection against earthquake, windstorm, flooding, and lightning, and can provide centralized safe water supplies and waste water removal and energy for heat, lighting, and cooking, most of the world's population does not yet have access to these technologies, and vast numbers live in self-constructed shelters that provide minimal protection from the elements and none from floods, windstorm, and earthquake. Heating and cooking may be done by open fires that distribute combustion pollutants throughout the dwelling. Even in advanced housing, indoor air pollution is a potentially significant health problem offering continuing challenges to environmental health. "Modern" indoor air pollution stems primarily from unvented cooking and heating appliances, pollutants of human origin such as from respiratory infection and cigarette smoking, and effluents from building materials including formaldehyde and radon gas. In addition, housing still offers a major site for personal injury or death from slips and falls in bathtubs, on stairs, and other slippery surfaces. Design, construction, and improved ventilation and air handling in domestic dwellings could have significant impacts in human health and safety.

Food safety is a second carry-over unresolved problem of the twentieth century which will continue to provide a challenge into the twenty-first. Refrigeration, chemical preservatives, and rapid transportation and distribution are the three technologies that have most contributed to the reduction of food spoilage and the enhancement of food safety. In vast areas of the world, however, these technologies are not yet available, and food continues to serve as a major vehicle for the spread of infections, diseases caused by fungal mycotoxins, and spoilage-induced generation of mutagens and carcinogens. The addition of chemical food additives either during plant growth as pesticides and herbicides, or during food processing to enhance flavor, appearance or other characteristics continues to disseminate potentially injurious chemical substances for human health. Extensive distribution systems and centralization of agricultural industries raise the potential for massive exposure from a contaminative accident at a central point in the production or distribution of food supplies. Finally, new technologies that offer substantial advances in food preservation and storage, such as irradiation, must overcome significant psychological barriers to achieve acceptance among consuming populations, even though the prospect for improving public health is great.

The third unsolved issue of the twentieth century is chemicals. The expansion of the synthetic chemical industry has introduced tens of thousands of new chemicals in vastly increasing quantities over the past forty years. Hundreds of new chemicals are added to the inventory each year. Experience has shown that quantity and use is proportional to exposure, as illustrated by the direct proportional relationship between the total amount of lead that is utilized and the average level of lead in human blood. The last third of this century has witnessed extensive legislation to provide the base for regulation and enforcement of standards for control of waste pollution, as evidenced by the series of legislative acts taken in the United States in the 1970's. These acts address air, water, natural resources, clean water supplies, waste water and solid waste disposal, and the use of chemicals in food and consumer substances. These regulatory activities have proven effective in stemming the growth of environmental pollution, and have even caused some decreases in levels such as in environmental and blood levels of lead and other air pollutants. However, the lack of basic toxicologic information on thousands of these chemical substances, even those that are used in million-pound quantities, limits the progress that can be made in chemical substance control and imposes a massive challenge to environmental health research in the twenty-first century. In addition, the lack of either regulation or enforcement infrastructure in many developing and industrializing nations around the world, with the resultant extensive contamination of agricultural land, food substances, and air and water supplies, poses a very significant challenge to environmental health on into the twenty-first century.

A fourth major unsolved problem in environmental health is in the area of industrial disasters prevalent in developed and developing countries alike. Whether stemming from deficiencies in design and location, as the massive environmental contamination with the nuclear power generating accident at Chernobyl, faulty maintenance of safety systems such as with the chemical release of methyl isocyanate in Bhopal, India, or with accidents in process control, as at Seveso, Italy, large populations may be either outright killed or threatened for lifetimes with chronic or latent injury or disease. The threat of such massive technological disasters has generated a reaction against the technology itself, as evidenced by the public perception that nuclear power is the greatest health risk to mankind. These accidents illustrate the reality that perhaps the greatest challenge of the twenty-first century to environmental health is the growing public perception that the economic benefits of industrial development may be overshadowed by the risk to human health and environment. The resolution of this challenge will call on disciplines and techniques addressing psychological and behavioral factors that are not now a customary armamentarium of the environmental health professions.

A fifth challenge of the twenty-first century to environmental health lies in the area of accidents. Technologically-based traumatic injury and death forms an increasingly high proportion of human health impact of technology, and in some age groups is the first cause of death. In contrast to chemically-induced risks, there appears to be a curious tolerance of accidental death and injury, perhaps in part because it is associated with human fault. The area of accidents offers the greatest opportunity for risk reduction and progress in life saved and morbidity avoided. To focus societal attention on the costs of accidental death and injury requires education in the concept of relative risk and in the need for acceptance of effort and resources directed toward the magnitude of risk and the feasibility of intervention. The field of environmental health has not generally accepted as a challenge accidental death and injury. As other areas of environmental health come under control, this technologically-based source of human death and injury will assume even greater significance and therefore greater challenge to the field of environmental health.

Finally, what as yet undiscovered health issues arising from new technologies of post-industrialization pose unsuspected challenges to environmental health in the twenty-first century? While this subject must of necessity be speculative, one can project five overriding technological forces that will become apparent by the end of this century or soon thereafter, and will have significance both to human health and to the field of environmental health. The first is the enhanced communication and artificial intelligence made possible by the continuing evolution of computer technology. Aside from the many effects on man's perception of himself and his relationship to others, this technologic evolution will allow fundamental changes to occur in the distribution of populations, in the utilization of urban centers, transportation, and energy, in the need for industrial concentration, and in the role of worker versus robot in high-risk exposures. These are all major contributors to environmental health problems in industrialized countries. Improved communication will have a counterbalancing effect on urbanization, will allow decentralization of the work setting and a decreased reliance on daily commuter travel between home and the work place, with major impact on the needs for the automobile, energy consumption, and thus air pollution. The reduction in these needs will also contain the demands on resources. There will be substantial changes in social relationships as people come together for recreation rather than for work. These social behavioral factors may in fact emerge as significant public health issues by the end of the century. Certainly de-urbanization, especially of heavy industry, will help to control problems of air pollution, water supply, and waste water management through distribution and other related public health issues. The spread of robot technology will greatly reduce worker exposure to occupational hazards and diminish the burden of currently recognized occupational disease. Cancer, chronic lung disease, and physical trauma will decline markedly in the twenty-first century as infectious and heart disease have in the twentieth century.

The second major trend with an influence on environmental health will be the increasing growth and utilization of new artificially synthesized materials that will both decrease the dependence on declining resources and increase energy conservation through improved properties of insulation, vehicle weight, durability, etc. Natural resource utilization is one of the major causes of problems of environmental health relating both to those who mine and process natural resources and to those engaged in the manufacture and use of such products. The decreasing use of natural resources and increased recycling of existing resources will decrease demands on waste disposal, a major problem with public health implications in the 1980's.

A third major trend by the end of the century is the projected perfection of applications of molecular biology and bioengineering. While there are dramatic examples of medical applications of molecular biology, the more significant direction for bioengineering will be in industrial uses for biosynthesis, biodegradation, mining and mineral extraction, and a host of other purposes that will revolutionize many of the industrial processes of today. Undoubtedly these technologies will add to the ability to develop new materials.

Fourthly, the next twenty-five years will see the colonization of near space for industrial purposes. The advantages will be not only in enhanced industrial capability and productivity but perhaps also the development of techniques for disposing of waste materials. Although the risk for near space contamination is significant, space technology and biology will be an important industrial activity by the year 2000.

Finally, perhaps the most significant trend of all is in the arena of energy where a breakthrough in fusion technology to provide limitless clean and safe energy at an acceptable cost is in the cards for the next twenty to fifty years. The timing of the breakthrough is uncertain, but that it will take place in the twenty-first century is almost certain. The impact of energy development of this magnitude on environmental health will be staggering. We have, of course, no concept of the health implications of the successor technologies.

MEETING THE CHALLENGE

The word, anticipatory, best sums up the recommended response for the field of environmental health during this period of explosive spread of twentieth century technology around the world and the revolutionary developments toward twenty-first century technology at home. Education programs should be designed to provide the disciplinary strengths in the component sciences that underpin the evolving technologies. There may need to be less emphasis on the health effects of chemical exposures but a new emphasis on the health implications of computer technologies and their application to process control, robot labor, communication, information management and artificial intelligence. We will need new programs to manage public health issues in near space environments. Practitioners of environmental health will evolve from emphasis on skills required to contain contamination to acquisition of predictive and preventive management skills through health risk assessment of emerging technologies.

The field of environmental health must become more adept at anticipating new frontiers of knowledge that form the base for new technological applications in the twenty-first century. Research must focus on the fundamental properties of biologic events to better predict risk through understanding of mechanism. More importantly, the field must become much stronger in the information, communication, and behavioral sciences that will dominate technological development and societal responses over the next twenty-five years. Research must develop not only the knowledge base for more adequate risk assessment but also the applied technology to make risk prediction a reality.

A final twenty-first century challenge to environmental health is to shift to professionalism in the resolution of environmental health problems, with greater involvement of better educated environmental health professionals in the recognition, characterization, analysis, study, and resolution of the modern problems of environmental health. This challenge presupposes the existence of knowledgeable professionals who are expert in the analysis and solution of problems of environmental health.

CONCLUSIONS

The field of occupational and environmental health is rapidly evolving from its retrospective orientation toward the clean-up of gross, unsightly pollution and acute accidental exposures associated with mortality. There is currently an increasingly important focus on the determination of acceptable risk levels of exposure for products of existing technologies with the emphasis on low-dose exposures, latency, and chronic effects manifested as morbidity and shortened lifespan. The challenge of the twenty-first century for environmental health is to develop prospective assessment of risk for new technologies based on structure activity relationships, biologic activity, and in vitro toxicology. New disciplines and techniques will be required to recognize and manage the risks of future technologies such as computer sciences, communication technologies, and prolonged occupation of space. The importance of this future vision is to prepare today's research for use as tomorrow's tools and to educate today's students and trainees as tomorrow's professionals and leaders. Rising world expectations, industrial competition, and economic pressures will increasingly give the advantage to the culture that not only protects its work force and population from the hazards of industrialization and technologies, but in addition seizes on the work place as a setting for creating positive health effects and enhancing public health. New technologies will not only make this goal more attainable but may in turn introduce new risks of a markedly different nature which must be envisioned and explored early in the development of these technologies. One can already witness the psychological stress of information overload as the individual is brought in close proximity to the rest of the world, creating not only psychological but economic and political stresses. These prospects have meaning not only for new research but for the content of professional and research education and training today. More attention to behavioral sciences, better knowledge and skills in the new technologies of communications, computer sciences, and new materials, evolution of the field of aerospace medicine, and specialization in physical and social sciences as they relate to occupational and environmental health are some of the necessary steps to prepare these future leaders to meet the challenges of the twenty-first century.

Biochemical and Cellular Indices of Human Toxicity

Edward A. Emmett – The Johns Hopkins University Medical Institutions, Baltimore, Maryland, U.S.A.

Vito Foà – Clinica del Lavoro Luigi Devoto, The University of Milan, Milan, Italy

We are at a very interesting and important time in the history of Biomedical Science, when a number of advances in various disciplines are converging to make possible major new approaches to the understanding and prevention of diseases caused by occupational and environmental chemical exposures. The Sciences of Molecular Biology, Immunology, and Chemistry have provided us with superior, hitherto unavailable, techniques such as gene sequencing, assays using monoclonal antibodies and ultrasensitive methods of spectroscopy. From Medicine and Public Health we have gained a greater appreciation that chronic diseases are generally the result of multiple risk factors acting to produce disordered function. From Biochemistry, Pathology, Medicine and Toxicology we have gained an understanding of the nature of the biochemical and cellular events and in many cases the receptors, which are altered as the initial events leading to disordered function and disease. We thus have a better basis for the mechanism of the disease. The discipline of Epidemiology is no longer the province only of the classical epidemiologist, but is incorporating the medical clinician into the new hybrid discipline, clinical biochemical epidemiology. It is from combinations of these advances that new sensitive and specific cellular and biochemical indices for the effects of hazardous chemical agents have become available. These indices are increasingly applicable for use in human subjects. The development of these novel markers is of profound importance for Occupational and Environmental Medicine, where our fundamental preoccupation is with the prevention of disease from toxic agents in the ambient or occupational environment. The sequence in the development of such a disease is schematically represented in figure 1. A source of a toxic substance, such as an industrial process, chemical storage vessel or hazardous waste site, contaminates air, water, food, or surfaces in contact with the skin, the amount of toxic agent in this media is evaluated through exposure monitoring. As a result of absorption, distribution, metabolism and excretion there is a certain internal dose of the toxic agent at the target organ. The dose sustained by either this target organ or by another surrogate tissue is measured in biological monitoring. As a result of interaction with a receptor, biochemical and cellular events occur, which may be measured using markers of toxicity. These markers may be direct measures of agent-receptor interaction, for example adducts between DNA and xenobiotics; biochemical or cellular events subsequent to agent receptor interactions, for example elevated levels of certain porphyrins following interaction of 2,3,7,8-tetrachlorodibenzodioxin with the biochemical mechanisms controlling porphyrin metabolism; or measures of repair such as sister chromatid exchanges reflecting repair of DNA damage. Biochemical and cellular indices of these types are the focus of this conference. These and other biochemical and cellular disturbances are early stages in the development of occupationally and environmentally induced diseases which will ultimately be detected by the physician and enumerated by the epidemiologist.

INDICES WHOSE PRIME USE IS IN OCCUPATIONAL AND ENVIRONMENTAL MEDICINE

The development of biochemical and cellular indices of human toxicity has given us, for perhaps the first time, medical tests which are rather specific to occupational and environmental health. Previously we have borrowed and relied upon tests which have been developed principally for diagnostic purposes or for following the progress of a patient. However, many of the tests now becoming available, for example assays of benzo(a)pyrene-DNA adducts or of neurotoxic esterase, have their prime use in prevention rather than treatment. As has happened in other disciplines, we must expect that the development and increasing use of these tools will greatly expand our body of knowledge and will accelerate the maturation of our specialty.

PROPERTIES AND ADVANTAGES OF BIOCHEMICAL AND CELLULAR INDICES

Several important properties of biochemical and cellular indices of human toxicity bear recounting. 1. They reflect the dosage of an agent to the appropriate internal target. Individual differences in absorption, distribution, metabolism and excretion of hazardous agents are thus taken into account. 2. They measure the results of all exposures regardless of source (occupational, environmental, hobbies, etc) or the route of absorption (inhalation, ingestion, percutaneous). 3. They measure the effect of an internal dose modified by the sensitivity of the individual. Differences in individual sensitivity may include differences in receptor composition or density, differences in repair or replacement of biochemical or cellular lesions and past functional alterations. In this sense biological and cellular indices reflect the effective dose. 4. They may integrate the effects of many different agents, which may have additive synergistic or antagonistic effects.

These properties of biochemical and cellular indices make them particularly useful in a number of situations of current interest. An excellent example is in the study of hazardous chemical dump sites where there is potential exposure by a number of routes to many different agents (whose identity is frequently unknown) at low levels of exposure which may be intermittent and may occur over periods of years. Moreover, the exposed population is often very heterogeneous in terms of age, sex, race, health status and other characteristics.

NECESSARY CHARACTERISTICS FOR TEST VALIDITY AND USEFULNESS

As for any clinical tests, there are certain conditions which biological and cellular indicators of toxicity must fulfill before they are suitable for use. The characteristics will be discussed in more detail in the section dealing with validation. A test must be valid, in that it is accurate, precise and reproducible in different laboratories. It should also be sensitive, specific and have predictive value for the development of the disease which we wish to prevent. In public health we are primarily interested in the mutant, not the mutation.

We have recently faced the issue of the predictivity of a biochemical marker in finding that workers exposed to a mixture of isopropyl triphenyl phosphates have lowered levels of non-specific esterase activity in circulating monocytes. This reduction seems to be a reliable and predictable indicator of exposure to the mixture. Yet the biological consequence, if any, of reduced non-specific esterase activity in the monocyte is quite unknown, so that the predictive value of the test is quite uncertain (Emmett et al 1985, Levine et al 1986). Useful biochemical and cellular markers should also be available for use in a number of laboratories and the cost should not be unreasonable.

In the past it was felt that tests used for human monitoring should detect damage at an early irreversible stage, but the biochemical and cellular lesions we are now able to detect (e.g. markers of genotoxicity) may not all be reversible. However, irreversible biochemical lesions should at least be detected when they are few in number so that we may take public health steps to prevent further damage. An alternate approach might be to monitor for such lesions but not to consider any increase in such lesions acceptable.

DIFFERENT MONITORING DATA FOR DIFFERENT PURPOSES

The relative importance of information relating to different steps in the chain of causation illustrated in figure 1 is of interest. Figure 2 shows the most useful information for the purposes of control of exposure, legal enforcement of standards, and protection of health. Although the fine nuances represented on this figure might be argued interminably, some broad distinctions can be made.

In order to control exposure, the data of most use to an engineer will either pertain to the source itself, or will quantitate the amount of toxic agent in the media downstream from the source. For both regulatory and legal purposes external exposure measurements have great importance since they can demonstrate downstream exposure from one or more sources, and unlike biological monitoring do not measure exposure from entirely different sources, a phenomenon which may complicate legal attribution of causation. For the protection of health, however, where there is an appropriate biochemical and cellular marker, it is of the greatest potential significance and is likely to be more useful even than measures of internal dose in deciding who needs protection. In this respect, the development of disease is a late outcome which we would like to prevent, and external exposure measurements are too often very poor surrogates for the measurement of dose.

To a considerable extent, the utility of the different methods of monitoring explains the selection of different monitoring tools in different areas of the world. In the United States the field of Occupational and Environmental Health has been staffed largely by engineers; the Federal Occupational Health and Safety legislation requires a safe and healthful workplace rather than worker health; legal confrontations between government, labor and business are common and the courts have usually been the final arbitrator of these confrontations. In this environment primary emphasis has been given to ambient air monitoring. In certain European countries by contrast, the field of Occupational and Environmental Health is staffed predominantly by physicians, emphasis is placed on prevention of disease in the exposed populations rather than merely a healthy workplace; and issues between government, labor and industry are generally resolved by negotiation and consensus. In this environment biological monitoring and monitoring of cellular and biochemical indices have become better developed.

These differences between National systems are rather well-explored in the book Regulating America, Regulating Sweden, (Kelman, 1981) to which the interested reader is referred.

PRACTICAL CONSIDERATIONS IN THE USE OF INDICES

At least five special practical considerations arise in situations where we use biochemical or cellular markers to assess potential toxicity. 1. We must have information on the "normal", or background levels in the general population or for any particular group that we are studying. This will necessitate studies in different communities and the use of comparison or control populations where reference information is not available. 2. We need to decide what level of any given effect constitutes an undesirable effect. Will we consider that any pertubation is undesirable, if not, at what level do we have an unacceptable adverse effect, an acceptable effect or merely a new level of adaptation. 3. The regulatory and legal implications of this type of information need to be evaluated. Since an unacceptable effect in a biochemical or cellular index reflects both exposure to an external causal agent, whose site of origin may not be directly determined, together with individual susceptibility, the contributory fault of any particular employer or polluter may be difficult or impossible to establish. However, in the absence of clear evidence of proximate fault or liability we should not deny individuals and populations the protection afforded by the use of biological and cellular markers, but we should seek societal discussion and consensus as to the ways in which these indices should best be used. 4. There need to be education of workers, communities and governmental and industrial authorities as to the meaning of results obtained using these indices, so that the public is neither unduly alarmed or complacent over this type of information. 5. There must be appropriate communication with the individual upon whom the test has been performed of the results and their interpretation. Further, wherever the use of such indices is experimental, this must be clearly conveyed to all participants.

There are some of the issues raised by these exciting new techniques which are the focus of this conference. We are about to enter a Brave New World of occupational and environmental health, one in which all the issues we and other authors have raised, will need to be considered very carefully, but one in which we should have much more sensitive, specific and predictive indices of human toxicity at our disposal.

REFERENCES

Emmett, E.A., Lewis, P.G., Tamala, F. et al (1985). Industrial exposure to organophosphorus compounds: Studies of a group of workers with a decrease in esterase-staining monocytes. J.Occup.Med. 27:905-914.

Kelman, S. (1981). Regulating America, Regulating Sweden: A Comparative Study of Occupational Safety and Health Policy, Boston, M.I.T. Press.

Levine,S., Fox, N.L., Thompson, B., et al (1986). Inhibition of esterase activity and an understanding of circulating monocytes in a population of production workers. J.Occup.Med. 28:207-211.

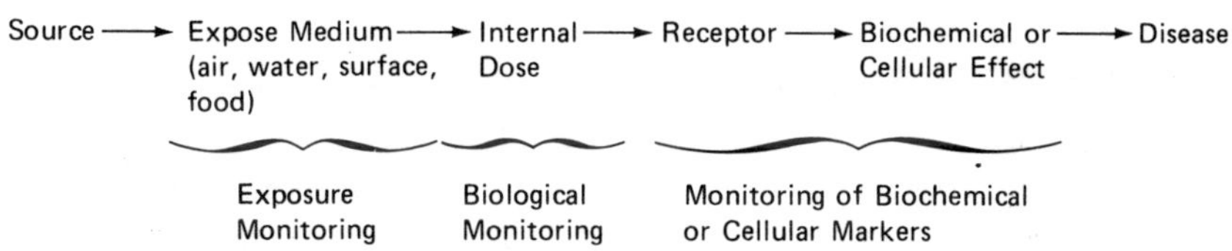

	Source →	Exposure → Monitoring	Biological → Monitoring (Internal dose)	Monitoring of → Biochemical or Cellular Markers	Disease Indentification
Control of Exposure	+++	+++	+		
Regulation and Legal Purposes	+	+++	+	+/–	+/–
Protection of Health	+	+	++	+++	+

Part I

Biological indicators of dose and effects: general concepts

Coordination:
A. Berlin (CEC, Luxembourg)
G. Chiappino (Milano, Italy)

1

Validation of biological monitoring: an introduction

Sven Hernberg, M.D. and **Antero Aitio**, M.D., Institute of Occupational Health, Helsinki, Finland

Biological monitoring was defined by a joint EEC/NIOSH/OSHA seminar (Berlin et al. 1984) as "the measurement and assessment of agents or their metabolites either in tissues, secreta, excreta, expired air or any combination of these to evaluate exposure and health risk, compared to an appropriate reference". Monitoring is a repetitive, regular and preventive activity; it should not be confused with diagnostic procedures. The use of the term biological monitoring unfortunately has not been consistent (Zielhuis 1985). Some authors have used this term to cover also physiological responses such as pulmonary or liver function tests and even tests for the early diagnosis of diseases. We would like to advocate the narrower definition and to stress the distinction between monitoring and health surveillance, which is a wider concept encompassing several aspects of health screening and early diagnosis (WHO 1975). The EEC/NIOSH/OSHA seminar defined health surveillance as "the periodic medicophysiological examinations of exposed workers with the objective of protecting health and preventing disease. The detection of established disease is outside the scope of this definition".

Biological monitoring can be divided into monitoring of uptake, sometimes also called exposure monitoring, and monitoring of effect. The principles for the validation of each category are quite different.

MONITORING OF UPTAKE

The purpose of monitoring of the uptake is to achieve an estimate of the biologically active body burden of the chemical in question. Its rationale is to ensure that worker exposure does not reach adverse levels.

Meaningful exposure monitoring presupposes the meeting of four conditions:

1. The chemical and/or its metabolites are present in some tissue, body fluid or excretion suitable for sampling;
2. Valid and practical methods of analysis and sampling are available;
3. The measurement strategy is adequate (the samples are representative);
4. The results can be interpreted in a biologically meaningful way.

Representativeness of samples and analytical accuracy, precision and sensitivity are addressed in other presentations. This paper discusses the biomedical interpretation of monitoring test results.

Reference values

Even perfectly sampled and measured values are meaningless if they cannot be compared with some standard. Ideally standards should be based on well-known dose- or exposure-effect and dose- or exposure-response relationships. Unfortunately such relationships are known for only a few substances, such as lead, cadmium and perhaps mercury. Whenever enough data is available, the reference point for a biological monitoring test should be defined as the no-detectable-adverse effect or response level. Provided exposure is continuously lower than the biological value corresponding to that level, adverse effects do not occur. Hence, in principle, there should be no need for further examinations. Today we are far from such an ideal situation, however. Thus biological monitoring must usually be combined with health surveillance, which relies on the examination of early adverse effects.

As the exposure-effect and exposure-response relationships are usually insufficiently known, they rarely provide a satisfactory basis for a reference value. In such instances the biological value must be compared with some other type of reference value.

There are two types of such values. One set represents concentrations of chemicals found in non-exposed persons, living in circumstances otherwise comparable to those of the studied populations. These reference values are often different for different populations as many (toxic) chemicals occur in the environment and have access to the body, e.g., via food and drink. Older reference values for trace elements are often higher than modern ones because of analytical improvements.

The other type of reference value, used for occupationally exposed workers, represents values regarded as "acceptable", although they exceed the levels found in the non-exposed. These reference values can also be subdivided into two types.

Health-based reference values are defined as levels not giving rise to detectable adverse toxic effects. They are based on medical considerations derived from exposure-effect and exposure-response relationships; they do not consider technological or economic feasibility. Due to the scarcity of data, only few have been proposed. WHO has recommended such values for lead, mercury, cadmium, carbon monoxide, and some pesticides. New toxicological information may change health-based values. For example, when the values for lead and cadmium were decided upon, their possible carcinogenicity was not known.

Administrative standards usually emerge as an incorporation of health-based values with technological, economic, administrative and other nonscientific considerations. Due to this compromise, they accept a certain risk for the most sensitive individuals. There are at present few administrative standards for biological monitoring. Lead is an exception; several countries base their surveillance mainly on blood lead monitoring. Recently, the Deutsche Forschungsgemeinschaft has published biological, administrative values for 21 chemicals, and ACGIH has signalled a notice of intent to establish biological reference values for 10 chemicals (DFG 1985, ACGIH 1985).

Biological monitoring has been and can be validated in an indirect way also (e.g., Lauwerys 1983). This type of validation involves comparisons with concentrations of the chemical in the ambient air and with effects corresponding to those concentrations. A toxicological data base exists for many chemicals, i.e., at least some kind of exposure-effect and exposure-response curves are available. However, the concentrations of a particular chemical in the air are not necessarily closely related to its actual uptake. Table 1 summarizes the most important reasons for this discrepancy.

Table 1: Factors affecting uptake of chemicals in the body, not likely to be accounted for by air measurements.

1.	Differences in the concentration of the chemical at different locations
2.	Variation in the concentration of the chemical at different points in time
3.	Particle size and aerodynamic properties of particles
4.	Solubility characteristics of the chemical
5.	Alternative absorption routes (skin, gastrointestinal tract)
6.	Protective devices and their efficiency
7.	Respiratory volume — work load
8.	Personal working habits
9.	Exposure outside the workplace

Considering all these factors, the indirect method of validating biological exposure tests over air measurements to adverse effects (or their absence) is far from ideal. Whenever a biological reference value is indirectly derived from administrative workroom hygienic standards (e.g., TLVs), the degree of uncertainty is high and even conceptual confusion may occur. Furthermore, <u>average</u>

concentrations in air and biological media are conventionally compared. If the average biological value corresponds to the hygienic standard, then about half of the workers have taken up more, and the other half less, than the amount "acceptable" according to the air standard. Thus, although the biological measurement usually reflects more accurately personal absorption, such indirectly derived reference values differ from other reference values. Their proper use requires that not the average, but the lower confidence limit should be used as the biological hygienic standard.

Carcinogens

While knowledge on dose-effect and dose-response relationships is necessary for the interpretation of the results of biological monitoring whenever toxic effects are concerned, the issue is fundamentally different when the aim is to prevent cancer. Toxic effects usually have a threshold, but no such level appears to exist for carcinogens — at least not for initiators. Therefore all exposure to carcinogens is undesirable in principle and, consequently, everything that exceeds "normal" levels should be interpreted as "excessive". The validation process consists of ensuring that the reference values are correct. Whenever a level is above zero, or alternatively, the population reference value defined as "acceptable", there is no longer a question of scientific decision but rather a technological-administrative-economic procedure, in which not merely medical aspects are involved.

Toxicokinetics

The validation of biological monitoring requires insight into the toxicokinetics of the chemical, especially its clearance rate is crucial.

For chemicals with long half times, such as lead, cadmium and mercury, the concentration in the blood or urine reaches a plateau, which reflects the equilibrium between daily intake and excretion. In conditions of stable exposure the day-to-day concentrations vary only little, and even single determinations of blood or urine levels give an accurate picture of the exposure. Chemicals with short half times are different in this respect and thus the validation of their biological monitoring is weakened. First, the concentrations — especially in blood — alter rapidly and thus only reflect very recent exposure. The sampling time must be strictly standardized to give meaningful results. Second, the representativeness of the sample may be poor whenever the concentrations of the chemical in the workplace air fluctuate. One single biological value cannot reflect these changes. It is therefore not feasible to monitor chemicals with half times of less than a few hours. However, many chemicals have several successive half times, reflecting their distribution in different compartments of the body. Thus lipid soluble organic solvents, which have a short half time (in the order of minutes only) in the blood and richly vasculated organs, tend to accumulate in fat deposits and are slowly released after exposure has ceased. Uptake over an entire day — and even several days — can be estimated from samples collected about 16-18 hours after the end of exposure. Considered in this way, the half times of, e.g., lipid soluble solvents may be in the order of 10 — 20 hours. Excretion in the urine may also function as an integrator; the half times are generally longer than those in the blood. Thus measurement of a metabolite in the urine is often a kinetically recommendable alternative for biological monitoring.

Separate reference values should also be calculated for chemicals in different physicochemical forms, because different compounds of a single element may behave completely differently in the organism. Thus the relationships between air and blood or urine concentrations may be drastically different for different compounds. Nickel and chromium are good examples of this.

EFFECT MONITORING

Effect monitoring means that instead of measuring the substance itself or its metabolite, some specific or semispecific effect of it is measured. Monitoring an early effect of the chemical would, in principle, be ideal for preventing adverse health effects; it would account for individual differences in susceptibility, not only in the uptake of the chemical. However, the very rationale of biological monitoring is to prevent effects; thus effect monitoring is conceptually contradictory. Some clarity would be provided by requesting that the effect on which the monitoring is based be non-adverse. Examples of effects which traditionally are considered non-adverse are the depression of 5-aminolevulinate dehydratase (ALAD) activity and the increase of zinc protoporphyrin (ZnPP) concentration in the erythrocytes. On the other hand, e.g., increased excretion of proteins such as beta-microglobulin and retinol binding globulins in the urine in exposure to cadmium or other nephrotoxins is probably not reversible. These effects should therefore be regarded as adverse; they do not fulfill the criteria for effect monitoring.

Some of the effects commonly used for biological monitoring also lack specificity. For example, erythrocyte ZnPP is elevated also in iron deficiency anemia. The problem becomes even more difficult as soon as one moves from biological monitoring in sensu strictu to the wider area of health surveillance. Here many completely non-specific tests are used. For example, liver function tests can be abnormal in many liver diseases and as a result of alcohol use, the blood picture reacts not only to benzene exposure but also to a variety of other exposures and a great number of hematological and other disorders.

Validation of tests used in health surveillance for measuring early effects of chemicals is a difficult process. Both the sensitivity and specificity of the test should be known. The more dangerous the chemical, the more important is the sensitivity of the test. However, increasing sensitivity usually leads to decreasing specificity. Two types of specificity should be considered. Diagnostic specificity means that an abnormal result can be referred to a certain condition. Anemia, for example, can be primary or secondary, and is therefore not "condition-specific". Chemical specificity means that an abnormal test specifically relates to a given chemical exposure. ALAD-depression is for all practical purposes specific for lead. In contrast, abnormally low conduction velocity in the peripheral nerves is a common manifestation of several toxic exposures (and other conditions as well) and is thus non-specific. The validation of tests used in health surveillance therefore alwaysinvolves an assesment of their sensitivity and specificity.

As already stated, lack of specificity renders most effect measurements unsuitable for biological monitoring in a strict sense. Such examinations should therefore explicitly be referred to the domain of health surveillance. In practice, however, biological monitoring and health surveillance form a continuum in workers' health protection; all sharp divisions are therefore artificial. Because biological monitoring tests generally are poorly validated, one cannot rely solely on them; health surveillance is therefore often needed in addition. Actually, one could say that biological monitoring is a component of a broader health surveillance program.

CONCLUSION

Before biological monitoring can be generally recommended as the primary method of assessment of uptake, more information on toxicokinetics and the metabolism of most chemicals is needed. Dose-effect and dose-response relationships have been clarified for only a few of the substances for which biological analyses are technically possible. Practical aspects must be further elaborated. Analytical sensitivity and accuracy must be improved and new methods must be developed for a great many substances. These considerations may suffice to illustrate the uncertain basis on which biological monitoring still rests. Therefore, alternative ways, i.e., environmental monitoring, for ensuring accurate enough assessment of exposure must always be considered.

REFERENCES

American Conference of Governmental Industrial Hygienists: Threshold limit values for chemical substances and physical agents in the work environment and biological exposure indices with changes intended for 1985-86. ACGIH, Cincinnati, Ohio, 1985.

Berlin, A., Yodaiken, R.E. and Henman, B.A. (eds): Assessment of toxic agents at the workplace. Roles of ambient and biological monitoring. (The Hague: Nijhoff Publ., 1984).

Deutsche Forschungsgemeinschaft: Maximale Arbeitsplatzkonzentrationen und Biologische Arbeitsstofftoleranzwerte 1985. Mitteilung XXI der

Senatskommission zur Prüfung gesundheitsschädlicher Arbeitsstoffe. VCH Verlagsgesellschaft, Weinheim, BRD, 1985.

Lauwerys, R.R.: Industrial chemical exposure: Guidelines for biological monitoring. Biomedical Publications, Davis, CA 1983.

World Health Organization (WHO): Early detection of health impairment in occupational exposure to health hazards. WHO Tech.Rep.Ser. 571. (Geneva:WHO, 1975).

Zielhuis, R.L.: Biological monitoring: Confusion in terminology. Am.J.Ind.Med. 8:515, 1985.

2

The current focus of biological monitoring

P. Schulte, W. Halperin, M. Herrick, B. Connally –
National Institute for Occupational Safety and Health (USA)

ABSTRACT

Biological monitoring is an increasingly used component of occupational and environmental health practice. We have surveyed recent biological monitoring research in order to characterize it, identify problems and gaps, and develop recommendations. The literature from nine occupational and environmental health journals for the period of 1981-85 has been reviewed and characterized according to trends by year and journal for effect parameters, substances monitored, study design, biological media sampled and goals of studies. We recommend that biological monitoring research shift to entail more studies on the relationship between internal dose and disease for established toxins and to develop new monitoring methods for the less studied toxins.

INTRODUCTION

Biological monitoring is a component in a continuum of preventive practices useful in the preservation of occupational safety and health (Halperin and Frazier, 1985). This continuum spans the primary, secondary and tertiary levels of prevention. Biological monitoring, the monitoring of a worker for absorption of an intoxicant or its metabolite, is a more accurate measure of exposure than environmental monitoring because it indicates the internal dose.

There is a gray zone between biological monitoring and medical screening. We define medical screening as the search for early stages of disease. Sometimes biological monitoring involves the measurement of altered physiological responses such as elevated protoporphyrin. Some would argue that this is an example of medical screening and not biological monitoring but the boundaries are not precise (Berlin et al. 1984, Zielhuis 1985, Goyer and Rogan 1986).

In recent years there seems to have been a proliferation of scientific journal articles on biological monitoring. Numerous books and conferences also have addressed the subject (Lauwerys 1983, Aitio et al. 1984, Berlin et al. 1984). We thought it useful to address the trends in the literature in order to localize areas needing augmented interest, and to identify generic problems in biomonitoring research.

METHODS

We reviewed nine environmental or occupational health journals, for the period 1981-1985, in order to identify and characterize the extent and nature of reports of biological monitoring research in humans. In this review, we have included articles if they involved analysis for the following:

1. The concentration of an intoxicant in biological media such as blood, urine, expired air, hair, adipose tissue, saliva etc; or

2. The concentration of metabolites in biological media; or

3. The determination of non-adverse biological changes (e.g., free erythrocyte protoporphyrin) that are the results of the reaction of the organism to exposure (Lauwerys, 1983).

Additionally, we included articles that involved the detection of potentially adverse biological effects, such as the increased levels of hepatic enzymes in plasma or abnormal urine cytologies, in exposed individuals.

The nine journals evaluated are: International Archives of Occupational and Environmental Health; the Scandinavian Journal of Work and Environmental Health; Journal of Occupational Medicine; International Journal of Epidemiology; British Journal of Industrial Medicine; American Journal of Industrial Medicine; American Industrial Hygiene Association Journal; American Journal of Epidemiology; and the Archives of Environmental Health. The journals were selected to be representative of the field and the breadth of biological monitoring research. Each issue of the journals was reviewed and articles pertaining to biological monitoring research in humans were evaluated. Studies of biological monitoring in animals or articles on monitoring methods were not considered.

We classified the articles according to the following categories of study design: (1) cross-sectional, (2) case-control, (3) cohort, (4) case-series, (5) case report, (6) industrial hygiene, or (7) review and commentary. The type of biological medium that was sampled was also identified, as was the substance that was monitored. All studies were characterized according to the goal of the study. The possible goals were stated as the following questions: (1) Did the study attempt to link exposure (historical or measured) with biological monitoring, (2) did the study attempt to link current biological monitoring of exposure with a health outcome, (3) did the study attempt to link a historical level of biological monitoring with a health outcome, (4) did the study attempt to use biological monitoring to confirm indirect data, such as exposure records? These evaluations were tabulated by journal and year. Tabulations were made for the various content categories and related to the denominator of total studies of biological monitoring. Results were evaluated yearly and for the period 1981-1985.

RESULTS

For the five-year study period, a total of 3,738 articles were reviewed of which 585 or 15.6% involved biological monitoring (Figure 1). This percentage was relatively

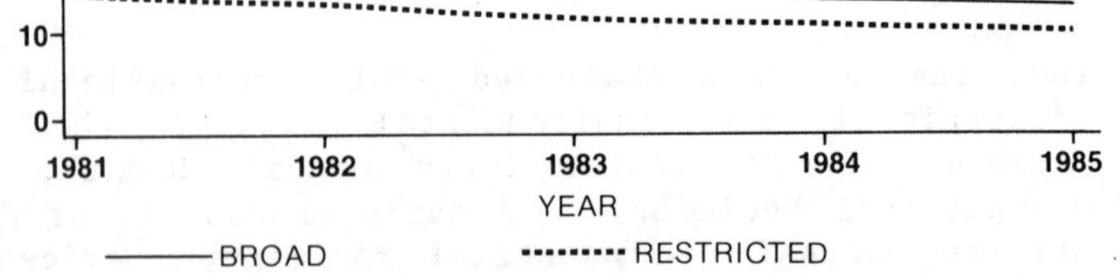

FIGURE 1. PERCENTAGE OF JOURNAL ARTICLES INVOLVING BIOLOGICAL MONITORING, 1981 — 1985

constant over the period with the highest percentage, 18% in 1983, and the lowest, 12% in 1985. Since there is some confusion in the literature regarding terminology, the percentage of articles was also plotted using a restricted definition of biological monitoring that excluded the category "potentially adverse health effects." This showed a similar trend.

As would be expected there was substantial inter-journal variation (Figure 2). The journals could be grouped into three categories according to percentages of articles involving biological monitoring. The high percentage group consists of the International Archives of Occupational and

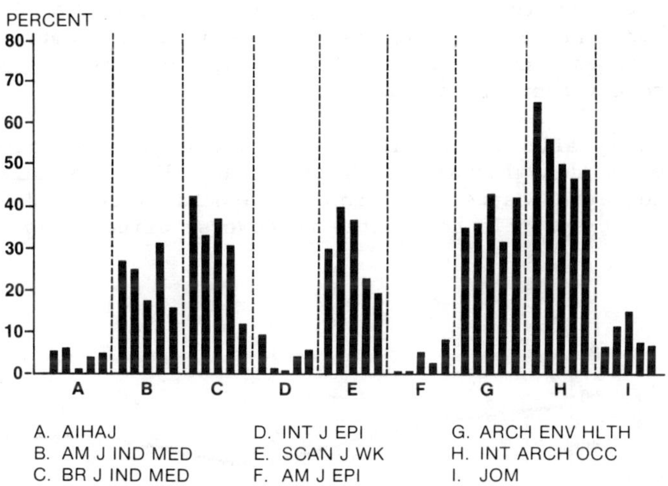

FIGURE 2. FREQUENCY OF BIOLOGICAL MONITORING ARTICLES BY JOURNAL, EACH FROM 1981 – 1985

Environmental Health which had an average of 59% of articles on biological monitoring. A second group consisting of the Scandinavian Journal of Work and Environmental Health, American Journal Industrial Medicine, Archives of Environmental Health and the British Journal of Industrial Medicine had an average of 30% and a final group consisting of the Journal of Occupational Medicine, International Journal of Epidemiology, American Journal of Epidemiology, American Industrial Hygiene Association Journal, had an average of 5.7%. In the lowest percentage group the two epidemiology journals generally showed an increasing trend throughout the study period, as exemplified by the American Journal of Epidemiology, which had only 0.6% in 1981 but 8% in 1985. Excluding 1981, due to a large number of studies of serum cholesterol and heart disease, a similar pattern was seen for the International Journal of Epidemiology. These increases may be indicative of a trend in epidemiology to use biochemical or molecular endpoints in studies. It may, however, represent a redistribution of biological monitoring articles among journals since, in the high percentage group, there was a general decrease from 65% in 1981 to 49% in 1985.

Effect Parameters

In 1985, Zielhuis reported on continuing confusion over the definition of biological monitoring. The confusion generally surrounds the distinction between biological monitoring of the intoxicant or its metabolites versus medical screening for signs and symptoms of early pathology. We grouped the articles into three categories suggested by Zielhuis (1984, 1985) and found that the measure of toxin comprised 69% of the studies, of a metabolite, 19%, and of an agent-specific nonadverse effect, 12%.

Using all effect parameters (six categories discussed below) as a denominator, we found that 41% of studies measured a toxic substance, 11% measured a metabolite, and 7% measured an agent-specific reversible nonadverse effect (Figure 3).

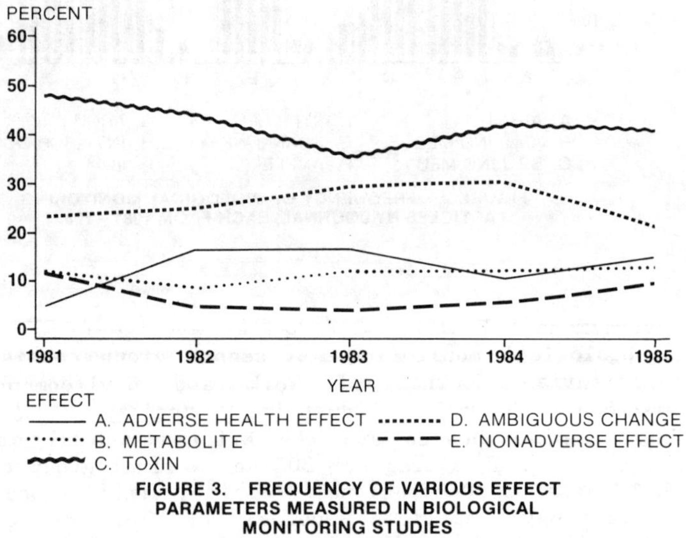

FIGURE 3. FREQUENCY OF VARIOUS EFFECT PARAMETERS MEASURED IN BIOLOGICAL MONITORING STUDIES

Alternately, we grouped the articles into six categories, the three previously mentioned and three others. Of these latter three categories, the most frequent were measures of biological substances that were not agent-specific but were nonadverse biological effects. This group, exemplified by clinical chemistries, could be termed ambiguous effects and comprised 26% of studies. Another group consisting of measures of health effects such as abnormal urine cytology, comprised 14% of studies, and a third group involved 31 (5.3%) studies involving genetic monitoring and 4 (0.6%) involving genetic screening.

Substances monitored

An average of 58 different chemical substances was monitored each year during the study period. The most frequent were lead, 23%; cadmium, 5.2%; toluene, 3.8%; asbestos and mercury 3.6% each, polychlorinated biphenyls (PCB), 3.8%; and styrene, 3.1% (Figure 4). No consistent patterns were observed over the study period except for a peak, in 1984, for 9% of studies involving PCBs. Most of these studies concerned Yusho disease that resulted from PCB contamination of rice oil.

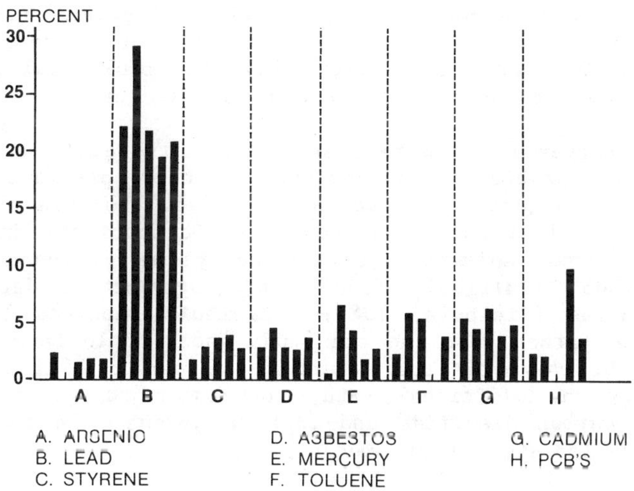

FIGURE 4. MOST COMMONLY MONITORED SUBSTANCES IN REVIEWED ARTICLES, EACH FROM 1981 – 1985

The frequency of substances that were included in published studies parallels the frequency with which substances are used in industry. We compared the ranking of substances obtained from the literature with the frequency with which they were found in a systematic sample of industries in the United States by the National Occupational Hazard Survey (NOHS) for 1972-74; the Spearman rank correlation coefficient yields an r value of 0.76. However, according to the NOHS data, other toxic substances, such as ethanol, silica, aromatic hydrocarbons, trichloroethylene, welding gases, products of combustion, and ethylene glycol, which occur in the workplace with great frequency, accounted for little or none of the research evaluated.

These findings appear to represent a dilemma in the field. For example, although research on lead now serves as a model for biological monitoring, the preponderance of studies on lead during the period 1981-1985 show a fair amount of redundancy. Moreover, despite this extensive number of studies, we agree with Hernberg (1980) who has concluded that there are still areas of knowledge about lead that are deficient. For example, there is clear need for an analysis of the relationship between lead, more or less at acceptable environmental levels, and the effects on sperm. At the other end of the spectrum there is a paucity of studies on other important toxins, such as, nickel, dioxin, ethylene oxide, aluminum, manganese, methyl bromide and various mutagenic and carcinogenic compounds. There are no studies that we could find on nitrosamines to which there are multiple possibilities for human exposure, and of which Bogovski (1984) believed biological monitoring should be of utmost importance. Finally, very few studies involved biological monitoring of combined exposures or exposure to mixtures.

We surveyed the data base of 585 studies to address some of these problems and evaluate whether previously identified research gaps had begun to be filled. For example, in 1980, a WHO Study Group on Health-Based Occupational Exposure limits for four solvents (trichloroethylene, toluene, xylene and carbon disulfide), found inadequate or no data from field studies (Zielhuis, 1984). Zielhuis reported that the data were rather poor and the study group could have discarded the whole body of data. During the period 1981-1985 there was only one additional study concerning trichloroethylene, five on carbon disulfide and four on xylene. In comparison there were 133 studies on lead.

Type of study

The cross-sectional design was used in an average of 73% of studies with a high of 83% in 1982, and a low of 63% in 1981 (Figure 5). Other study types were far less prevalent: experimental studies, 9.0%; case reports, 5.2%; case-series, 4.6%; cohort studies, 3.4%; and case-control, 2.7%. The use of an experimental design showed the most dramatic fluctuation. In 1981 it comprised 13% of the studies, dropped to 3% in 1982, and increased to 11% in 1985. The most consistently increasing type of design is the case-control design that rose from less than 1% of studies in 1981 to more

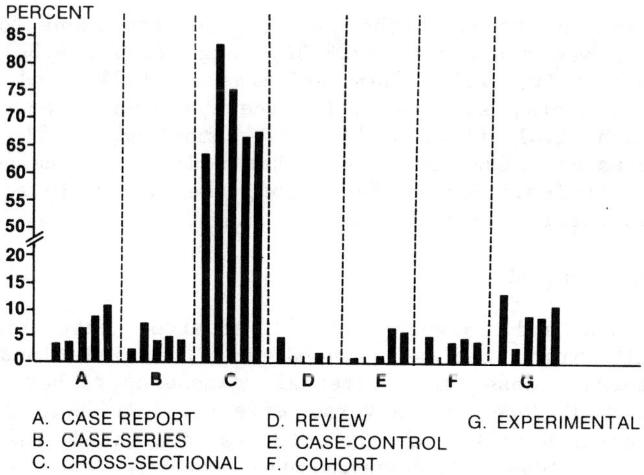

FIGURE 5. TYPE OF STUDY DESIGN, EACH FROM 1981 – 1985

than 6% in 1984 and 1985. On the average, it was the sixth most common design.

Biological materials

Approximately half of the studies involved sampling blood for biological monitoring (Figure 6). There appears to be an

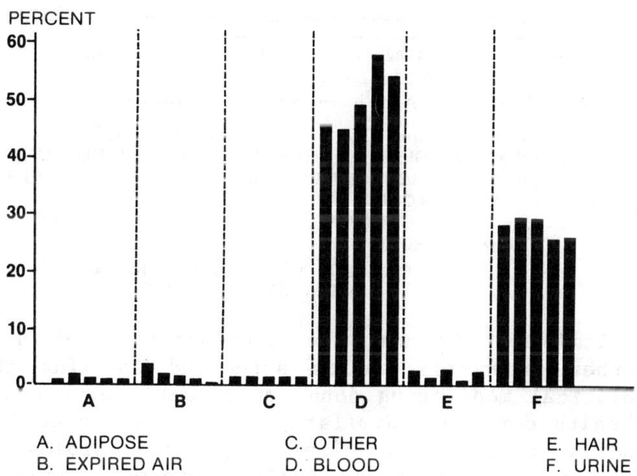

FIGURE 6. BIOLOGICAL MATERIALS SAMPLED, EACH FROM 1981 – 1985

increasing trend in the use of blood specimens for monitoring. Urine was monitored in 28.3%; lung tissue, 4.3%; expired air, 2.3%; hair, 2.2%; adipose tissue, 1.2%; and saliva 1.1%. Other media, such as, milk, teeth, semen, bone, feces, liver, conjunctival fluid and skin, comprised 14.9% of the media evaluated. Each of these other media was examined in 1-2% of the studies. Except for blood, the trends in all other media were fairly constant.

Purpose or goal

Bernard and Lauwerys (1984) observed that most biological monitoring studies focused on the relationships between internal dose and external exposure rather than between internal dose and adverse effects. We have confirmed this observation--74% of the studies evaluated the relationship between level of environmental exposure and biological level of the intoxicant or metabolite (Figure 7). Typical among this type of study was the correlation of measurement

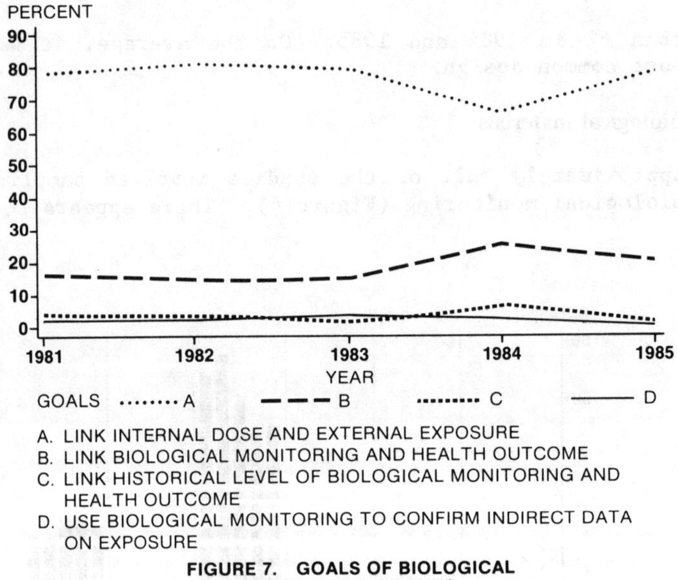

FIGURE 7. GOALS OF BIOLOGICAL MONITORING STUDIES

of lead in air and the concentration of lead in blood. Another group of studies attempted to link the results of biological monitoring done concurrently with the assessment of a health outcome. Similarly, but to a lesser degree, 3.4% of studies attempted to link historical biological monitoring data with a subsequent health outcome. An example linking historical levels of biologicial monitoring to a health problem is a study that links past levels of mercury in urine with current evidence of neurotoxicity (Williamson and Teo 1982).

In 2% of the studies biological monitoring was used to supplement or confirm indirect data on exposures. For example, several studies, most involving biomonitoring of heavy metals, relate blood and urine levels to job classifications and length of employment or to historical air monitoring records in company personnel records (Hesley and Wimbish 1981, Hassler et al. 1983, Piikivi et al. 1984).

CONCLUSION

Although biological monitoring is increasing in use in industry (Ratcliffe et al. 1986), the amount of research in the last five years has stayed relatively constant. The type of research is still predominantly the measurement of toxin in biological media and the correlation with environmental exposure. There is little research on the relationship between internal dose and adverse effects. Aside from lead and cadmium there appears to be too little research related to agents that are either common in industry, or that have unusual toxicity. Even with agents that are commonly subject of research reports, there is no assurance that efforts are appropriately targeted to the gaps in our knowledge or the needs of the workforce. Other important substances merit investigation.

We believe that it would be useful to conceptualize research on biological monitoring into three phases. The first phase is development of adequate methods. The second phase is field testing of those methods. The third phase is routine use of those methods in practice in industry or as part of medical research that attempts to guage the relationship between exposure and adverse outcome. Although we recognize that science progresses through trial and error, aided by chance findings, we recommend that scientists interested in biological monitoring establish a list of agents that should be prioritized for research and field validation.

ACKNOWLEDGMENTS

The authors thank Clorinda Battaglia and Sylvia Morgan for their assistance in the development of this manuscript.

REFERENCES

Aitio, A.; Riihimaki, V.; Vanio, H., editors. Biological monitoring and surveillance of workers exposed to chemicals. Washington: Hemisphere Publishing Co., 1984. 403.

Berlin, A.; Yodaiken, R.E.; Henman, B.A., editors. Assessment of toxic agents at the workplace. Boston: Martinus Nijhoft Publishers, 1984: XIII.

Bernard, A.; Lauwerys, R. Present status and trends in biological monitoring of exposure to industrial chemicals. (Presentation in Cincinnati, Ohio, 1984) J Occup Med. (in press)

Bogovski, P.: N-Nitrosocompounds in biological monitoring and surveillance of workers exposed to chemicals. Aitio, A.; Riihimaki, V.; Vanio, H., eds Washington: Hemisphere Publishing Corp., 1984:193-207.

Halperin, W.E., Frazier, T.: Surveillance for the effects of workplace exposure. Annu Rev Public Health. 6:419-32; 1985.

Hassler, E.; Lind, B., Piscator, M. Cadmium in blood and urine related to present and past exposure. A study of workers in an alkaline battery factory. Br J Ind Med. 40:420-425; 1983.

Hernberg, S. Lead. Occupational Medicine. C. Zenz, editor. 715-769, 1980.

Hesley, K.L.; Wimbish, G. Blood lead and zinc protoporphyrin in lead industry workers. Am Ind Hyg Assoc J. 42:42-46; 1981.

Lauwerys, R.R. Industrial chemical exposure: guidelines for biological monitoring. Davis, CA: Biomedical publication; 1983:4.

National Occupational Hazard Survey, National Institute for Occupational Safety and Health, DHEW Publications No. (NIOSH) 74-127, May 1974.

Piikivi, L.; Hänninen, H.; Martelin, T.; Mantere, P. Psychological performance and long term exposure to mercury vapors. Scand J Work Environ Health. 10:35-41;1984.

Ratcliffe, J.M.; Halperin, W.E.; Frazier, T.M.; Sundin, D.S.; Delaney, L.; Hornung, R.W. The prevalence of screening in industry: a report from the NIOSH National Occupational Hazard Survey. J Occup Med. (in press).

Williamson, A.M.; Teo, R.K.C. Occupational mercury exposure and its consequences for behavior. Int Arch Occup Environ Health. 50:273-286; 1982.

Zielhuis, R.L. Approaches in the development of biological monitoring methods: laboratory and field studies in biological monitoring and surveillance of workers exposed to chemicals. Aitio, A.; Riihimaki, V.; Vanio, H. eds. Washington: Hemisphere Publishing Corp; 1984:373-385.

Zielhuis, R.L. Biological monitoring: confusion in terminology. Am J Ind Med. 8:515-516; 1985.

3

The use of biological indicators in the surveillance of groups or individuals

P.A. Bertazzi and **P.G. Duca** – Institute of Occupational Health, Clinica del Lavoro L. Devoto and Institute of Medical Statistics and Biometry of the Universita' Degli Studi, Milan.

INTRODUCTION

A monitoring/surveillance programme of a population exposed to environmental or occupational toxic agents is aimed at preventing or identifying at the earliest possible stage, effects of exposure. In our domain, when a surveillance programme is undertaken, certain evaluations have already been made, either implicitly or explicitly: for example, that the complete removal of exposure from the environment is not feasible or is insufficient to the scope (e.g. late effects); that adequate technical methods and instruments and sufficient scientific knowledge exist to guarantee the reliability of the data collected and the sound interpretation of the results; that the programme is effective, so that those who undergo it receive considerable benefit to their health in comparison to those who, given comparable conditions, do not. In these evaluations, ethical, economic, and scientific issues are entangled.

Different types of surveillance activities are currently adopted. Environmental monitoring is aimed at the detection of chemicals in the ambient air and water, in food, soil, etc.. Biological monitoring is used to measure the intake of chemicals by the human body. Health surveillance is targeted to the identification of early adverse effects (reversible or curable) of chemicals to human health. Epidemiological surveillance is based on the exploratory analysis of possible associations between disease incidence or mortality and different occupations or exposures. All apply to the identification and evaluation of the risk at the group level. Thanks to the use of biological indicators, biological monitoring and health surveillance also allow, at least under certain circumstances, evaluation at the individual level. The biological and health significance of the numerous indicators of toxicity, mutagenicity and carcinogenicity currently available or being studied, and the problems related to their measurement methods have been recently reviewed [1,2,3]. Here we limit ourselves to addressing some issues arising from their use in practice, especially with regard to the evaluation, at a group or individual level, of the collected information.

REFERENCE VALUES

Biological indicators are used at the individual level in order to guarantee adequate protection even to persons with multiple exposure or to hypersusceptible subjects; at group level, the objective is to improve safety of the work environment. Both aims can be accomplished provided that appropriate reference values are available for a comparative evaluation.

Reference values for the individual and for the group can be either the values measured in the same subjects before the beginning of exposure (internal reference), or the values measured in a sufficiently numerous and appropriately selected group of non-exposed subjects. (external reference). In both cases, the role of confounding factors must be taken into account and controlled.

External reference

The external reference value is the one characteristic of a non-exposed population, assumed as representative of the background risk. The 'non-exposure' of the reference population must be carefully ascertained. Sources of exposure can, in fact, be multiple and different. Thus, the food chain might carry unsuspected exposures which are not always differentiable by the physico-chemical nature of the measured compound, as it is the case, for example, with arsenic. Exposures from industrial processes can contaminate the surrounding area, thus rendering the resident population unfit as a 'non-exposed' reference population; this is the case, for example, of metal smelting plants.

In addition, the reference population must have some comparable characteristics with the exposed group: social class, for example, is associated with a number of factors, e.g. hygienic conditions, life style, health status, and access to health care services, which might differentiate the potential for exposure of different subjects; the area of residence may also have notable importance, since, as has been demonstrated for cadmium, different towns or regions may be characterised by quite different background levels. The value of an indicator can then be influenced by personal characteristics (age, sex), health experience (disease and drug consumption), and life style (alcohol consumption, smoking, diet). Therefore, a reference population aside from being 'non-exposed' to the factor in the study, must be comparable to the one exposed and surveilled for the series of factors listed above since they are capable in themselves of producing changes in the indicator values.

A third element of concern is the size of the compared groups, which is a function of: - minimum difference between the populations that is considered relevant; - significance level, α; -power of the test (1-β). Let us assume that the relevant biological variable be a continuous one, and have a Gaussian distribution with the same dispersion and a different mean in the study and reference group. If the indicator is such that an increasing risk corresponds to its increasing values, we can reasonably suppose that, for example, the risk among the exposed subjects is doubled when a doubled proportion of them exceeds a given value. Let us then consider two hypothetical situations, one in which the 50th percentile in the exposed population has the same value as the 75th in the reference population, and a second one in which the 90th percentile among the exposed corresponds to the 95th among the reference. Given these situations, a doubled risk, as defined above, is associated with a departure of the exposed population mean from the reference population mean equal to 0.675 (first situation) and 0.363 (second situation) standard deviation (Fig.1). In testing the null hypothesis (exposure

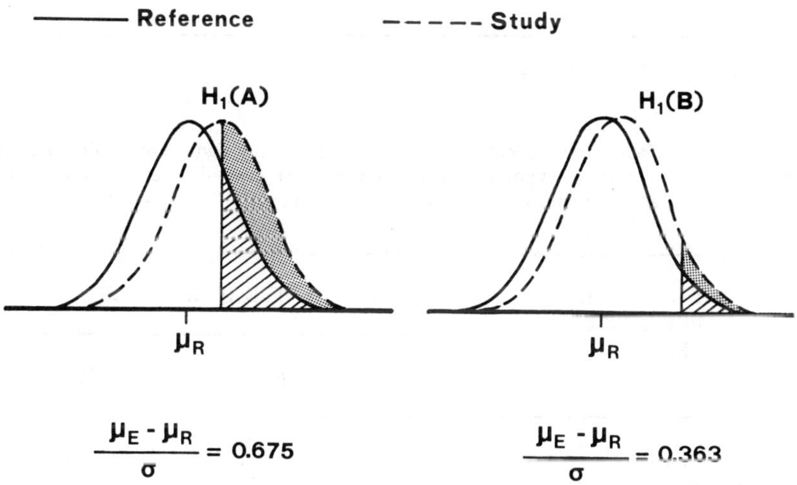

Fig. 1. – Hypothetical examples of the distribution of the values of an indicator in the study and reference populations. The 75th percentile in the reference corresponds to the 50th in the study population (left); the 90th to the 95th (right). μ=population mean; σ=standard deviation.

does not increase the risk) vs. the alternative hypothesis (exposure causes a doubling of the risk) the size of the exposed and reference groups clearly affects the power of the test. Tables 1 and 2 show the probabilities of obtaining a statistically significant result when the alternative hypothesis H1(A) or, respectively, H1(B) is true. It is apparent from these examples that in order to have a powerful tool for detecting risk, the size of the groups needs not only to be adequate but also balanced: thus, for example, 25 subjects in both groups ensure a higher power than 10 vs. 100 subjects.

Table 1: Probabilities to obtain a statistically significant result when the alternative hypothesis is true (power; 1-β), given different sizes of the exposed and reference group. α=0.05, one tailed. H1(A): $(\mu E - \mu R)/\sigma$=0.675

SIZE OF THE REFERENCE GROUP	SIZE OF THE STUDY GROUP		
	# 10	# 25	# 100
# 10	.44	.56	.65
# 25	.56	.77	.92
# 100	.65	.92	.999
# ∞	.69	.96	1.000

Table 2: Probabilities to obtain a statistically significant result when the alternative hypothesis is true (power; 1-β), given different sizes of the exposed and reference group. α=0.05, one tailed. H1(B): $(\mu E - \mu R)/\sigma$=0.363

SIZE OF THE REFERENCE GROUP	SIZE OF THE STUDY GROUP		
	# 10	# 25	# 100
# 10	.20	.25	.29
# 25	.25	.36	.49
# 100	.29	.49	.82
# ∞	.31	.57	.98

It follows that differences between the study and the reference group which are modest but meaningful from a biological

point of view might turn out to be statistically not significant merely because of the low power of the test, thus causing the failure of the surveillance programme especially in groups of limited size.
Comparison between means yields information on the exposed group as such. In order to examine a departure from background risk for the individual, tolerance limits are used. In this case, we need to identify the interval which covers, with a specified confidence, at least a stated percentage of the individual values in the population from which the sample has been drawn.These intervals are known as 'tolerance intervals' and their end point as 'tolerance limits'. In the application we are concerned with, a one sided tolerance limit seems most appropriate: it divides the population into two parts, so that at least the specified proportion of values will fall below this limit with a stated probability [4]. Table 3 shows how the size of the reference group can affect the estimate of tolerance limits.

Table 3: Expected values for one-sided upper tolerance limits.

INTENDED VALUES OF THE LIMIT	SIZE OF THE REFERENCE GROUP			
	# 10	# 25	# 100	# ∞
75th	91	86	81	75
95th	99.5	98	97	95

Only when the size of the reference group is=∞ can we expect to estimate, with confidence of 95%, the intended tolerance limits (respectively, 75th and 95th).Given n=25, and that the intended values for the one-sided tolerance limit are the 75th or 95th, what we can expect are to estimate are, instead, the 86th and 98th; etc.

Internal reference

In this case, the individual or group is characterized by a baseline value, and after commencement of exposure, possible deviations from it are looked for.
In this way the component of interindividual biological variability is eliminated. However, the intraindividual variability component remains. In addition, age or other characteristics of the subject(s) inherently changing as time passes may influence the value of the indicator, independently from exposure, thus causing confounding.
The examination of the indicator values in an individual at successive times, can be properly accomplished by

constructing an appropriate control, chart particularly when the program includes repeated measures at close time intervals. The approach based on the cumulative sum of successive deviations can also be useful: (CUSUM).
At group level, a collective control chart can be constructed analogously, reporting mean or median values (according to the more or less symmetric shape of the distribution of the variable). Comparison between mean values in the same group at different times requires methods suitable for paired data. Interpretation of results yielded by these methods is far from straightforward when the number of observations increases.

INTERPRETATION AND ACTION

Assume that we have determined reference values in an appropriate manner. The results of the comparison of the values among the exposed subjects with reference values lead to classifying the observed values in distinct categories or regions, in short defined as "under control","alarm", "out of control". Each category calls for a distinct type of action: respectively: - reevaluation according to the schedule of the original surveillance protocol; - reevaluation at closer time intervals; - removal of the subject(s) from exposure, adoption of environmental measures, implementation of secondary prevention measures. Such an evaluation implies different problems for the individual or the group, depending on which reference is being used (internal or external).

Individual safety

When an external reference is used, it is advisable not to base the evaluation of the observed values on the most extreme percentiles of the reference distribution (e.g., 95th, 98th), but to be prudent, for example on the 75th. In this way, the second type of error (i.e., not identifying an effect due to exposure when in fact it exists) is reduced: the very nature of a surveillance activity seems to allow such a procedure, admittedly with the increased yield of false alarms. In defining the region of action, the intraindividual variability of the indicator and the quality of the measurement play a major role. Given this variability, it is important not to limit the "alarm region" only to a range of values above the chosen percentile, but to all the values which fall around it, for example, to those lying within an interval of 1 - 2 standard deviations. Values which depart from the reference to an extent which is not compatible with a random fluctuation (e.g., >2DS) will be interpreted as out of control, and values falling below the alarm region (e.g., <2DS), in control.
With an internal reference, values which depart - as an example - 2 standard deviations from the initial value can be considered in the alarm region; it is worth mentioning that a

sequence of 5-7 values in increasing order should also be considered an alarm signal. Out of control could analogously be considered a value which swerves more than 3 standard deviations from the initial value. The interpretation of these results requires a careful consideration of intraindividual biological variability and analytic variability: in particular, the analytic method should be maintained constant during the entire duration of the programme, to avoid false or missing alarms.

Group safety

When the comparison is with an external reference two facts must be taken into particular consideration: the quality of measurement (analytic variability) and the random sampling fluctuation, due to the interindividual biological variability. The size of the compared groups, as discussed above, may influence results and their interpretation decisively. For this reason, a possible suggestion, especially in the case of small study groups (e.g., less than 30 subjects), is to lower the significance level protecting against first type error in order to gain statistical power. A resulting strategy could be to consider out of control, values departing more than 1.645 standard error from the reference mean, and in the alarm region, those falling in the region included within 1.00 and 1.645 standard error. Such a strategy will, of course, result in an elevated number of false alarms.

The comparison of a group with an internal reference involves evaluation criteria analogous to those of the individual values. The difference is that, having to do with means (or medians), their dispersion will be measured by the appropriate standard error. In this evaluation, interindividual and technical variability are more important than the intraindividual one. The larger the surveilled group, the less - however - will be the importance of the variability component bound to analytic imprecision.

EFFECTIVENESS OF PREVENTION

Correctly established reference values provide us with an important tool for surveillance, but they do not ensure that prevention or protection is effective. The definition of the 'action regions' as indicated above, has not yet undergone practical application, and an in-field evaluation is still required.

In addition, the adopted values do not necessarily correspond to those of a comparable non-exposed population: economic and social reasons often lead the interested parties to agree on different limits.

The biological indicators employed are not always completely reversible and without significance to health: this is the case not only in the domain of 'health surveillance', but also for some markers proposed for 'biological monitoring'.

For this reason, along with the cross-sectional comparison with the reference value, longitudinal studies are required to evaluate the prognostic value of the end points of the surveillance in respect to long term health effects. Such an evaluation needs to be both qualitative and quantitative. In the first case, the aim is to verify whether the indicator should be considered 'adverse'or 'non-adverse'; the second evaluation can bring to light the possible existence of dose-response relationships.

Finally, the problem of an overall evaluation of the surveillance programme as such exists. Its effectiveness needs to be established in relation to short, medium, and long-term health effects. No doubt, several traditional chronic intoxications have almost disappeared in many areas or countries. Surveillance programmes have certainly contributed to this, but it is obvious that numerous different reasons and conditions concurred in producing improvements in work places. Problems still remain open, however. For example, the prevention of effects other than those of traditional intoxications and the prevention of subtle effects, appearing after a long time, without a detectable early toxic stage.

To verify how effective a surveillance programme is to these ends, appropriate epidemiological investigations should be conducted: they can be based on two types of comparison, i.e., between the surveilled group and a non-exposed population or, more properly, between an exposed and surveilled group and an exposed but not surveilled group. The latter approach carries self-evident ethical problems. Its adoption would be greatly facilitated should sufficient information already exist to carry out historical investigations. In areas where good recording systems of diseases or deaths exist, along with available documentation on work exposures and surveillance programmes performed, an evaluation based on a case-control design could be envisaged.

REFERENCES

1. LAUWERYS R.R. (1983): Industrial chemical exposure: guidelines for biological monitoring. Biomedical Publications, Davis, CA, USA.

2. BERLIN A., DRAPER M., HEMMINKI K., VAINIO H. (1984): Monitoring human exposure to carcinogenic and mutagenic agents. IARC Scientific Publications No. 59, Lyon, France.

3. BERTAZZI P.A., ALESSIO L., DUCA P.G., MARUBINI E. (1984): Monitoraggio biologico negli ambienti di lavoro. Principi, metodi, applicazioni. Franco Angeli, Milano, Italy.

4. BLISS C.I. (1967): Statistics in biology. Vol. 1, pp 197-198. Mc Graw Hill Co., New York, NY, USA.

4

Influence of factors other than exposure on the levels of biological indicators

L. Alessio – Associate Professor of Occupational Health, University of Brescia
A. Berlin – Commission of the European Communities, Luxembourg;
V. Foà – Associate Professor of Occupational Health, University of Milan

The use of biological indicators to assess exposure and early effects in workers exposed to industrial toxics is now a common feature, and it can be foreseen that it will be possibile to extend this method to a larger number of substances in the near future.

Correct use of biological indicators requires a thorough knowledge of how these tests behave in relation to exposure. However, this information alone is not sufficient and due consideration should also be given to all situations which, although independent of exposure, may neverthless effect the biological indicator levels. There are many such situations; the most important are: physiological factors; personal habits; use of drugs; genetic disorders; acquired diseases; physiological and pathological factors that can alter the characteristics of the biological medium used for determination of the indicators.

We will now try to illustrate the classification given above with some practical examples taken from our own experience and from literature.

Physiological factors include diet, age and sex.

To highlight the importance of dietary factors, we have the example of the influence of consumption of fish and crustaceans on the levels of urinary arsenic in non-occupationally exposed subjects (7). After a meal of 100 g of crab meat, the levels of total urinary arsenic in non-occupationally exposed subjects showed a marked increase (30-550 ug/l), reached a peak after 10 h, and returned to base line values afer 25 to 30 h.

This example shows that routine determination of total urinary arsenic can be used for periodic monitoring of occupationally exposed subjects, provided they are instructed to avoid eating sea-food in the 48 h prior to collection of the urine sample.

The biological indicators may differ according to sex. For example, in femal subjects, at the same PbB levels, the erythrocyte protoporphyrin values are significantly higher compared to those of male subjects (2).

An example of the influence of age on the levels of biological indicators is found in the increase that may occur in the levels of urinary cadmium with increase in age (3). This phenomenon is probably due to the fact that CdU is influenced by the body burden.

Of the <u>personal habits</u> that can interfere with the indicator levels, smoking and alcohol consumption are particularly important.

Smoking may cause direct absorption of substances naturally present in tobacco leaves, or of pollutants present in the working environment that have deposited on the cigarettes, or of combustion products.

An example of the first type of interference is the influence of smoking on Cd levels in biological media. Blood Cd levels in non-occupationally exposed subjects are higher in smokers and increase in proportion to the number of cigarettes smoked (3). Moreover, where workers smoke at the workplace or anyway before washing their hands properly, cigarettes may also act as a vehicle for pollutants present in the environment. This is particularly significant in the case of certain metals with low melting point, like lead, for example. The combustion temperature of cigarettes exceeds 700°C, so that a good part of the deposited metal can be transformed into vapour.

Lastly, toxic substances, like carbon monoxide, can be generated during the combustion of cigarettes: in fact, the levels of carboxyhaemoglobin are higher in smokers and it is common knowledge that workers must be instructed absolutely not to smoke for some hours before the blood test if reliable data are to be obtained.

Another major personal habit is alcohol which may influence the biological indicator levels. The consumption of alcohol

can cause intake of substances naturally present in alcoholic beverages, can be the direct cause of metabolic changes similar to those caused by the industrial toxic in question, and can interfere with the biotransformation and elimination of industrial toxics.
It is a known fact that wine has a "natural" lead content, so that in heavy drinkers the quantity of lead ingested daily with the diet can increase significantly; it has, in fact, been demonstrated that heavy drinkers show higher PbB levels than control subjects.
Ethanol can also directly inhibit erythrocyte ALAD activity, and it was found that the levels of this enzyme in alcoholics were markedly more reduced than could have been expected from the PbB levels. Moreover, ethanol can alter the metabolism of various organic solvents and consequently the behaviour of their biological indicators.
Ingestion of alcohol in a single dose inhibits the metabolism of trichloroethylene, xylene, styrene and toluene, probably due to competition for alcoholdehydrogenase, an enzyme essential for the breakdown of both ethanol and solvents (6, 10, 12, 13).
Regular alcohol ingestion can affect the metabolism of solvents in a totally different manner; for example, in regular drinkers occupationally exposed to toluene, the blood levels of toluene were lower than those of a reference group. This mean that regular drinking accelerates toluene metabolism, presumably due to induction of the microsome oxidising system (12).
Less information is available on the possible interference of drugs on the levels of biological indicators.
It was demonstrated that phenylsalicylate, a drug widely used as an analgesic, can significantly increase the levels of urinary phenols. This phenomenon was observed in subjects working with benzene who had disproportionately high levels od the urinary metabolite compared to the degree of exposure (9).
Administration of cimetidine and propanol, which are well known inhibitors of microsomal drug metabolism, have a negligible influence on toluene kinetics in man, which is assessed by determining the levels of hippuric acid and-cresol in urine (6).
Since toxicological studies on animals and casual observations in man clearly point to the possibility that commonly used drugs can influence solvent metabolism, and vice versa, we believe that this question warrants more in-depth study.
Some data are available on the influence that genetic factors

may have on the levels of biological indicators. For example, an hereditary ALAD deficiency was found in a case of lead poisoning which was much more severe than would have been expected from the degree of exposure.

The frequency of the heterozygote form of hereditary ALAD deficiency is about 1% in the general population and in such subjects the levels of the enzyme are reduced by about 50% (5).

In lead-exposed subjects a study was made to ascertain whether erythrocyte protoporphyrin behaviour varies according to the degree of exposure in subjects with G6PD deficiency and in individuals with thalassemia minor (4).

Study of the relationship between PbB and EP levels showed that in subjects with G6PD deficiency, the behaviour of EP was similar to that observed in exposed workers not suffering from genetic disorders. However, in thalassemic patients, at the same PbB levels, the EP values were significantly lower (4).

Another situation of interest is that of the considerable individual differences in excretion of toluene metabolites observed in 24 workers who collected their urine samples 4-8 times during the 8 h workshift urinary hippuric acid and o-cresol were determined on the samples and the relationship between the two metabolites was studied. While the correlation coefficient of the regression lines for individuals were all significant, the slopes of the regression lines varied widely. In the Authors' opinion, these findings are due to the fact that an individual difference exists in the pattern of toluene metabolism and the ratio between aliphatic and aromatic oxidation is presumably established congenitally (8).

Acquired pathological states can influence the levels of biological indicators. The critical organ can undergo alterations both due to the specific action of the toxic agent and due to other reasons. An example of situations of the first type is the behaviour of urinary cadmium levels. In prolonged occupational exposure this test permits evaluation of exposure; however, when tubular disease due to cadmium sets in, urinary excretion increases markedly and the levels of the test no longer reflect the degree of exposure (9). Examples of the second type of situations are the increase in EP observed in sideropenic subjects who did not show any abnormal lead absorption, and the "normal" erythrocyte ALAD levels found in subjects with high reticulocytosis, even when there is high lead exposure (2).

The last topic we shall deal with concerns the physiological and pathological changes in the biological media, on which the biological indicators are determined, that can influence the test values.

Determination of biological indicators on urine poses a recurrent problem in occupational medicine. For practical purposes, only spot samples can be obtained from individuals during work, and the varying density of these samples means that the levels of the indicator can fluctuate widely in the course of the same day.

In order to overcome this difficulty, it is advisable to eliminate over-diluted or over-concentrated samples according to selected specific gravity or creatinine values. Several Authors also suggest adjusting the values of the indicators according to specific gravity and expressing the values according to creatinine (9).

Without entering into the discussion as to whether or not such adjustments are of use for all the indicators used in industrial toxicology, some observations that are closely connected with the topic of this paper are appropriate.

We know in fact that there is a high inter-individual variability of creatinine levels. For example, the values of the metabolite are lower in females than in males, and higher in young subjects than in the elderly (1). Therefore, if the values of the indicators are corrected for creatinine, reference and limit values which take account of these differences will be necessary.

Pathological factors can also considerably alter the biological media on which the indicators are determined. For example, in anaemic subjects exposed to metals (mercury, cadmium, lead, etc.) the blood levels of the metal may be lower than what would be expected on the basis of exposure; this is due to the reduction of red blood cells, which are the cells that transport the toxic in circulation. Therefore, when determinations of toxics or metabolites that are bound to red blood cells are made on whole blood it is always advisable to determine the haematocrit.

To conclude, it appears appropriate to emphasize that inadequate knowledge or insufficient assessment case by case of the situations that can affect the levels of biological indicators can lead to incorrect evaluations with false positive or false negative results.

It is a well known fact that in carrying out biological monitoring programmes it is indispensable that the percentage of false negatives be minimal; however, a high percentage of false positives leads to a high number of errors in evaluation which can create undue alarm and even lead to unnecessary measures being taken, such as removal from exposure, which have a negative effect on work organization and worker competence.

In this paper we have discussed only those tests that are fairly specific for the toxic under study.

Factors of non-occupational origin can have an even greater effect on the levels of indicators that are not specific. For example, the tests which are capable of revealing an influence on the microsomal enzyme functions, which thus indicate, in a non-specific manner, that an effect is taking place. On the other hand, special attention should be addressed to the study of the non-occupational factors that can influence tests which will probably be widely used in the near future. Of these, particular importance will be assigned to the indicators which have the potential of detecting subtle alterations induced in the target molecules by mutagens and carcinogens, for example, by measurement of DNA adducts.

REFERENCES

1. ALESSIO L., BERLIN A., DELL'ORTO A., TOFFOLETTO F., GHEZZI I. (1985): Reliability of urinary creatinine as a parameter used to adjust values of urinary biological indicators. Int. Arch. Occup. Envir. Health 55: 99-106.
2. ALESSIO L., FOA' V. (1983): Lead, in: Human Biological Monitoring in Industrial Chemicals Series. Alessio L., Berlin A., Roi R., Boni M. eds. Commission of the European Communities, EUR 8476 EN, Luxemburg, 105-106.
3. Castoldi M.R., Calzaferri G., Odone P., Dell'Orto A., Zocchetti C., Alessio L.(1983): Behaviour of cadmium biological indicators in subjects living in Milan area. Med. Lav. 74: 442-452.
4. Cherchi P., Carta P., Anni M.S., Alessio L., Giacomina C., Casula D. (1985): Monitoraggio biologico di lavoratori esposti al piombo portatori di alterazioni genetiche eritrocitarie. Atti del 48° Congresso Nazionale della Società Italiana di Medicina del Lavoro e Igiene Industriale, Pavia settembre 1985, Monduzzi ed., Bologna, 203-213.
5. Doss M., Laubenthal F., Stoeppler M. (1984): Lead poisoning in inherited δ-aminolevulinic acid dehydratase deficiency. Int. Arch. Occup. Environ. Health, 54-55.
6. Dossing M., Baelum J., Hansen S.H., Lundqvist G.R. (1984): Effect of ethanol, cimetidine and propanolo on toluene metabolism in man. Int. Arch. Occ. Env. Health 54:309-315.

7. Foà V., Colombi A., Maroni M. (1984): The speciation of the chemical forms of arsenic in the biological monitoring of exposure to arsenic. Science Tot. Env. 34: 241-259.
8. Hasegawa K., Shiojima S., Koizumi A., Ikeda M. (1983): Hippuric acid and o-cresol in the urine of workers exposed to toluene. Int. Arch. Occup. Env. Health 52:197.
9. Lauwerys R.R. (1983): Guidelines for biological monitoring, Biomedical Publications, Davis, California, p. 17-21.
10. Riihimaki V., Salvalainen K., Pfaffli P., Peakari K., Sippel H.W., Laine A. (1982): Metabolic interaction between m-xilene and ethanol. Arch. Toxic. 45: 253-263.
11. Sato A., Nakajima T., Koyama Y. (1980): Effects chronic ethanol consumption on hepatic metabolism of aromatic and chlorinated hydrocarbons in rats. Brit. J. Med. 37:383-386.
12. Waldron H.A., Cherry N., Johnston J.D. (1985): The effects of ethanol on blood toluene concentrations. Int. Arch. Occup. Environ. Health 51: 365-369.
13. Wilson H.K., Robertson S.M., Waldron H.A., Gomperez D. (1983): Effect of alcohol on the kinetics of mandelic acid excretion in volunteers exposed to styrene vapour. Brit. J. Med. 40: 75-80.

5

An x-ray fluorescence, XRF, technique to assay *in vivo* body burdens of heavy metals

P. Bloch (1), I.M. Shapiro (2) – University of Pennsylvania, School of Medicine, Department of Radiation Therapy, (on Sabbatical at Oxford University, UK. Physical Chemistry Laboratory) (1), School of Dental Medicine, Department of Biochemistry (2), Philadelphia, Pa 19104

ABSTRACT

The usual measure of heavy metal exposure of an individual is by assessing its concentration in body fluids, hair, nail clippings, excreta or by evaluating specific proteins affected by the metal contaminant. All of these assays depend on the mode of intake and excretion, its rate of absorption by various tissues, and the time elapsed since the heavy metal exposure. A determination of the cumulative concentration of the heavy metal in select tissues such as teeth, bone and brain may offer a measure of the total body burden. We have developed an XRF technique that permits assaying multiple heavy metals in these tissues in situ.

Using the XRF technique it was found that the tooth lead levels in 300 children from an urban setting (Phila.) were significantly higher than the levels found in 200 children from a more rural setting (Bennington, Vt). The XRF technique was also used to assay mercury levels in head and bone tissues in situ in approximately 300 dentists and 200 dental auxiliaries who were involved with Hg containing amalgams in their dental practice. This assay demonstrated that approximately 15% of the dentists had detectable Hg in brain tissue (head) whereas the dental auxiliaries had much lower mercury levels. Recent improvements in mercury hygiene and the younger population of dental auxiliaries compared to the dentists could account for these differences.

X-ray fluorescence can also be used to investigate the clearance from the blood of an administered iodine contrast agent such as iothalamate. Iothalamate is filtered from the blood by the renal glomeruli, thus its clearance can be used to assess the glomerular filtration rate, GFR, of the kidneys. This may be a useful procedure to be used on individuals suspected of exposure to nephrotoxic agents.

INTRODUCTION

Exposure to high concentration of heavy metals such as lead, mercury, arsenic etc are known to cause acute clinical disease. At low exposure to these metals more subtle chronic long term health deficits appear. For example, low levels of airborne lead exposure result in changes in renal, hematologic, and neurological systems.[1,2] The clinical symptoms express themselves in irritability, fatigue, skeletal pain, colic, headache, depression and apathy. In addition, chronic low level mercury in dentistry has been demonstrated to be associated with increased incidence of polyneuropathies and, mild visuographic dysfunction.[3,4]

The diagnosis of disease from chronic low level toxic metal exposure is difficult, both because an accurate assessment of the heavy metal burden is lacking, and the relationahip between body-burden of metals and various organ function have not been evaluated with precision.

The body-burden of heavy metals in vivo is generally deduced indirectly. For example, the concentration of mercury in blood or urine is accepted as a measure of mercury exposure. The measure of lead exposure is usually obtained from either the blood lead levels or deduced from lead-sensitive metabolic intermediates associated with hemoglobin synthesis and degradation such as free erythroporphyrins or zinc protoporphyrins.[5,6] The body fluid concentration and/or the appropriate porphyrin concentration give information on the circulating heavy metal concentration in the body at the time of measurement, which is important for evaluating acute metal toxicity. However little information is obtained from these assays concerning total cumulative metal exposure, which is necessary for evaluating low level chronic metal toxicity. To obtain the cumulative history of a metal exposure such as lead or mercury, it is necessary to sample tissues that both accumulate and permanently store the metal.

Both primary and permanent teeth have been shown to sequester lead with the total tooth lead concentration being proportional to the quantity of ingested lead.[7-10] In addition, a close correlation between bone and tooth lead levels _in situ_ has been observed[11] indicating that metals can be immobilized for long periods of time in the bone matrix. There are also indications that neurological or brain tissue could become a reservoir for some heavy metals, for example, mercury.[12,13]

An X-ray fluorescence (XRF) technique has been developed to measure heavy metal concentrations in teeth, bones and soft tissues _in situ_.[13-15] In addition, the XRF technique permits simultaneous evaluation of multiple heavy metals. This could be important information to gather since an exposure to a number of toxic metals may produce a greater deleterious health effect than an exposure from a single metal. To study this type of synergism, information needs to be accumulated on the body burdens of multiple heavy metals in individuals occupationally or environmentally exposed.

The XRF method is both an important research tool for determining body burdens of metals as well as providing data to assist clinicians in evaluating individual cases suspected of metal toxicity. In addition, when chelation therapy is required, the XRF technique permits a fast non-invasive method of monitoring the reduction of the body store of the heavy metal.

The XRF technique can also be used to measure the clearance of an administered non-radioactive iodine labelled substance. In studies of patients with chronic renal disease it has been demonstrated that the clearance of radioactive iodine labeled iothalamate provides information on the glomerular filtration rate, GFR, of the kidneys.[16] The clearance of non-radioactive iothalamate has been measured in rabbits using the XRF technique[17]. The assessment of glomerular function by the clearance of non-radioactive iothalamate, which is a contrast media used in many radiographic studies, is a relatively simple technique for measuring the GFR. Thus it could be a useful procedure for evaluating kidney clearance capability in individuals exposed to nephrotoxic agents. Data will be presented showing that changes in kidney clearance rate are detectable within 72 hours, using XRF, after the administration of a nephrotoxic dose of the chemotheraputic agent cisplatin[18]. In addition preliminary data on individuals with histories of elevated lead blood levels who also had diagnosed kidney disease showed prolonged clearance rates of iothalamate[19].

Because of the multiple applications of X-ray fluorescence the mechanism and technique will be described.

X–RAY FLUORESCENCE TECHNIQUE

X-ray fluorescence occurs when an inner shell electron vacancy is filled by the transition of a higher order orbital electron. The X-ray energy emitted corresponds to the difference in binding energies of the inner and outer shell electrons. This energy difference is unique for an element. For example, for lead, the filling of a K shell vacancy by L shell electrons results in two characteristic X-ray energies at 72.8 and 75.0 Kev, whereas for mercury the two characteristic X-ray energies for the same transition are 68.9 and 70.8 Kev. To produce an inner shell vacancy the atom is irradiated with X-rays of sufficient energy to knock out the tightly bound K-shell electron. This is referred to as the photoelectric effect. For lead, the K shell binding energy is 88 Kev. To supply this energy the tissue is irradicated with 122 and 136 Kev gamma rays from a 10mCi Co-57 point source. The sealed Co-57 source is housed in a cylindrical tantalum shield 16 mm in diameter and 15 mm long. A borehole 4 mm in diameter and 6 mm deep in the tantalum shield serves to collimate the gamma ray source. At 2 cm from the source, the radiation beam is approximately 12 mm in diameter.

Since the concentration of heavy metals in tissue is small compared to the low atomic number elements, H, C, N and O, most of the incident gamma rays will interact with the outer shell electrons of these elements to produce a broad energy spectrum of Compton scattered X rays. Figure 1 shows the measured XRF spectrum between 25-31 and 68-76 Kev from both a 3 ml water sample and the same volume of water containing 500 ppm of lead, mercury and iodine. The spectra demonstrates that the XRF from the metal atoms can be separated from the broad Compton scattered spectrum due to water. To achieve the required separation of Compton scattered X-rays, from the XRF fluorescence from the metal, a detector with good energy resolution is required. We have used an intrinsic germaniumm detector having a cross-sectional area of 110 mm^2 and a thickness of 5 mm (Princeton Gammatech Model IG 110-7). The detector is operated at liquid nitrogen temperature. The magnitude of the pulse from the Ge-detector is proportional to the energy of the X-ray absorbed. The pulse height distribution from the irradiated sample is obtained by sorting the signals from the detector in a 8192 channel pulse height analyzer (Tracor Northern).

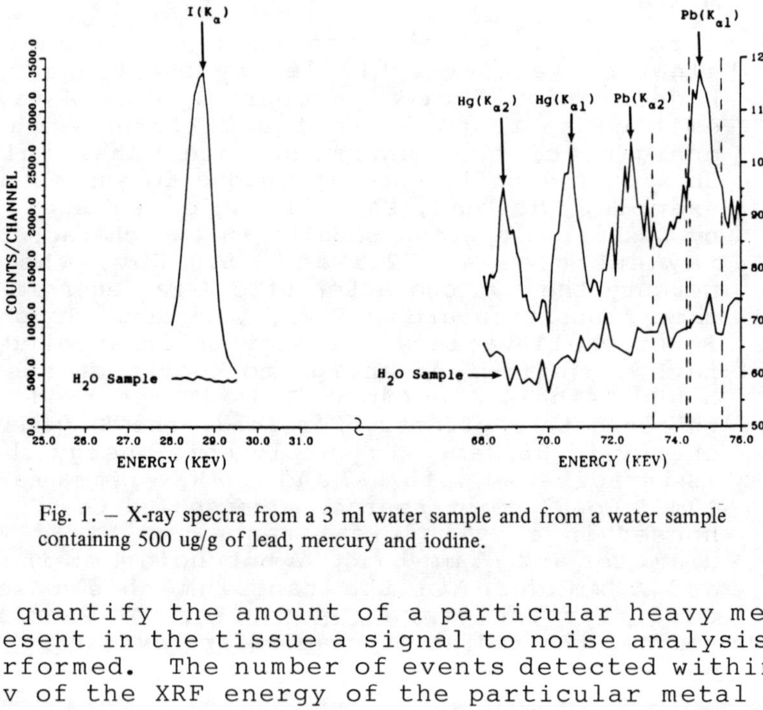

Fig. 1. – X-ray spectra from a 3 ml water sample and from a water sample containing 500 ug/g of lead, mercury and iodine.

To quantify the amount of a particular heavy metal present in the tissue a signal to noise analysis is performed. The number of events detected within .5 Kev of the XRF energy of the particular metal for example for lead (74.5 - 75.5 Kev) is taken to correspond to the signal (XRF from the metal) plus noise (Compton scattered X-rays from tissue). The number of scattered X-rays (noise) recorded in the fluorescent peak are estimated by simultaneously recording the number of events in an adjacent energy interval (73.3 - 74.3 Kev). It has been shown that the signal to noise ratio is directly proportional to the concentration of heavy metals present.[13,14]

The sensitivity of the XRF technique in detecting a small concentration of a heavy metal in tissue is statistically limited by the number of events recorded in the fluorescent window. For tissue samples irradiated in a test tube a large number of events are recorded. It has been demonstrated that an *in vitro* XRF assay results in sufficient precision to detect, with 70% confidence, the presence of 3 ug/g of mercury.[13] For measurement of heavy metal content in tissues *in situ* however the number of counts recorded in the fluorescent window is much reduced in order to minimize the gamma ray exposure to the tissue. For example, we irradiate, using the Co-57 source approximately 1 cm^2 of area with an exposure of approximately 60 mR to assay lead or mercury levels *in situ*. This exposure is less than 1/10 of a dental radiograph. Using this exposure permits detection with approximately 70% and 95% confidence the presence of 20 ug/g and 30 ug/g of mercury respectively.

The XRF technique has been used to evaluate in tissues *in situ* the mercury exposure in the dental profession due to handling mercury containing amalgam and the lead burdens in children. These are briefly summarized below.

MERCURY EXPOSURE IN THE DENTAL PROFESSION

The XRF technique was used to evaluate the mercury content in the temporal region and the wrist of 298 dentists[13] and 207 dental auxiliaries. It was found that 70% of the dentists and 89% of the dental auxiliaries had mercury burdens below the detection limit (20 ug/g) of the XRF. Thirteen per cent of dentists and 6% of the dental auxiliaries had temporal mercury levels in excess of 40 ug/g. The lower concentration of mercury in the dental auxiliary group (mean age 34) compared with the dentists studied (mean age 59) might be explained by the greater number of years that dentists had handled the mercury containing amalgams. The lower mercury concentration in the dental auxiliaries may also represent recent improvements in mercury hygiene.

LEAD BURDENS IN CHILDREN

Lead is accumulated by calcified tissues, thus bone or tooth lead concentration can be used to indicate lead intake Since the lead levels in shed teeth are elevated in children living in urban lead polluted environments[7] we selected to assay lead levels in teeth *in situ*. Recent data indicates a close correlation between tooth and bone lead levels.[11] For ease of setup we now routinely evaluate lead levels in bone in preference to teeth.

The XRF technique was used to assay tooth lead levels in 300 urban Philadelphia children and 200 children from a more rural setting in Bennington Vt. Six per cent of the children tested in Philadelphia had in excess of 20 ug/g[14] while no children from Bennington Vt. had detectable levels (less than 20 ug/g). This result might be expected, since the airborne lead, due to the use of leaded gasoline, would be higher in an urban setting.

POTENTIAL APPLICATION OF XRF TO MEASURING RENAL DYSFUNCTION

The XRF technique can be used to measure the clearance from the body of an administered iodinated molecule. The clearance of iothalamate, creatinine and/or inulin are cleared from the blood by the renal glomeruli with little tubular re-absorption. Thus clearance of the iothalamate is a measure of the glomerular filtration rate, GFR, of the kidneys. Fig.2 shows the clearance curve of iothalamate in a dog after administration of 2cc of iothalamate -29% iodine. The detector was placed on the dog's nose and the iodine content measured as afunction of time. The GFR determined using inulin was within 5% of that obtained from the measured clearance curve for the dog data shown in the Figure. The XRF technique did not however require the collection of urine as is necessary when creatinine or inulin clearance is used.

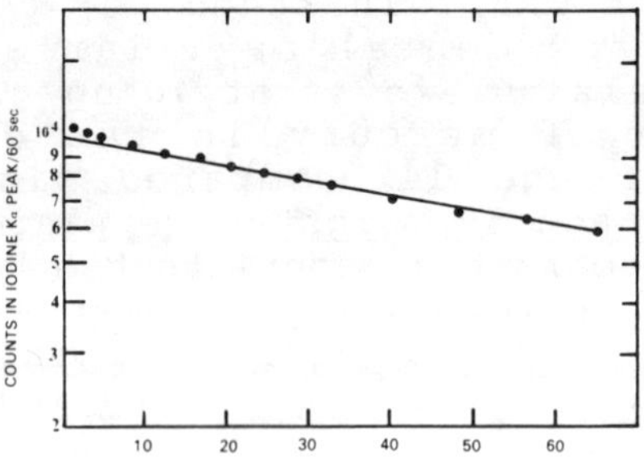

Fig. 2. – Clearance of iodine from a dog after the administration of 2 cc of iothalamate – 29% iodine.

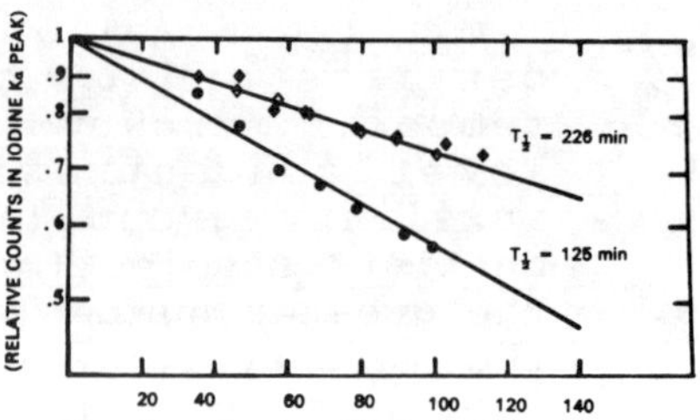

Fig. 3. – Clearance of iothalamate in a rat after administration of 5 mg/kg cisplatin. ● Pre-crisplatin cisplatin injection, ♦ 3 days post and ◇ 6 days post cisplatin.

The change in renal function in a rat could be demonstrated, after the administration of cisplatin which is a chemotherapeutic agent known to be nephrotoxic at high doses.[18] Approximately 0.5 cc of iothalamate was injected in a rat's tail vein and the clearance of the iodine contrast agent detected by positioning the tail near the XRF probe. The rat was then given cisplatin, 5 mg/kg, and the study repeated 3 and 6 days post injection. Figure 3 shows that the kidney clearance rate changed from a half time of 122 minutes prior to cisplatin administration to 226 min. 3 days after its administration. No further reduction was detected 6 days post cisplatin administration.

As a final example ten patients with long histories of elevated blood lead levels and known kidney impairment were administered 3-10 ml of 600 mg/ml iothalamate solution, approximately 10-20% the quantity used for an IVP. Rather than measure the iodine content <u>in situ</u>, the patients were catherized and 1 ml blood samples obtained at 15 minute intervals up to 90 minutes post injection. Whole blood and/or plasma was analyzed for iodine using XRF. The half life for clearance was found to range between 143 and 833 minutes.[19] Normal clearance is approximately 90 minutes.

More elaborate nuclear medicine techniques can be used to measure kidney function. For example, the clearance rate of injected radioactive Tc99m DTPA (diethylenetriamine-pentaacetic acid) is measured and differential information on the glomerular filtration rate for each kidney determined.[20,21] The XRF technique gives more restrictive information, however because it is cheaper to perform it might have application for screening of individuals suspected of exposure to nephrotoxic agents.

SUMMARY

An X-ray fluorescence technique has been described that can be used to measure multiple heavy metal concentration in tissue <u>in situ</u>. With an X-ray exposure of 60 mR to an area of approximately 1 cm^2 the presence of 20 ug/g of lead or mercury could be detected. The technique has been used to determine the body burden of lead in over 500 children. It has also been used to assess the mercury burden in nearly 500 dentists and dental auxiliaries who handled mercury containing amalgams in their dental practice.

The XRF technique can also be used to measure the clearance from the blood of an iodine labelled substance. Preliminary data was presented that indicated that the glomerular filtration rate of the kidneys could be deduced from the measured clearance of iothalamate. This technique may have potential in evaluating the renal function of individuals suspected of exposure to nephrotoxic agents.

REFERENCES

1) Batumen, V., Landy, E., Maesaka, J.K., et al. (1983) Contribution of lead to hypertension with renal impairment. N. Engl. J. Med. 309: 17-21

2) Lilis, R., Gavrilescu, N., Nestorescu, B., et al. (1968) A nephopathy in chronic lead poisoning. Br. J. Ind. Med. 25: 196-202

3) Shapiro, I.M., Cornblath, D.R., Sumner, A.J., et al. (1982) Neurophysiological and neuropsychological function in mercury exposed dentists. Lancet 8282: 1147-50

4) Iyer, K., Goodgold, J., Eberstein, A., Berg, P. (1976) Mercury Poisoning in a Dentist. Arch. Neurol. 33: 788-790

5) Sassa, S., Granick, J.L., Granick, S., et al. (1973) Studies in lead poisoning. Microanalysis of erythrocyte protoporphyrin levels by spectrofluorometry in the detection of chronic lead intoxication in the subclinical range. Biochem. Med. 4: 119-123

6) Lamola, A.A. and Yamane, T. (1974) Zinc Protoporphyrin in erythrocytes of patients with lead intoxication and iron deficiency anemia. Science N.Y. 186: 936-938

7) Needleman, H.L., Tuncay, O.and Shapiro, I.M. (1972) Lead levels in deciduous teeth of urban and suburban american children. Nature, Lond. 235: 111-112

8) Strehlow, C.D. (1972) The use of deciduous teeth as indicators of lead exposure. (Ph. D. Thesis) New York University

9) Needleman, H.L., Davidson, I., Sewell, E.M. and Shapiro, I.M. (1974) Subclinical lead exposure in Philadelphia school children: Identification by dentine lead analysis. New Engl. J. Med. 290: 245-249

10) Shapiro, I.M., Mitchell, G., Davidson, I. and Katz, S.H. (1975) The lead content of teeth: Evidence establishing new minimal levels of exposure in a living pre-industrialized human population. Arch. Environ. Health 30: 483-486

11) Bloch, P. and Shapiro, I.M. (1986) An X-ray fluorescence technique to measure in situ the heavy metal burdens of individuals exposed to these elements in the workplace. Journal of Occupational Medicine. IM Press

12) Shapiro, I.M., Cornblath, D.R., Sumner, A.J., Uzzel B., Spitz, L.K., Ship, I.I. and Bloch, P. (1982) Neurophysiological and neutropsychological function in mercury-exposed dentists. Lancet 1147-1150
13) Bloch, P., Shapiro, I.M. (1981) An X-ray fluorescence technique to measure the mercury burden in dentists _in vivo_. Med. Phys. 8: 308-311
14) Bloch, P., Garavaglia, G., Mitchell, G. et al. (1976) Measurement of lead content of children's teeth _in situ_ by X-ray fluorescence. Phys. Med. Biol. 20: 56-63
15) Ahlgren, L., Mattson, S. (1979) An X-ray fluorescence technique for the _in vivo_ determination of lead concentration in a bone matrix. Phys. Med. Biol. 24: 136-145
16) Tessitore, N., Schiairo, C.L., Corgnati et al. (1979) I^{125} - Iothalamate and creatinine clearances in patients with chronic renal disease. Nephron. 24: 41-45
17) Gronberg, T., Almen, T., Golman, K., Liden, K., Mattson, S. and Sjoberg, S. (1981) Non-invasive estimation of kidney function by X-ray fluorescence analysis. Method for _in vivo_ measurements of iodine containing contrast media in rabbits. Phys. Med. Biol. 26: 501-506
18) Safirstein, R., Miller, P., Dikman, S., Lyman, N., Shapiro, C. (1981) Cisplatin nephrotoxicity in rats: defects in papillary hypertonicity. Am. J. Physiol. 241 (2): F175-185
19) Bloch, P. and Baron, J. (1985) An X-ray fluorescence technique to determine glomerular filtration rate by clearance of iothalalmate from plasma. Med. Phys. 12: 512
20) Gates, G.F. (1982) Glomerular filtration rate: Estimate from fractional renal accumulation of Tc^{99m} DTPA (Stannous) Am. J. Roentgenal 138: 565-570
21) Gates, G.F. (1983) Split renal function testing using Tc^{99m} DTPA: A rapid technique for determining differential glomerular filtration. Clin. Nuc. Med. 8: 400-407

6

The immune response as a biological indicator of exposure

M.H. Karol, P.S. Thorne, J.A. Hillebrand – Department of Industrial Environmental Health Sciences, Graduate School of Public Health, University of Pittsburgh, Pittsburgh, PA 15261

ABSTRACT

The immune response is known for its qualities of exceptional sensitivity and specificity. These characteristics can be exploited by using the response as a biological indicator of exposure. The potential of this procedure was explored using an animal model in which exposure to the environmental agent was either through the inhalation route, or by topical exposure. No adjuvants, injections or other invasive means were used to stimulate the immune response. For inhalation exposures, toluene diisocyanate (TDI) was used as the hapten. Groups of English smooth-haired guinea pigs were exposed for 3 hours per day on 5 consecutive days to TDI vapor in concentrations of 0.12 to 10 ppm. The antibody response, measured twenty-two days following the initial exposure, indicated a concentration-dependent response. For dermal exposures, doses of TDI or diphenylmethane diisocyanate (MDI) were applied to the abdomen of BALB/cBy mice. Four days later, contact sensitivity was assessed by ear challenge. For both isocyanates, a dose-dependent increase in ear thickness was noted at 24 hours. These results indicate that several immunologic responses would be appropriate for biologic monitoring to indicate extent of exposure. The responses are highly specific, being based upon immunologic recognition. It is recommended that studies by conducted to determine the sensitivity of the immunologic response compared with other biologic measures of exposure to highly reactive industrial chemicals.

INTRODUCTION

Biologic monitoring encompasses both analytic measurement and measurement of the biologic response to known amounts of an agent. In recent years, interest has increased in the area of response to a chemical. In this situation biologic monitoring would represent a warning level of exposure to the chemical whether the exposure was by inhalation, ingestion or absorption through the skin. Currently, biologic exposure indices have been proposed by the American Conference of Governmental Industrial Hygienists for ten chemicals. In each case, analytical measurement is recommended, not the biologic response to the chemical.

With certain highly reactive organic chemicals, determination of the amount of chemical in a biologic fluid would be extremely difficult. An example of such a chemical is toluene diisocyanate a component of polyurethanes. TDI is not only self-reactive but may also hydrolyze, or react readily with hydroxyl, sulfhydryl or amino groups on proteins. Efforts were undertaken recently to determine if biologic monitoring for exposure to TDI could be accomplished using an immune response. Established animal models were utilized for inhalation (Karol *et al.*, 1978, 1980; Karol, 1983) and dermal exposure (Stadler *et al.*, 1985), routes appropriate to workplace or environmental situations.

INHALATION EXPOSURE

Groups of guinea pigs were exposed to TDI vapors for 3 hours per day on 5 consecutive days in concentrations ranging from 0.12 ppm to 10 ppm. On day 22, blood was drawn from animals and evaluated for anti-TDI antibodies using the passive cutaneous anaphylaxis (PCA) technique with a 6 hour latent period between intradermal injection of serum and intravenous administration of antigen (TDI-guinea pig serum albumin, TDI-GSA). This procedure detects mainly the IgG_1 class of cytophilic antibody.

As indicated in Figure 1, the antibody response was concentration-dependent. At concentrations of 0.25 - 1 ppm, the response increased with increasing TDI concentration. A threshold level was observed at 0.12 ppm since at this concentration or lower (0.02 ppm, data not shown) no response was obtained. Additionally, a plateau of responses was observed between 1-10 ppm.

The potential of this system for biologic monitoring is apparent. However, to be of value, the range must be determined over which the response occurs in humans. Early studies with workers (Karol, 1981) indicated that antibody production occurred following high industrial exposures, as a result of spills or splashes. In those studies, the antibody response was observed to decrease with time from exposure. A similar finding was obtained in the animal system. This method of biologic monitoring, therefore, shows promise as being able to document an exposure to a highly reactive chemical when contact has occurred via the respiratory tract.

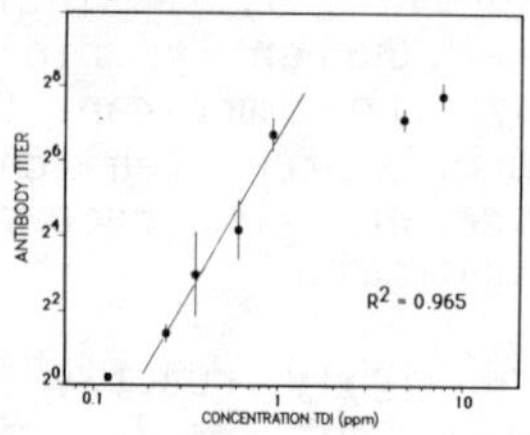

Fig. 1. – Antibody response in guinea pigs following inhalation of TDI. Animals were exposed to concentrations of TDI for 3 hr/day on 5 consecutive days. On day 22, sera were analyzed for cytophilic antibody using the PCA method with TDI-GSA as the antigen. Points represent the mean + S.E. of responding animals in each group (n = 5 to 12/group). (Data from Karol, 1983).

DERMAL EXPOSURE

Another frequent route of exposure to industrial and environmental agents is via skin contact. The immune response was explored for biologic monitoring following dermal contact to reactive industrial chemicals.

Dermal contact with small chemicals has traditionally been associated with development of contact sensitivity, a Type IV immunologic response. We have described a simple murine model for detection of contact sensitivity (Stadler and Karol, 1985). In the model, the immunologic response, measured as ear swelling 24 hours following challenge, was found to be dependent upon the exposure dose. Results obtained in this system using TDI and MDI are shown in Figures 2 and 3.

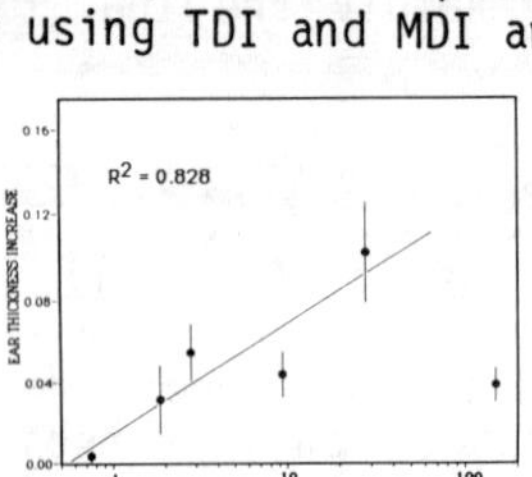

Fig. 2. – Mouse ear swelling test (MEST), Groups of BALB/cBy mice, 4 per group, were exposed on the abdomen to TDI. Four days later, contact sensitivity reactions were elicited by application of 100 ug TDI to the ear. Points indicate the mean ± S.E. for each group.

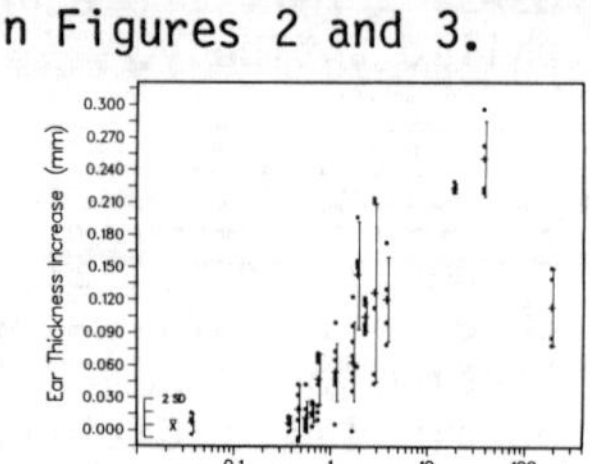

Fig. 3. – MEST for MDI. BALB/cBy mice were exposed on the abdomen to MDI. Individual responses are indicated to ear challenge performed 4 days later. Bar at lower left indicates response of control animals, exposed to acetone (vehicle and challenged on the ear with 200 ug MDI.

For TDI, the immunologic response was found to be dose dependent over the range 1-30 mg/kg. Lower doses failed to produce a response, whereas higher doses, in excess of 30 mg/kg, caused a submaximal response. The same pattern of response was observed using MDI (Figure 3), i.e., a threshold effect, a region of dose-response reactions, and reduced responses at high doses.

Both the contact sensitivity following dermal exposure and the antibody response which followed inhalation exposure demonstrated a temporal characteristic. Skin reactions were readily detected at 4-5 days following exposure, but could not be elicited 30 days after the exposure (Stadler and Karol, 1985).

DISCUSSION

The two models described above indicate that measurement of immunologic response offers promise as a tool for biologic monitoring of exposure to reactive chemicals, agents which typically pose difficulty for routine analytical procedures. The utility of the approach derives from the demonstrated concentration (dose)-dependence of these immunologic responses. The methods appear applicable for measurement of exposure to other chemical haptens, in addition to isocyanates. As indicated in Table 1, evidence has been obtained for dose-dependent immunologic responses to several other industrial chemicals.

Table 1: Evidence for a dose-dependent immunologic response to chemicals.

Chemical	Experimental Model	Response	Year
Picryl Chloride	Mouse	CS	1982
Kathon^{R} Biocide	Guinea Pig	CS	1983
Formaldehyde	Guinea Pig	CS	1984
Phenylenediamine	Human	CS	1974
Glutaraldehyde	Human	CS	1974
Dinitrochlorobenzene	Human	CS	1983
Dicyclohexylmethane Diisocyanate	Guinea Pig, Mouse	CS	1984
TDI, HDI, MDI	Mouse	CS	1986
Toluene Diisocyanate	Guinea Pig	Ab	1983
Subtilisin	Guinea Pig	Ab	1986

Abbreviations: CS, contact sensitivity; Ab, antibody production; TDI, toluene diisocyanate; HDI, hexamethylene diisocyanate; MDI, diphenylmethane,4'4'-diisocyanate.

Some caveats are required, however, before widespread adoption of this biologic monitoring procedure is encouraged. First, the dose-dependent response occurs over a limited range of doses (concentrations). Second, the response is temporal in nature. Third, individual variations exist in responses. Nevertheless, the benefits to be derived from the proposed monitoring scheme also deserve recognition. They are:

- o Standard procedures are used.
- o The response is highly specific.
- o The sensitivity appears to be equal to that of other biologic responses, i.e., irritation.

In conclusion, measurement of immunologic responses to reactive chemicals should be considered for purposes of biological monitoring. In animal systems, the responses are dose-dependent and occur following exposure protocols typical of those which occur in the occupational/industrial setting. Clinical reports, although limited, support the development of immunologic responses following chemical exposures.

ACKNOWLEDGEMENT

The studies were supported by grant ES01532 from the National Institute of Environmental Health Sciences.

REFERENCES

Karol, M.H. (1981). Survey of industrial workers for antibodies to toluene diisocyanate. J. Occ. Med. 23: 741-747.

Karol, M.H. (1983). Concentration-dependent immunologic response to toluene diisocyanate (TDI) following inhalation exposure. Toxicol. Appl. Pharmacol. 68: 229-241.

Karol, M.H., Dixon, C., Brady, M. and Alarie, Y. (1980). Immunologic sensitization and pulmonary hypersensitivity by repeated inhalation of aromatic isocyanates. Toxicol. Appl. Pharmacol. 53: 260-270.

Karol, M.H., Ioset, H.H., Riley, E.J. and Alarie, Y. Hapten-specific respiratory hypersensitivity in guinea pigs. Amer. Ind. Hyg. Assoc. J. 39: 546-556.

Stadler, J.C. and Karol, M.H. (1985). Use of dose-response data to compare the skin sensitizing abilities of dicyclohexylmethane-4,4'-diisocyanate and picryl chloride in two animal species. Toxicol. Appl. Pharmacol. 78: 445-450.

7

A quality control programme for biological monitoring of organic compounds in urine

S. Valkonen, A. Palotie, A. Aitio – Institute of Occupational Health, Laboratory of Biochemistry, Arinatie 3, SF–00370 Helsinki, Finland

BACKGROUND

Several studies have indicated that the quality of the analyses of toxic chemicals in body fluids is only poor to fair (see Aitio 1981). This is at least in part due to the fact that no continuous programmes exist for the quality control of such analyses. In recent years, the situation has improved markedly with regard to toxic heavy metals since several international programmes are now in operation. In addition, commercial reference materials for internal quality control are now available for trace element analyses.

However, there has been no permanent programme for the analyses of toxic organic chemicals in body fluids. In trying at least to some extent answer to this need a quality control programme for organic compounds in urine was started in 1979 by the Institute of Occupational Health in Finland.

Today altogether 34 laboratories in 13 different countries participate in the programme. So far 14 quality control rounds have been performed on five organic compounds in urine.

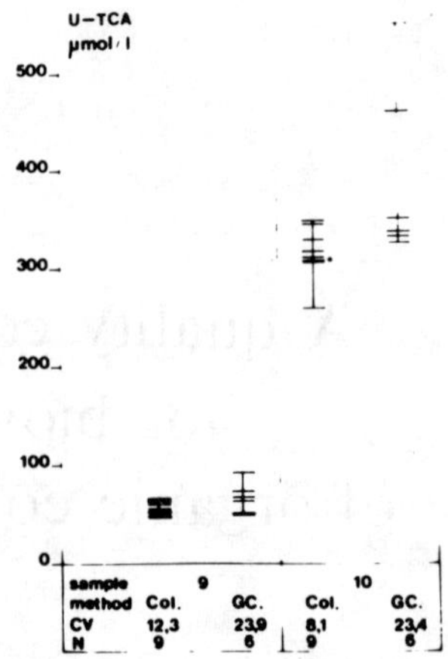

Fig. 1. – Coefficient of variation of the analysis of trichloroacetic acid in urine at two concentration levels in laboratories using colourimetric and gas chromatographic methods. The results of individual laboratories are indicated by horizontal bars.
The plots (.) beside the lines indicate the number of identical laboratories that have result.

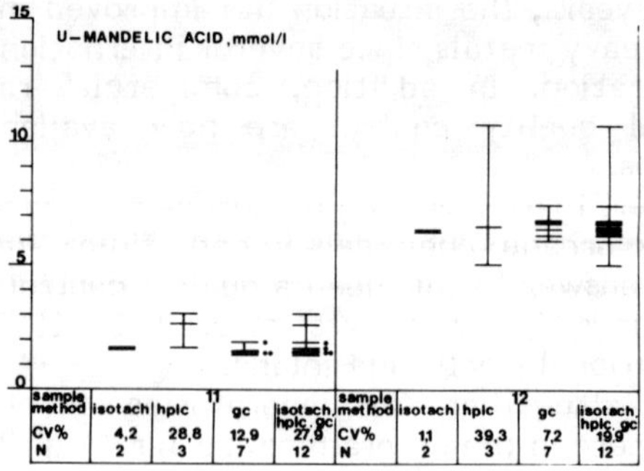

Fig. 2. – Coefficient of variation of the analysis of urinary mandelic acid in laboratories employing different methods. The results of individual laboratories are shown as horizontal bars.
The plots (.) beside the lines indicate the number of indentical laboratories that have result.

METHODS AND RESULTS

Until now, 4 rounds for urinary mandelic acid, 2 for urinary trichloroacetic acid, 6 for urinary chlorinated phenols and 2 rounds for urinary phenol have been performed. With the exception of the first phenol round, natural specimens pooled from exposed workers have been used.

The interlaboratory coefficient of variation has usually been between 10 - 25 % and sometimes even higher than this (fig. 1 - 4).

Since the number of laboratories performing any single analysis is small, few methodological comparisons have been possible.

In the case of trichloroacetic acid about two thirds of the laboratories used the colourimetric Fujiwara method and one third gas chromatographic methods. In the most recent trichloroacetic acid comparison round (performed in fall 1985), 15 laboratories in 9 countries participated (fig. 1). The coefficient of variation (CV %) of the laboratories using the colourimetric method (N=9) was 8.1 and 12.3 %, whereas the CV of the laboratories using gas chromatographic methods (N=6) was 23.9 - 23.4 % at concentration levels of about 60 and 340 µmol/L, respectively.

For the analysis of mandelic acid in urine, isotachophoretic, liquid chromatographic and gas chromatographic methods have been used (fig. 2) The interlaboratory coefficient of variation of the laboratories using gas chromatographic methods (N=7) was 12.9 % at lower concentrations and 7.2 % at higher concentrations. The overall coefficient of variation of all laboratories (N=12) was 28 % and 20 %, respectively. However, the number of laboratories in each subgroup is very small.

The main aim of external quality control is to improve the quality of the analyses. Unfortunately, no clear signs for such a tendency has been observed in this quality control programme. As an example the rounds of urinary mandelic acid are presented in figure 3. The coefficient of variation at different concentration levels has remained essentially the same, 10-25 % during the last three rounds. However, the values were over 60 % in the first round performed in 1979.

Two rounds performed with analysis of urinary phenol could serve as an exemple of possible improvement in analytical procedures achieved by external quality control (fig. 4).

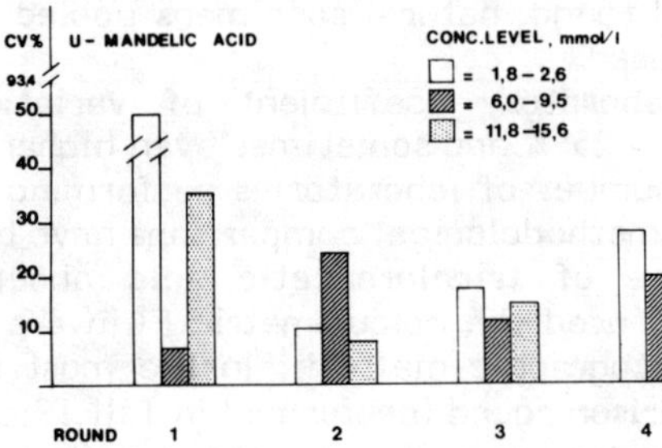

Fig. 3. – Coefficient of variation of the analysis of mandelic acid in urine in four consecutive quality control rounds, at 3 different concentration levels.

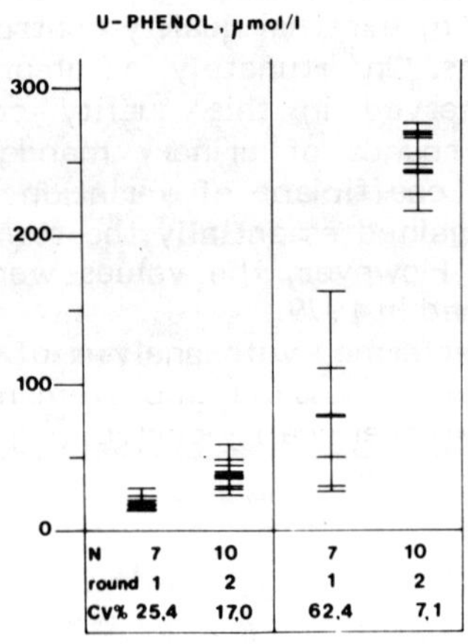

Fig. 4. – Coefficient of variation of the analysis of urinary phenol at two concentration levels in two consecutive quality control rounds. The results of individual laboratories are indicated by horizontal bars.

In the first phenol round where the urine samples were spiked with 3,7 µmol/L or 148 µmol/L phenyl-ß-D-glucuronide, the variation between the results achieved in different laboratories was very large. Later some laboratories discovered that the hydrolysis procedure they used probably was not quite severe enough. An incomplete hydrolysis evidently would result in erroneously low phenol concentrations. In the next round, when more drastic hydrolysis methods were used the coefficient of variation decreased markedly. However, use of natural samples obtained from exposed workers in the second round might also contribute to the difference: conjugates other than glucuronide, probably present in workers' urine might be more susceptible to hydrolysis. Anyway, this example emphasizes the importance of use of natural - or as close to that as possible - specimens in external (and internal !) quality control programmes.

CONCLUSIONS AND FURTHER PERSPECTIVES

These experiences clearly demonstrate the need for a permanent quality control programme for toxic organic compounds in urine. The interlaboratory variation coefficient is high and better uniformity in the analytical results is required. In spite of the small number of laboratories participating, some methodological comparisons and improvements have been possible. The quality control programme will be continued for urinary phenol, mandelic acid and trichloroacetic acid. In the near future also methyl hippuric acid will be included in the programme. More laboratories are wellcome to the programme.

REFERENCES

Aitio, A. (1981) Quality control in the occupational toxicology laboratory, WHO Regional Office for Europe, Copenhagen, 1 - 49.

8

Animal data and human solvent biomonitoring connected by a mathematical model

L. Perbellini, P. Mozzo*, F. Brugnone, A. Zedde – Institute of Occupational Medicine;* Physical Health Service, Policlinico Borgo Roma – 37134 Verona, Italy

Various mathematical-models simulating the uptake, distribution and excretion of volatile chemical products in the human body have been proposed on many occasions: to define the kinetics of gaseous anaesthetics (Fiserova-Bergerova 1983), to analyse the body distribution of antineoplastic drugs (Bischoff et al. 1971) and to explain the toxicokinetics of organic solvents used in industrial processes (Sato et al. 1977, Fernandez et al. 1977).

However, the results suggested by the mathematical models have seldom been confirmed by the results of human or animal experiments.

Recently, Droz (1985) suggested that mathematical pharmacokinetic models can be very useful when the "Biological Exposure Index" of solvents is to be established.

We confirm that these models give precise information on the concentration of solvents in the blood, in alveolar air, in fatty tissue and in the main body tissues, both during and after various levels of exposure. Moreover, interesting data on the kinetics of the solvent metabolites can be obtained; these products are widely used during the biological monitoring of occupational exposure to solvents.

We prepared a physiological-mathematical model with 8 compartments aimed at studying the pharmacokinetics of an organic solvent in the 5 main body compartments and

simultaneously the synthesis, distribution and urinary excretion of its metabolites.
A complete description of the model is published elsewhere (Perbellini et al. in press).
Studies on the connections between animal and human exposure to solvents have been carried out using mathematical models only on few occasions (Ramsey et Andersen 1984). It is possible that the pharmacokinetic differences between various animal species and between the human and animal species can be ingestigated using these models, the results of which could be very useful in occupational medicine.
Figure 1 reports the results of the styrene concentrations in arterial blood of rats if an exposure to 336 mg/m^3 for 6 hours is simulated. The solid line represents the styrene concentrations indicated by the mathematical model and the black dots the experimental data reported by Ramsey and Andersen (1984), when the rats were actually exposed to styrene. The data calculated using the mathematical model is similar to data obtained in the laboratory experiments.
Figure 2 reports the results of a simulated and a real exposure:the rats were exposed to 2520 mg/m^3 for 6 hours, the physiological parameters used in the mathematical model were the same as those used for exposure to 336 mg/m^3. The difference between the mathematical and experimental data is marked: this discrepancy is explained by the styrene biotransformation rate, which cannot be the same at very different exposure concentrations.
In rats, Hildebrand and Andersen (1981) found that the biotransformation rate of many industrial solvents was similar when the exposure is not excessive (near or lower than T.L.V.). In these conditions the metabolic capacity of the body cannot be saturated.
Figures 3 and 4 show the styrene kinetics in blood during 2 different human experimental exposures (Vigaeus et al. 1983, Ramsey and Andersen 1984). The mathematical simulation gives results similar to those obtained experimentally. It is therefore clear that we have inserted in the model the exposure parameters and the physiological conditions of the tested subjects reported by the Authors.
Furthermore the venous blood concentrations of n-hexane found experimentally by Veulemans et al. (1982) were well calculated by the mathematical model (Fig. 5).
Fig. 6 reports the n-hexane concentrations in venous blood and in alveolar air suggested by mathematical analysis when an exposure to 180 mg/m^3 for 8 hours is simulated. The alveolar

ventilation ranges between 6 and 15 l/min and the cardiac output ranges between 5 and 10 l/min. The data reported in Fig. 6 are confirmed by our previous investigations and provide interesting information on blood and alveolar limits of n-hexane during or after an exposure to the T.L.V.
The urinary kinetics of the solvent metabolites can be described in the same way.
The reported examples confirm that the physiological mathematical models provide much useful information about exposure to single industrial solvents. By studying the blood, alveolar, fat and tissue concentrations of the solvents and/or their metabolites during and after exposure, it is then possible to analyze the varying exposure levels, the effects of the workload and the body burden during the "ceiling" exposure. In addition, the mathematical models can be used for determining individual exposure to solvents: the main physiological parameters, solvent exposure and solvent concentrations in some biological samples permit analysis and correlation of all data collected on individual pharmacokinetics of the solvent.

REFERENCES

Bischoff K.B., Dedrick R.L., Zaharko D.S., Longstreth J.A.: Harm Sci 1971; 60: 1128-1133.
Droz P.O., Int. Symp. on Occup. Exposure Limits. A.C.G.I.H. Cincinnati, Ohio, 1985.
Fernandez J.G., Droz P.O., Humbert B.E., Caperos J.R. J. Ind.: Med. 1977; 34: 43-55.
Fiserova-Bergerova V.: Vol. I-II, 1983, CRC Press, Inc. Boca Raton, Florida.
Hildebrand R.L., Andersen M.E.: Toxicologist 1981; 1:86-87.
Perbellini L., Mozzo P., Brugnone F. Zedde A.: Br. J. Ind. Med. (in press).
Ramsey J.C., Andersen M.E.: Toxicol Appl. Pharmacol., 1984; 73: 159-175.
Sato A., Nakajima T., Fujiwara Y., Murayama N.: Br. J. Ind. Med. 1977; 34: 56-63.
Veulemans H., Van Vlem E., Jansses H., Masschelein R., Leplat A.: Int. Arch. Occup. Environ. Health 1982; 49: 251-263.
Vigaeus E., Lof A., Bjurstrom R., Byfalt Nordquist M.: Scand. J. Work Environ. Health 1983; 9: 479-488.

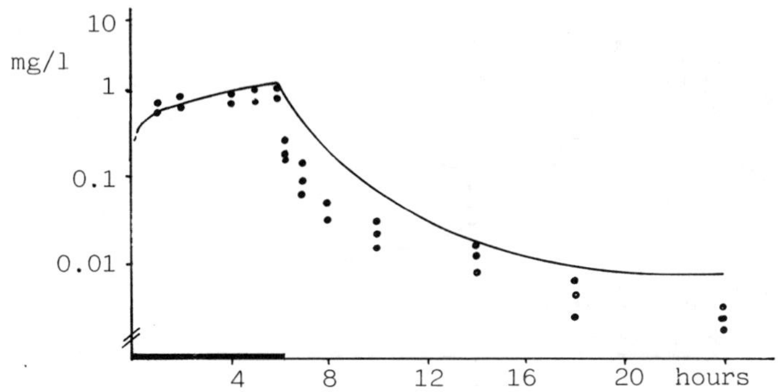

Fig. 1. – Styrene concentrations in arterial blood of rats exposed to 336 mg/m^3 for 6 hours. (Data from Ramsey and Andersen 1984). Solid line: data given by the mathematical model.

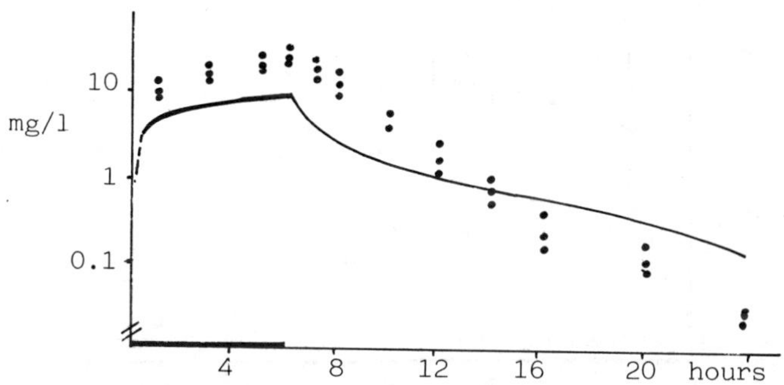

Fig. 2. – Styrene concentrations in arterial blood of rats exposed to 2520 mg/m^3 (see Fig. 1 and text).

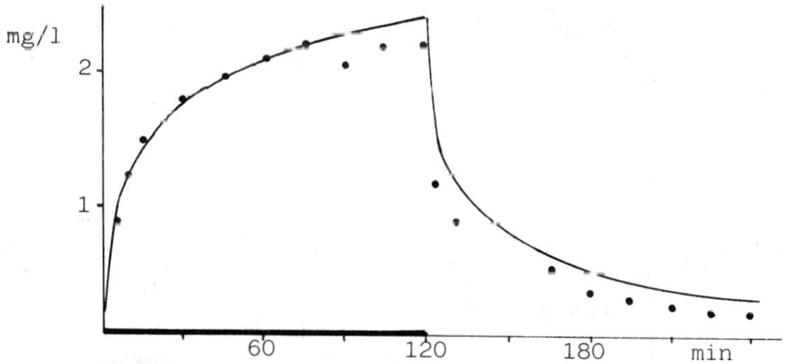

Fig. 3. – Styrene kinetics in human arterial blood during and after an exposure to 300 mg/m^3 for 120 min. (Data from Vigaeus et al. 1983). Solid line: data given by model.

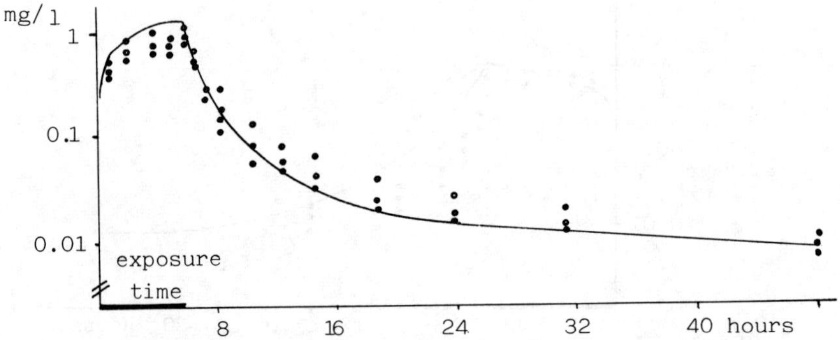

Fig. 4. – Styrene kinetics in human arterial blood during and after an exposure to 336 mg/m^3 for 360 min. (Data from Ramsey and Andersen 1984). Solid line: data given by model.

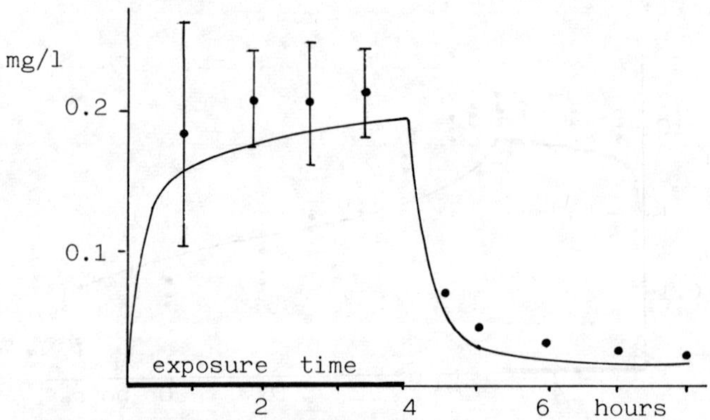

Fig. 5. – n-Hexane venous blood concentrations during and after an exposure to 360 mg/m^3 for 4 hours. (Data from Veulemans et al. 1982). Solid line: data suggested by the model.

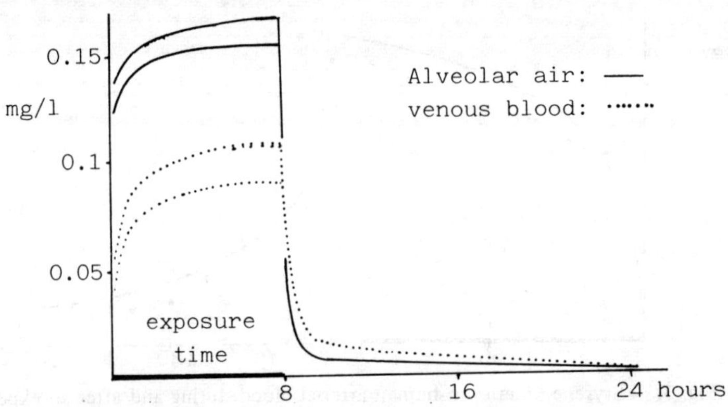

Fig. 6. – Alveolar and venous blood concentrations of n-hexane during and after a hypothetic exposure to n-hexane (180 mg/m^3 for 8 hours). See text.

9

Change of ceruloplasmin during silicosis

Y.R. Li, L.F. Liu, G.J. Chen, G.P. Cai, B.C. Liu, N.M. Chen – Institute of Occupational Medicine, Chinese Academy of Preventive Medicine.

INTRODUCTION

It has been reported that serum ceruloplasmin (Cp) content was elevated during silicosis (1). In China, serup Cp contents were estimated among a great number of silicotic patients (2). Values higher than those of normal persons were also obtained. It is of interest to clarify the cause of its elevation. In this paper, results of several studies on change of Cp in the process of experimental silicosis were reported: (I) Cp was found to be increased in the alveolar macrophage (AM) of silicotic rats as shown in the gel electrophoretic profile and from the results of radioimmunological assay. (II) By using indirect fluorescent immunological technique, distribution of Cp in the normal lung was observed only in the alveolar wall, while in the silicotic rat lungs, it spread along with the collagen fibers throughout the nodules. (III) Stimulation of collagen synthesis in fibroblasts by Cp was proved in 2BS diploid cells. The results indicated that Cp is an AM derived factor which may stimulate the process of fibrosis.

1. IDENTIFICATION OF Cp IN SUSPENSION OF AM

(a) Electrophoresis and immunodiffusion methods

Materials and methods

Human ceruloplasmin (Serva product)
Rat ceruplasmin, prepared with Jamieson's method (3)
Rabbit anti human Cp serum
Rabbit anti rat Cp serum
Alveolar macrophages from normal and silicotic rat lungs.
Rats were injected intratracheally with 40 mg of silica suspended in 1 ml of saline. After 20 days, the rats were killed. Lung washing cells were harvested by repeated washing of the lung with saline, centrifuged and homogenized. From normal lung, the washing contained 90% AM and from rats injected with silica, it contained 80% AM.
SDS-PAGE Neville's method (4)
PAGE Davis' method (5)
Immunodiffusion method.

Results

A protein band (B band) which was the main difference observed between the profiles of normal and silicotic cells, was appeared in SDS-PAGE profiles of suspension protein from AM of silica dusted rats (Fig. 1,2). For identification of B band protein, its relative mobility (Rm), staining characteristics and immunological properties were compared with those of Cp (Serva). There were 6 bands shown in the SDS-PAGE profile of standard Cp (Fig. 1). This was similar to those reported by Prozorovski (6). Rm of B band protein was 0.50 (M.W. 66KD) which corresponded to the F_3 band in the Cp profile. Both B band and F_3 of Cp can be shown when stained with p-phenylene diamine or O-anisidine which are specific dyes for Cp. P-phenylene diamine staining also revealed a protein band with Rm of 0.08 (M.W.20KD) from extraction of B band and from standard Cp (Fig. 2).

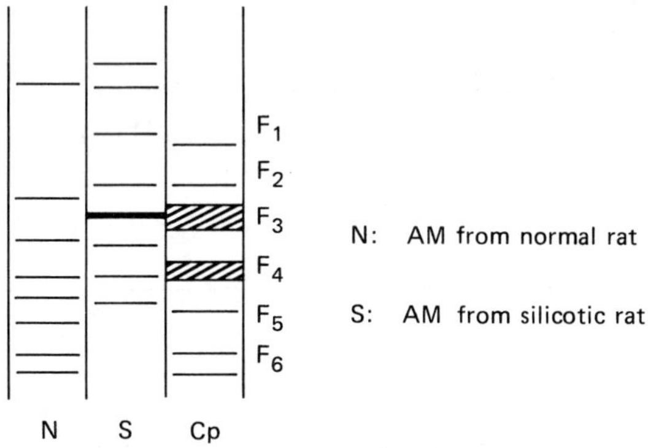

Fig. 1. – SDS–PAGE profile of rat AM.

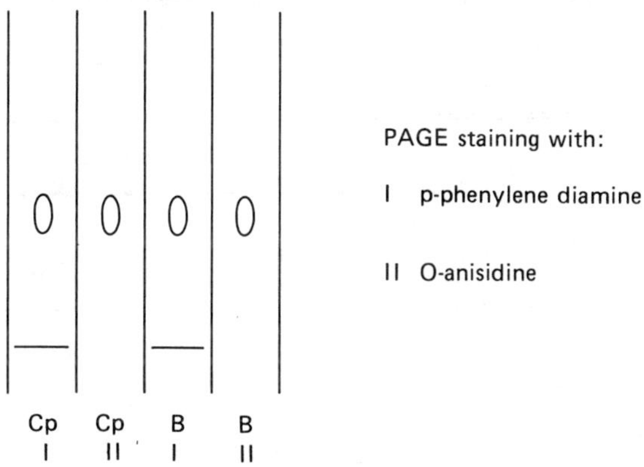

Fig. 2. – Specific staining for Cp and B Band.

In the immunodiffusion tests, B band protein was eluted from the gel by dissolving in Tris buffer and was allowed to react with the antibodies. We found that there was common antigenic property between rat Cp and human Cp, between human Cp and B band protein (Table 1 and 2).

Based on the results of electrophoretic behavior and immunological reaction of B band, B band protein was considered as a main part of Cp.

Table 1: Results of immunodiffusion test between rat or human Cp and the antibodies.

Antigen	Antibody	
	Rabbit anti rat Cp	Rabbit anti human Cp
rat Cp	+	+
human Cp	+	+

Table 2: Results of immunodiffusion test between Cp or B band protein and antibodies.

Antigen	Antibody	
	Rabbit anti human Cp	Rabbit anti B band protein
human Cp	+	+
B band protein	+	+

(b) Radioimmunological test for the prescence of Cp in AM

Materials and method

^{125}I labelled rabbit anti human Cp antibody was prepared with purified antibody and $Na^{125}I$ using chloramine T oxidation method. The labelled antibody was incubated with AM obtained from normal and the silica-injected rats. After 1 hour, the unreacted part of the labelled antiserum was washed out and centrifuged. CPM values of the cells were determined.

Results

In this experiment, CPM values was related to the amount of Cp antibody attached to the macrophages and was also related to the amount of Cp in the macrophages. The results showed that in normal AM, it contained Cp and in silicotic AM, Cp was increased to twice as much as those in the normal AM (Table 3). The increase of Cp was considered to be induced by the ingestion of silica dusts. The increase of Cp in silicotic AM has never been reported.

Table 3: Reaction of AM with ^{125}I-labelled anti-human Cp anti-body.

Kind of cell	Number of cell	CPM(average)
Normal AM rat	5×10^6	3002
AM from silicotic rats	5×10^6	6644

II. DISTRIBUTION OF Cp IN NORMAL AND SILICOTIC LUNG SPECIMENS

Indirect fluorescent immunological method

Rabbit antihuman Cp antibody was allowed to react with fresh lung specimens of normal and silicotic rats separately. Then FITC-labelled goat anti-rabbit IgG was added to perform the indirect fluorescent immunological reaction. After staining, the lung specimen was examined under a fluorescence microscope (Olympus). The prescence of green illumination of FITC would indicated the distribution of Cp.

Results

In normal AM, Cp was found to locate only in the alveolar wall, together with the layer of connective tissue. In silicotic rats, it was present along with the collagen fibers distributed throughout the silicotic nodules (Fig. 3, 4). The results confirmed that Cp was connected with collagen fibers structurally and was increased in silicotic lung.

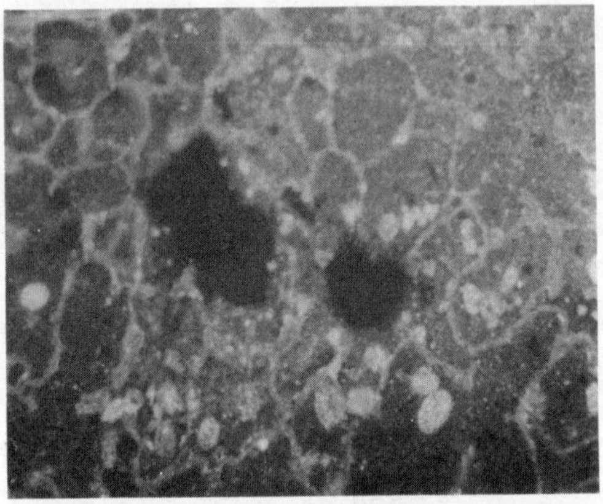

Fig. 3. – Cp in normal lung.

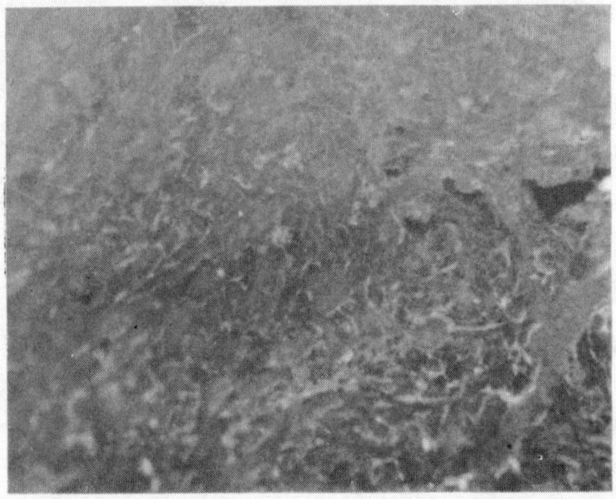

Fig. 4. – Cp in silicotic lung.

III. STIMULATION OF COLLAGEN SYNTHESIS IN FIBROBLAST BY Cp

Method

1. Cultivation of fibroblast

2BS cells was used. It was a strain of diploid cells obtained from embryonic lung by the Beijing Institute of Biological Products. The cells were subcultured for three days at 37°C in E 70 medium (MEM containing 10% FBS).

2. Addition of Cp and incorporation of ^{3}H-proline or ^{3}H-thymidine

On the 4th day of culturing, E 70 medium was discarded and replaced by fresh MEM medium which contained 10, 20, 30 or 60 mg/dl of Cp respectively. A control group of 2BS cells was cultured at the same time in a medium without Cp. 30 minutes later, 1 uc of ^{3}H-proline or ^{3}H-thymidine was added and incubated for another 24 hours. After incubation, the cells and the medium were separated by centrifugation. Noncollagenous proteins in the medium were eliminated by digestion with pepsin. ^{3}H-proline (including ^{3}H-hydroxyproline) in the cells and in the medium were estimated with FJ-353 G scintillator (Jackanan's method (7)). ^{3}H-TDR in DNA of the cells were also counted according to Aalto's method (8).

Results

Incorporation of ^{3}H-proline into cell proteins and collagen secreted in the media were examined. The results showed that incorporation of ^{3}H-proline into the cell proteins was 24.96% higher than in the control group (without Cp in the culture). Incorporation into collagen in the media was 102.8% higher than in the control (Fig. 5, 6).

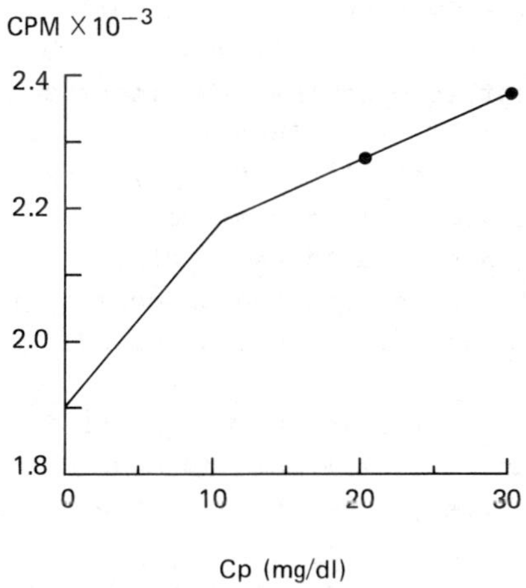

Fig. 5. – Incorporation of ^{3}H-proline into 2BS cells.

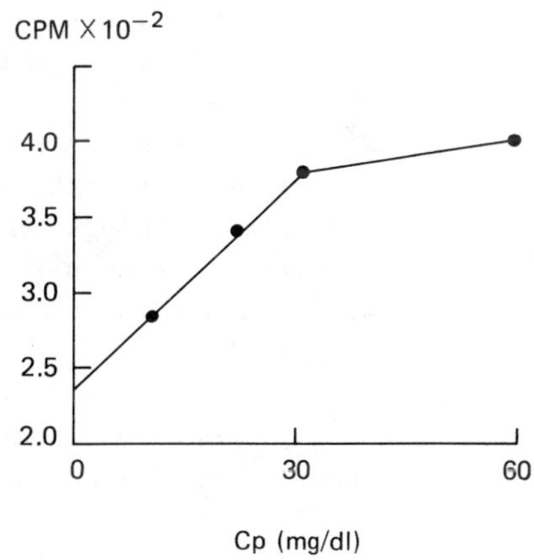

Fig. 6. – Incorporation of ^{3}H-proline into collagen in media.

CPM of ^{3}H-TDR incorporated into DNA was also raised with the increase of concentration of Cp. In the group with 60 mg/dl Cp, the result was 105.0% higher than in the control (Table 4).
The increase of collagen and proliferation of the cells were observed after stimulation with Cp in the fibroblasts.

Table 4: Incorporation of ^{3}H-TDR (CPM).

Group	CPM	Average	%
Cells without Cp	15321		
	12134		
	13452	13636	100
Cells with 60 mg/dl Cp	27850		
	26430		
	29571	27950	205

DISCUSSION

The role of ceruloplasmin has been known as an antioxidant agent (9). In this study, ceruloplasmin was found to be increased in AM after ingestion of silica. It may react as an antioxidant to inhibit the peroxidation caused by silica in AM. We also noticed that its distribution was closely connected with collagen fibers and it can stimulate the synthesis of collagen in fibroblasts. The results of our experiments proved that the increase of Cp was obviously related to the silicotic fibrosis process.

Cu^{++} has been known to be a cofactor of lysyl oxidase in catalyzing the cross-linking of collagen molecules. Cp is a copper containing enzyme which reacts as a diamine oxidase, therefore we considered that its role in the process of fibrosis may be mainly due to the Cu^{++} contained in its molecule and its oxidase activity.

According to Prozorovski (6), the molecular weight of Cp was 132 KD. It contained two subunits of M.W. 22 KD and 18 KD. The former formed a covalent polymer, therefore in the SDS-PAGE profile of Cp 6 bands could be seen (Fig. 1, F_{1-6}). B band from silicotic AM corresponded to F_3 band in Prozorovski's result. Rm was 0.50 and M.W. was 66 KD. It is a trimer to the subunit. An additional fragment from B band with Rm 0.80 and M.W. about 20 KD was usually noticed after storing. This fragment corresponded to F_6 of standard Cp (Fig. 5).

ABSTRACT

Ceruloplasmin was found to be increased in alveolar macrophages of silicotic rat lung both from the profile of SDS-PAGE and by radioimmunological assay. Using an indirect fluorescent immunological method, Cp was observed to distribute along with collagen fibers both in normal rat lung and in silicotic rat lung. Stimulation on synthesis of collagen caused by Cp was also proved in this study. On the basis of the above results, we considered that Cp is one of the alveolar macrophage derived factor, with fibrogenic effect.

REFERENCES

1. Fati S. et al. (1961): Behavior of Ceruloplasmin and Serum Copper in Pulmonary Silicosis and other Respiratory Diseases, Folia Med. 44: 731
2. Li Y.R. et al. (1972): Changes of Serum Ceruloplasmin and Lysozyme in Silicosis, Hygienic Research, 2:62 (in Chinese)
3. Jamieson G. A.(1972): Glycoprotein Part A, 672 Elsevier Publ. Co.
4. Neville Jr. D.M. et al. (1971): Molecular Weight Determination of Protein Dodecyl Sulfate Complexes by Gel Electrophoresis in a Discontinuous Buffer System, J. Biol. Chem., 246: 6328.
5. Davis B. J. (1964): Dis Electrophoresis, Annals N.Y. Acad. Sci 121:404.
6. Prozorovski V.N. et al. (1982): Evidence that Human Ceruloplasmin Molecule Consists of Homologus Parts, Int. J. Peptide Protein Res. 19:40
7. Jalkanen M. et al. (1980): A Rapid Assay to Measure Collagen Synthesis in Cell Cultures, J. Biochem. Biophys. Method, 2:331
8. Aalto M. et al. (1979): Fractionation of Connective Tissue Activating Factors from The Culture Medium of Silica Treated Macrophages, Acta Path. Microbiol. Scand. C. 87:241
9. Goldstein I. M. et al. (1979): Ceruloplasmin, J. Biol. Chem. 254:4040.

10

Biological tracers of environmental exposure to trichloroethylene

G. Ziglio, G. Beltramelli, F. Pregliasco, G. Ferrari – Istituto di Igiene, Università degli Studi di Milano, Italy

ENVIRONMENTAL EXPOSURE

Trichloroethylene (TRI) is a widespread environmental contaminant. WHO EHC 50 on Trichloroethylene (WHO, 1985) reports that there is clear evidence of the carcinogenic properties of TRI in mice and some evidence of it in rats. The build up of TRI in the human body occurs through water consumption and inhalation. Desorption of TRI from water may increase levels in indoor air (Andelman, 1985), while use of home-products represents a frequent short-term, high-level exposure. In polluted areas, typical concentrations may range from 1-2 to hundreds ug/L in drinking water, and from 2-5 to 20-50 ug/m^3 (24h average) in urban atmospheres; generally, concentrations do not exceed 1-2 ug/Kg in foods (Bauer, 1981; Ziglio, 1982; Pearson, 1982).

BIOLOGICAL MONITORING

An estimate of the total uptake of TRI, starting from exposure levels, is hampered by: the multiple-source exposure pattern, the spatial, temporal and biological variability of individual exposure and the possible chronic health effects not related to parent compound. Consequently, biological monitoring has to be used for exposure assessment and, possibly, for risk evaluation.

PREVIOUS STUDIES Previous cross-sectional studies aimed to evaluate and compare the percentile distributions of plasmatic trichloroacetic acid (TCA) in subject groups differently exposed to TRI at known environmental level (Ziglio et al, 1982; Ziglio et al, 1983; Ziglio et al, 1984a). A gas-chromatographic procedure (GLC-ECD) was used (Ziglio, 1979; Ziglio et al, 1984b).
A summary of the major results are reported in Table 1.

Study design	TRI levels Water ug/L(sd)	Air ug/m^3(sd)	TCA percentile distribution 10	50	90
City of Milan 197 adults (M+F) June 1980	80(32)	7,6(2,7)	19,5	38	81
City of Milan 68 adults (M+F) July 1980	12(1,7)	7,7(3,3)	8	19	36
City of Trent 89 adults (M+F) July 1980	<0,5	not meas	4	7,5	39
Town of Sesto S.G. 114 school-children November 1980	87(19,9)	8,6(3,7)	15	36	80
City of Milan 335 samples from 136 blood-donors (M+F) April '84 to March '85	50(21,3)	8,5(7,7)	11	23	48

Table 1: Summary data on TCA percentile distributions (ug/L) found in different studies

Plasmatic TCA was found in all subjects, including those exposed at background levels. TCA levels did statistically differ among subjects served by drinking water contaminated at different levels. However, TCA levels did not differ whether or not contaminated water was consumed. Plasmatic TCA distributions were not related to sex, age, body structure or chronic use of common drugs (Ziglio et al, 1982).

RECENT STUDY A new study was conducted to evaluate the contribution of ingestion and inhalation to total TRI exposure. From a cohort of 136 preselected blood-donors (M-F, aged 18-62), a total of 335 biological samples were collected for a year. Plasmatic TCA (GLC-ECD), urinary TCA and Trichloroethanol (TCE-glu hydrolysis plus TCA esterification; extraction; HRGLC-ECD) were measured. TRI concentrations in tap water and atmosphere were evaluated simultaneously. Questionnaires were submitted individually on personal habits and microenvironmental characteristics. Details of study design, analytical procedures and complete results are reported elsewhere (Ziglio et al, 1986).
Plasmatic TCA concentrations distribution were comparable with those previously found (see table 1). Plasmatic and urinary metabolites were intercorrelated (r values ranging from 0.48 to 0.55). Multiple regression analyses were performed between individual data of exposure and metabolite levels. The r values were statistically significant and ranged from 0.45 to 0.48 (see Table 2).

$(TCA)pl = 1,32C_A + 0,14Cw + 10,25$ $\quad F = 42,41\ (P < 0,01) R = 0,45$

$(TCE)ur = 1,85\ C_A + 0,15Cw + 10,52$ $\quad F = 41,76\ (P < 0,01) R = 0,45$

$(TCE+TCA)ur = 3,37C_A + 0,37Cw + 19,01$ $\quad F = 53,12\ (P < 0,01) R = 0,49$

C_A = Concentration of TRI in atmosphere (mg/m^3)
Cw = Concentration of TRI in water supply (ug/L)
Urinary metabolites as ug/g creatinine

Table 2: Parameters of the regression analyses between individual exposure data and metabolite levels.

For each subject, TRI contents in tap water were assumed constant. Data from the questionnaire did not allow an accurate evaluation of individual water consumption. TRI was measured at a single fixed point before specimen collection.
From the coefficients of the regressions it can be calculated that 1 ug/m^3 of TRI in the atmosphere produces the same metabolite body burden as a water supply containing 10 ug/L TRI. Theoretical calculation leads to a similar relative contribution (Trouwborst, 1982). Water consumption or time

spent at home did not further explain total variability. All the regressions showed a non zero value for the intercept. A possible explanation for this may be: food (small contribution); same metabolites derived from other chlorinated solvents (small contribution, considering the low level measured); inaccuracy in evaluating environmental exposure; late effects of occasional home exposure.

CONCLUSIONS

TRI metabolites are ubiquitous in population groups exposed at ppb levels. Sufficiently adequate procedures exist for collecting, storing and analysing plasmatic and urinary metabolites. Urinary TCE may be used as a tracer for evaluating environmental exposure to TRI even in concomitant exposure to other volatile chlorinated compounds. Plasmatic TCA may be used as an alternative approach (less influenced by variability of exposure levels). However, Methylchloroform and Tetrachloroethylene, when present at high levels, may give positive interference.

Some recommendations for further research are indicated: extension of biological monitoring in different population groups; metabolite distribution in populations not exposed (reference populations) or exposed at the lowest compatible levels; biological monitoring in school-children; possible biochemical or cellular indices of early effects in hypersusceptible subjects.

REFERENCES

Andelman, J.B. (1985) Inhalation exposure in the home to volatile organic contaminants of drinking water. Sci. Total Environ., 47: 443-460.

Bauer, U. (1981) Human exposure to environmental chemicals. Investigations on volatile organic halogenated compounds in water, air, food and human tissues. Results of investigations. Zbl. Bakt. Hyg. I. Abt. Orig. B., 174: 200-237.

Pearson, C.A. (1982) C_1 and C_2 halocarbons. In: The handbook of environmental chemistry (Ed. by O. Hutzinger). Vol 3 part B: Anthropogenic compounds. Springer Verlag, Berlin. 69-88.

Trouwborst, T. (1982) Een vergelijkende analyse van de opname van (vluchtige) organische stoffen door de mens als gevolg van lucht-respectievelijk drinkwaterverontreiniging en de factoren die de lichaamsbelasting bepalen. H_2O, 15: 208-215.

WHO (1985) Environmental Health Criteria 50. Trichloroethylene. Geneva.

Ziglio, G. (1979) Determinazione gascromatografica del livello ematico di acido tricloroacetico in soggetti non professionalmente esposti a tricloro e tetracloroetilene. Ig. Mod., 72: 876-903.

Ziglio, G. (1982) Tricloroetilene e tetracloroetilene nell'atmosfera urbana di Milano. Ing. Amb., 9: 482-490.

Ziglio, G., Beltramelli, G., Pregliasco F., Arosio, D., Ciarambino, A. e Skouse, D. (1982) Indagine epidemiologica sulla diffusione e gli effetti della trielina nell'acqua potabile di Milano. Quaderni Istituto di Igiene della Università di Milano. N° 11, Giugno.

Ziglio, G., Fara, G.M., Beltramelli, G. and Pregliasco, F. (1983) Human environmental exposure to trichloro- and tetrachloroethylene from water and air in Milano, Italy. Arch. Environ. Contam. Toxicol., 12: 57-64.

Ziglio, G., Beltramelli, G., Pregliasco, F., Arosio, D. e De Donato, S. (1984a) Esposizione ambientale a solventi clorurati in popolazioni studentesche di un comune del Nord Italia. Ig. Mod., 82: 133-161.

Ziglio, G., Beltramelli, G. and Pregliasco, F. (1984b) A procedure for determining trichloroacetic acid in human subjects exposed to chlorinated solvents at environmental levels. Arch. Environ. Contam. Toxicol., 13: 129-134.

Ziglio, G., Beltramelli, G., Pregliasco, F., Ferrari, G. e Filipponi, S. (1986) Esposizione ambientale a trichloroetilene. Interim Report. Istigiene, Milano. 04-28-86.

11

A mathematical model of the fall of zinc protoporphyrin levels in workers

Stephen M. Hessl, M.D., M.P.H., **Daniel Hryhorczuk**, M.D., M.P.H. – Division of Occupational Medicine, Cook County Hospital, Chicago, Illinois; University of Illinois at Chicago.

Charles Tier, Ph.D., Department of Mathematics, Statistics, and Computer Sciences, University of Illinois at Chicago.

INTRODUCTION

We recently observed a temporal fall of zinc protoporphyrin (ZPP) levels in whole blood in 51 patients with occupational chrónic lead intoxication who were removed from exposure, treated with intravenous calcium disodium edetate (EDTA), and followed for periods up to 2,273 days (Hryhorczuk, 1985). ZPP levels fell with a mean half-life of 68 days, to a mean baseline level of 36 mcg/dl of whole blood. The baseline ZPP was reached in an average of 143 days following removal from lead exposure. These data suggested that the fall of ZPP levels was largely a function of red blood cell (RBC) turnover. We propose the following mathematical model for quantitative description of ZPP kinetics in lead workers.

METHODS

The following terms were defined as essential variables for the model:

a	=	age of RBC
m	=	parameter which describes the random destruction (hemolysis) of RBC's.
T	=	potential life span of RBC's.

Φ (a)	=	density function for the age distribution of red blood cells.
Z (t)	=	concentration of ZPP in cells at age = 0.
C (t,a)	=	concentration of ZPP in RBC's of age a at time t.
D (t)	=	total concentration of ZPP over all ages of RBC's.

Several assumptions were made to limit the number of variables in the model and to remain consistent with contemporary descriptions of the life cycle of red blood cells (Berlin and Berk, 1975) and with evidence that ZPP remains in the red blood cell for the life span of the cell (Piomelli, 1975). These assumptions were:

A. Life cycle of red blood cells

1. Stationary age distribution (the age distribution of RBC's does not fluctuate in time).
2. Constant number of RBC's of a = 0 for each t (there is a constant production rate of RBC's).
3. Random rate of destruction of RBC's = m (the hemolysis of RBC's is independent of age and concentration of ZPP).
4. Finite potential mean life span = T (the life span is approximately 120 days and independent of ZPP at blood lead levels below 70 ug/dl).

B. Action of ZPP

1. There is no random loss of ZPP in RBC's with a>0.
2. ZPP has no effect on density function of RBC's (life span or random destruction).
3. The concentration of ZPP in new cohorts of RBC's (age = 0) can be tracked.

RESULTS

Red blood cell survival can be described by the following equations:

Equation 1

$$\phi(a) = \begin{cases} \dfrac{me^{-ma}}{1-e^{-mT}} & ,0<a<T \\ 0 & a>T \end{cases}$$

Equation 2

$$\int_0^T \phi(a)\, da = 1$$

As described in an earlier model of red cell survival, it is assumed that the age distribution of red blood cells is regular and stable, that a constant number of cells is produced per unit time, that all RBC's have a finite life span (T) and that there is a steady risk of destruction (m) (Dornhorst, 1951). Under these conditions the age distribution would be as follows:

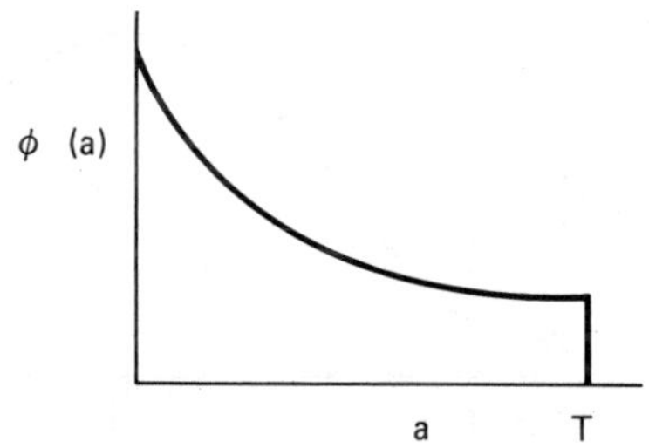

Fig. 1. Age distribution of red blood cells.

The concentration of ZPP in whole blood at time t (D(t)) can then be described by the following equations

Equation 3

$$C(t,a) = \begin{cases} Z(t-a) & ,t>a \\ 0 & ,t<a \end{cases}$$

Equation 4 $$D(t) = \int_0^{\infty} C(t,a)\,\phi(a)\,da$$

Equation 5 $$D(t) = \int_0^{t} Z(t-a)\,\phi(a)\,da \quad 0<t<T$$

$$= \int_0^{T} Z(t-a)\,\phi(a)\,da \quad t>T$$

The concentration of ZPP in blood at any point in time is related to the concentration of ZPP in the RBC's at age = 0 ($Z(t)$), the distribution of cells by age and other assumptions described above. Equation 5 should then predict the change of concentration of ZPP over time.

DISCUSSION

Two recent studies provide empirical data on the temporal change in ZPP levels in whole blood, relative to occupational lead exposure. One study describes the rise of ZPP in workers who returned to lead exposure after a ten-weeks strike (Lerner, 1982). The other study describes the fall of ZPP after removal from lead exposure (Hryhorczuk, 1985). Figure 2 illustrates these observations. Note that a new, higher baseline is expected after exposure.

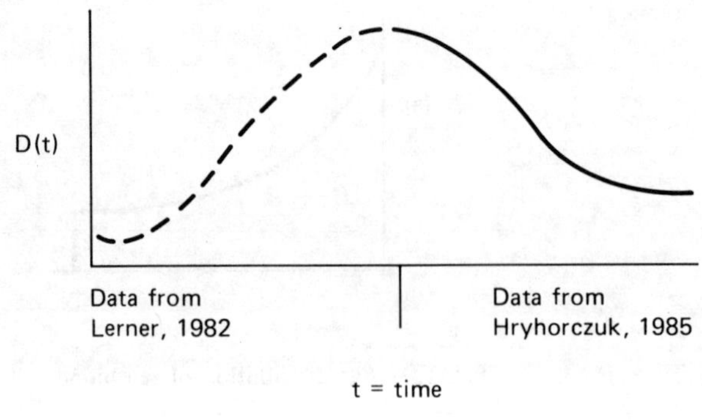

Fig. 2. – Changes in concentration of ZPP over time.

Equation 5 models the behavior of ZPP over time, as described in figure 2 provided that Z(t) varies over time and exposure shown in figure 3.

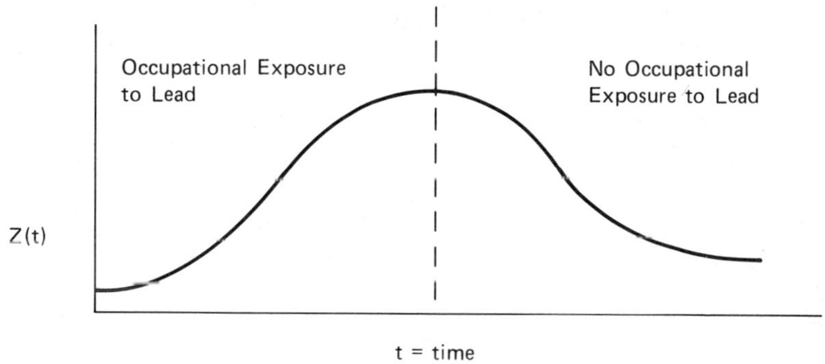

Fig. 3. – Concentration of ZPP in newly-formed RBC's.

Further studies are needed to demonstrate Z(t) in bone marrow and to apply empirical data to equation 5 to validate this hypothesis.

REFERENCES

1. Berlin, N.I. and Berk, P.D. in The Red Blood Cell vol 2, Surgenor, D.M., ed. Academic Press (1975).

2. Dornhorst, A.C. "The Interpretation of Red Cell Survival Curves" Blood 6:1284-1292 (1951).

3. Hryhorczuk, D.O.; Hogan, M.M.; Mallin, K.; Hessl, S.M.; and Orris, P. "The Fall in Zinc Protoporphyrin Levels in Workers Treated for Chronic Lead Intoxication" Journal of Occupational Medicine 27:816-820 (1985).

4. Lerner, S.; Gartside, P.; and Roy B. "Free Erythrocyte Protoporphyrin, Zinc Protoporphyrin and Blood Lead in Newly Re-Exposed Smelter Workers: A Prospective Study" American Industrial Hygiene Association Journal 43:516-519 (1982).

5. Piomelli, S.; Lamola, A.; Poh-Fitzpatrick, M.; Seaman, C. and Harber, L. "Erythropoietic Protoporphyria and Lead Intoxication" J. Clin. Invest. 56:1519 (1975).

Report of session 1

Sven Hernberg – Institute of Occupational Health, Helsinki, Finland.

The papers presented in the first session addressed general aspects of biological monitoring. It became apparent that there still prevails a conceptual confusion regarding the very meaning of the term, although a joint EEC/NIOSH/OSHA seminar in 1984 tried to give biological monitoring a standardized definition. Especially the distinction between monitoring and health surveillance was not enough appreciated by all speakers. Much of the discussion on the validation of the tests focussed rather on surveillance than on monitoring in a strict sense.

Most of the tests used both for biological monitoring and for health surveillance, suffer from lack of validation. Especially the lack of knowledge on exposure-effect and exposure-response relationships complicates the meaningful evaluation of the results of many monitoring tests. The use of non-specific tests in health surveillance creates other types of problems, because non occupational causes for abnormal results must be ruled out. Non-specific tests are most useful for groups of individuals, as group differences can be interpreted better than individual values.

Correct study design is crucial whenever validation of biological monitoring tests or, alternatively, tests for health surveillance is planned. The use of unvalidated tests is not recommended. Consequently, the validity and efficiency of every monitoring or surveillance program, before initiated, must be evaluated in the light of scientific, ethical, economic and feasibility criteria.

Many otherwise potentially useful examinations do not fulfill the criteria of feasibility and economy, and therefore can not be recommended for routine use in occupational health practice. However, they may be highly relevant and useful in validation research.

Studies aimed at the validation of monitoring or surveillance tests must be staightforward; they must clearly define the surveyed population, the effect to be studied, and the appropriateness of the dose estimate. The methods as well as sampling schedule must be carefully standardized and reference values are needed for the interpretation of the results. Attention should also be given to factors relevant for interindividual variation such as genetic differences, interactions in mixed exposure, etc. For example, the use of drugs, alcohol and tobacco can explain variations; consequently, such factors should be accounted for. It is important to clarify the significance or lack of significance for any given endpoint. Moreover, the significance should be agreed upon by groups of experts.

Everybody involved in a monitoring program should be made aware of its significance and benefits. It is important that the occupational health physician, and not a hygienist or a technician, interprets the results and informs the workers participating in the program.

Analytical quality control programs are important for improving the accuracy and precision of the tests; some are in progress and more should be initiated. New methods, e.g., immunological monitoring, await their role in routine programs.

A list of the gaps in our knowledge would be extensive. Perhaps the most urgent need is to study direct dose-effect and dose-response relationships between biological values and the effects in question. In this effort one can, and should, use much more sophisticated and expensive effect indicators than the ones used in routine monitoring programs. Once the non-effect level is known, there is no longer need for a whole battery of effect indicators in routine programs, because as long as the dose indicator remains under that level, no effects should occur.

The development of new methods for biological monitoring and health surveillance should be encouraged. Both the development of new methods and the improvement of older ones still require considerable analytical progress. A strategy for validating biological monitoring or health surveillance tests should include the following steps:

1. Define priorities for xenobiotics to be studied (based on, e.g., toxicity, type of effects, population at risk, trends in use, etc.)

2. Agree on what is the critical effect (requires numerous independent studies and a large spectrum of examinations).

3. Relate critical effects to a dose estimate. Define the non-effect level and its confidence intervals.

4. Develop precise, accurate, sensitive, specific and repeatable methods for the dose estimate.

5. Determine the factors that influence the dose estimate (e.g., kinetic variation, physiological factors, interactions, etc.).

6. Study reasons for variations in sensitivity for critical (and other early) effects (e.g., age, genetic factors, diseases, etc.).

It is obvious that the validation of each test for biological monitoring or health surveillance is a formidable task, requiring a large number of separate studies. But why not make a worldwide effort of concentrating resources on this exercise, rather than producing the 100th or 200th study to show once again that asbestos exposure causes mesothelioma or that lead exposure inhibits ALA-dehydratase?

Finally, we should always remember that the ultimate goal of occupational hygiene is to create conditions so safe that at least health surveillance programs, aimed at the early detection of toxic effects, become unnecessary.

Part II

Biochemical indicies of liver toxicity: enzyme induction , lipid peroxidation and other indices relevant to liver function changes

Coordination:
N. Dioguardi (Milano, Italy)
D. Henschler (Würzburg, FRG)

12

Biological indices of enzyme induction as markers of hepatic alterations

R.J. Rubin – Department of Environmental Health Sciences,
The Johns Hopkins University

INTRODUCTION

Human exposures to low levels of many environmental and occupational chemicals (xenobiotics) can result in adaptive changes in the liver that can serve as biological markers of the degree of exposure. These adaptive changes take the form of stimulation or induction of de novo synthesis of enzymes that are involved in or related to the metabolism of the xenobiotic to which exposure has occurred. The resulting enhancement of the metabolism serves to increase the clearance of the xenobiotic from the body. From this point of view, the adaptive response is viewed as beneficial and is not to be considered a toxicologic response as defined as a frank clinical disease. Thus, evidence of enzyme induction, per se, is not necessarily to be interpreted as a toxicologic response but merely as an indicator of exposure. Assessment of enzyme induction is useful for (1) understanding drug interactions; (2) evaluating chemical exposures; and (3) defining occupational and environmental populations for epidemiological studies.

PATHWAYS OF XENOBIOTIC METABOLISM

The metabolism of the vast majority of xenobiotics has been subdivided into 2 phases, referred to as Phase I and Phase II. Phase I metabolism involves the oxidation of a lipophilic xenobiotic, which because of its lipophilicity is only poorly excretable by the kidney, to a somewhat more polar oxidized intermediate. Phase I metabolism is catalyzed by a family of isozymes, referred to as cytochrome P450. While somewhat more excretable than the parent compound, the oxidized intermediate is still relatively lipophilic. But it is now susceptible to Phase II metabolism which serves to conjugate the intermediate with bulky water-soluble molecules, such as glucuronic acid, glutathione, or sulfate, thus imparting a high degree of water-solubility and therefore excretability to the final metabolite.

As will be discussed below, a number of methods have been developed to assess cytochrome P450 activity _in vivo_ in humans. In addition, glucaric acid excretion in the urine can be used as a measure of the induction of the glucuronidation pathway while serum γ-glutamyl transferase (or transpeptidase) (γ-GT or γ-GTP) can be used to assess the induction of the glutathione adduct pathway. Little, if anything is known about the induction of the sulfation pathway. It should also be pointed out that cytochrome P450, being a hemoprotein, requires not only the synthesis of protein but also of heme. Thus, during the induction of cytochrome P450, there must be an increased supply of heme arising from porphyrin metabolism. For this reason, induction of cytochrome P450 has been associated with induction of porphyrin synthesis.

IN VIVO METHODS OF DETECTING ENZYME INDUCTION

In experimental animals it is a relatively easy matter to assay enzyme induction. One simply sacrifices the animal, removes the tissue of interest (in this case, most frequently the liver) and performs an _in vitro_ assay. Except for a few instances where liver biopsies can be taken from humans for legitimate medical purposes, this approach is not useful for humans. For this reason, a variety of indirect methods have been developed to assess enzyme induction in humans. A widely used method is to administer a test agent (most frequently a well known drug which has only inconsequential pharmacologic activity but which is known to be cleared from the body primarily by cytochrome P450-mediated metabolism) and to follow its rate of disappearance from the blood from which either the half-life (t 1/2) or its rate of clearance can be calculated. The most widely used of these tests is the antipyrine t 1/2 or clearance assay. Since this procedure is still invasive, requiring the drawing of sequential blood samples over an extended period of time, a modification has been developed in which the rate of disappearance from saliva is determined. An additional development of a non-invasive technique has involved the administration of a radiolabelled drug and the monitoring of expired $^{14}CO_2$. This procedure carries the disadvantage of requiring the administration of a radioactive compound to the human, a procedure that has become less acceptable in recent years.

The ultimate non-invasive technique involves the measurement of the urinary excretion of an administered test agent and/or its metabolite. In order to avoid even the necessity of administering a test agent to the human, a variety of techniques have been developed to measure the urinary excretion of endogenous molecules known to be associated with xenobiotic metabolism. These include the measurement of (1) urinary 6β-hydroxycortisol, which is an otherwise minor P450-mediated metabolite of cortisol; (2) D-glucaric acid an end-metabolite of glucuronic acid which is utilized in the glucuronidation Phase II pathway; and (3) porphyrins which respond to the need for an increased supply of heme during cytochrome P450 induction.

Other approaches involve the assay of serum enzymes which arise from the liver following enzyme induction. These include γ-GT and GSH-S-transferase both of which are involved in the glutathione Phase II adduct pathway.

Finally there are several miscellaneous techniques which are experimental and are currently undergoing evaluation. These include measurement of (1) hepatic blood flow; (2) serum lipids and lipoproteins and (3) what might appear to be the most promising new approach, specific isozymes of P450 in human lymphocytes by radioimmunoassay with monoclonal antibodies.

ROLE OF ENZYME INDUCTION IN DRUG INTERACTIONS

The first description of induction of the cytochrome P450 system in experimental animals was published in the mid 1950's by A.H. Conney working with the Millers [Conney et al, 1956]. It wasn't until almost 10 years later, following the description of the inductive potency of phenobarbital in animals that the first experiments were carried out to show that the same phenomenon occurs in man. Conney, then working in the laboratory of J.J. Burns, showed that phenobarbital administered to humans decreased the concentration of an anticoagulant, bishydroxycoumarin, in the plasma and reduced its pharmacologic action [Cucinell et al, 1965]. They showed that combined therapy with a coumarin anticoagulant and a stimulator of drug metabolism can be hazardous if the enzyme stimulator is withdrawn but therapy with the anticoagulant is continued without an appropriate decrease in the dose. Measurement of the coumarin level in blood was, in a way, the first described method for assessing enzyme induction in man.

In 1971 Hunter et al showed that another potent barbiturate, phetharbital, markedly enhanced the urinary excretion of glucaric acid and that the effect was relatively rapidly reversible and dose-dependent [Hunter et al, 1971]. In 1973 Whitfield showed that the decrease in plasma warfarin levels induced in humans by treatment with a number of hypnotics and sedatives (dichloraphenazone, various barbiturates and antipyrine) was inversely related to plasma γ-GT levels [Whitfield et al, 1973]. As plasma warfarin levels fell plasma γ-GT levels rose sharply to levels as high as 3-5 times that seen prior to treatment. Since the decrease in plasma warfarin levels had been shown to be related to the induction of P450, then plasma γ-GT levels began to be appreciated as an in vivo measure of induction. In 1979 Ohnhaus and Park showed that rate of antipyrine clearance from the blood following treatment with several known inducers was highly correlated with the urinary excretion of 6β- hydroxycortisol [Ohnhaus and Park, 1979]. Saenger in 1983 not only confirmed that phenobarbital in humans was associated with an enhanced excretion of 6β-hydroxycortisol but, showed that there were marked circadian rhythms in the excretion of this steroid metabolite. Since the rhythms were mirrored by equal changes in the parent molecule, cortisol, he concluded that by reporting the ratio of 6β-hydroxycortisol to cortisol, he was able to assess the extent of enzyme induction by using spot collections of urine samples, thus avoiding the necessity of requiring a 24-hour collection.

NON-THERAPEUTIC CHEMICAL EXPOSURES

Through the 1970's the use of indirect assays of enzyme induction in humans was being well established, particularly for a variety of drugs known to be inducers in experimental animals. There was little doubt that the process of enzyme induction could occur in humans. With the advent of massive environmental exposures to xenobiotics and increased attention to the role of occupational expsoures in worker's health, attention began to turn to the use of these assays as markers of human exposure. However, a number of studies enphasized the need for extreme care and caution in using such assays. They appeared to be sensitive to a number of what might be referred to as "life-style" factors, including smoking, diet, and alcohol consumption. Such influences required extreme precaution in designing large scale studies to detect enzyme induction as a marker for human exposure to environmental and occupational chemicals.

In 1976, Conney, using phenacetin clearance as the marker, showed that a diet high in charcoal-broiled beef content (known to contain high concentrations of polycyclic aromatic hydrocarbon inducers) resulted in a marked lowering of phenacetin levels in plasma [Conney et al, 1976]. Interestingly enough kinetic analysis of their data suggested that the induction had occurred in the gastro-intestinal tract itself and/or during its first pass through the liver. Additional studies revealed that a diet containing large amounts of protein enhances the metabolism of antipyrine and theophylline in man, whereas a high carbohydrate diet depresses the rate of metabolism of these drugs [Alvares et al, 1976; Kappas et al, 1976].

In 1979 Dollery et al showed that cigarette smoking enhanced both antipyrine and acetaminophen clearance [Dollery et al,

1979]. The effect was shown to be related to the number of cigarettes smoked per day with the maximum effect seen at between 20-40 cigarettes per day. Goldberg et al in 1980 showed the effect of chronic alcoholism on tolbutamide, warfarin, and phenytoin half-lives, all of which were decreased by approximately 50% [Goldberg et al, 1980].

In the meantime (in 1963) Hart and Fouts had made the accidental discovery that the chlorinated insecticide, chlordane, sprayed in their animal quarters to control an insect infestation, had resulted in a marked stimulation of the cytochrome P450 system [Hart et al, 1963]. It was only a short time later that evidence for enzyme induction in humans exposed to chlorinated hydrocarbon insecticides was obtained. Kolmodin et al reported a marked decrease in antipyrine half-life in workers exposed to a mixture of lindane and DDT [Kolmodin et al, 1969]. In confirmation of these results, Poland et al reported that workers in a DDT-factory had both a decreased phenylbutazone half-life and an increased 6β-hydroxycortisol excretion [Poland et al, 1970]. At the present time there are data indicating significant degrees of enzyme induction in workers exposed to chlorinated hydrocarbons following aerosol spraying, halothane used in anesthesia, PCBs, organochlorine pesticides such as chlordecone (kepone), petroleum fuels, and phenoxyacids and chlorophenols [Dossing, 1984].

POPULATION STUDIES

In recent years there have been a number of massive, accidental exposures of large geographical areas to a variety of man-made chemicals. These exposures have involved virtually thousands of human beings. Examples include dioxins in Sevesco, Italy, kepone in the James River region of Virginia, and the PCB episodes in both Japan and Taiwan. Much effort has been involved in assessing the human health impact of these exposures. While it is not the intention of this paper to review all the medical data accumulated in these episodes, a number of representative studies on assessment of the extent of enzyme induction resulting from the exposures will be presented. On the one hand, these studies serve the purpose of acting as markers for identifying exposed populations, but on the other hand they highlight the subtleness of the response in humans. In no way can enzyme induction be interpreted to be a toxicologic response; it is, as stated, simply an adaptive response and as such serves merely as a marker of exposure.

Ideo et al have observed a statistically significant increase in urinary D-glucaric acid excretion in adults living in the Sevesco area as compared to a control population living in an area which although geographically near Sevesco was quite remote from the site of exposure [Ideo et al, 1985]. It is interesting to note here that the assays were carried out years after the date of exposure and yet the evidence was strong for a chronic inductive effect.

More recently, Guzelian has shown an increase in D-glucaric acid excretion in individuals exposed to high levels of kepone [Guzelian, 1985]. His data further shows a return to control levels some 2 years later following cessation of cholestyramine treatment. Also at this time PCB, DDT, lindane, and hexachlorocyclohexane have been shown to cause enzyme induction in humans as assessed by significant decreases in antipyrine half-life.

INDUCTION OF PORPHYRIN PATHWAY

As stated above, induction of the hemoprotein, cytochrome P450, requires a commensurate induction of the heme moiety. In the cell the synthesis of heme is controlled by the first enzyme in the heme pathway, aminolevulinic acid (ALA) synthetase. Under normal conditions this enzyme is repressed by the end product of the pathway, i.e. heme. A number of studies have shown that as heme becomes depleted due to its enhanced incorporation into cytochrome P450, ALA synthetase is induced. Thus, induction of ALA synthetase activity can also be a measure of the induction of the xenobiotic metabolizing enzyme system. As with measures of cytochrome P450 activity itself, a number of indirect assays have been developed for assessing induction of ALA synthetase. The most widely used method involves measurement of the urinary excretion of heme intermediates, i.e. uroporphyrin and its decarboxylated product, coproporphyrin. Under normal condition, the upper limit for total urinary porphyrins is approximately 95 μg/ml and the ratio of coproporphyrin to uroporphyrin ranges from 2 to 5, with 6 being the upper limit of normal. In 15 farmers from Michigan who had been exposed to polybrominated biphenyls (PBB's), Strik et al found that 6/15 (40%) had normal total urinary porphyrins and normal coproporphyrin to uroporphyrin ratios while 5/15 (33%) had levels of total urinary porphyrins that exceeded the upper limit of 95 μg/ml [Strik et al, 1979]. These elevated levels were associated with normal levels of uroporphyrin and moderately elevated levels of coproporphyrin (copro/uro ratio = 7.3). This condition is referred to as coproporphyrinuria and it most likely reflects the induction of the heme biosynthetic pathway by PBB. The remaining 4/15 subjects had not only elevated levels of total porphyrins but also elevated levels of both copro- and uroporphyrin. This state has been defined as Type A Chronic Hepatic Porphyria and it most likely represents a stage in a continuum that ends in a state of frank hepatic disease referred to as Type D or Porphyria Cutanea Tarda. Thus, it may be concluded that the induction of the porphyrin pathway (as reflected by coproporphyrinuria) reflects an early adaptive response to the PBB exposure which, if high and/or long enough, converts to a disease state. The role of pure coproporphyrinuria as a marker of enzyme induction requires further investigation.

RECENT DEVELOPMENT OF MARKERS OF ENZYME INDUCTION

Recently Ohnhaus et al have shown that inducers that significantly shorten antipyrine half-life, and elevate serum γ-GT levels as well as D-glucaric acid and 6β-hydroxycorticol urinary excretion also increase hepatic blood flow in humans [Ohnhaus et al, 1985]. While this methodology has only limited value in large scale field studies because of the technical complexities involved in measuring hepatic blood flow, it would seem to have a role in well-controlled laboratory studies. At the very least these results suggest caution in concluding that any given exposure results in enzyme induction when the latter is assessed by the rate of clearance (or t 1/2) of a chemical whose intrinsic hepatic metabolism is flow-dependent. Fortunately, antipyrine clearance (the most widely used method for determining enzyme induction in humans) has been shown not to be flow-dependent [Ohnhaus et al, 1985].

Cytochrome P450 is a membrane-bound enzyme which is embedded in the endoplasmic reticulum (ER) of the cell. Induction of P450 is, in fact, associated with proliferation of the ER and a generalized increase in protein synthesis. Recent evidence [Luoma et al, 1982] has shown that in patients who were on inducing drug therapy (phenytoin, phenobarbital, or primidone) and who had significantly elevated levels of liver P450 and antipyrine clearance also demonstrated significantly elevated levels of apo-lipoproteins AI and AII in serum. These are the lipoprotein carriers for the high density lipoproteins (HDL) of serum. Associated with these increases were elevated levels of HDL-cholesterol. Furthermore, the increase in HDL-cholesterol was significantly correlated with the markers of enzyme induction, i.e. P450 levels and antipyrine clearance. These results suggest the judicious use of serum apo-lipoproteins and HDL-cholesterol as additional markers of enzyme induction. Of particular interest in this study is the fact that serum triglycerides were significantly lowered and inversely related to the markers of enzyme induction. This would suggest that the elevation of serum triglycerides seen in humans environmentally exposed to PCB's is unrelated to the process of enzyme induction and is most likely due to a direct effect on lipid metabolism.

As referred to above, cytochrome P450 is not a single enzyme but a family of closely related isozymes. Each isozyme has a somewhat limited specificity for specific substrates. Modern molecular biology with its methods of gene cloning and production of monoclonal antibodies which are specific for any given isozyme of P450 has begun to yield methods for assaying specific P450's. Song et al [Song et al, 1985] have recently used monoclonal antibody technology to show that human smokers have significantly elevated levels of the specific P450 associated with the metabolism of aromatic hydrocarbons. This assay was performed on lymphocytes isolated from peripheral blood. Earlier work had attempted to relate the occurence of lung cancer in humans to the inducibility of aromatic hydrocarbon hydroxylase activity (AHH) in isolated lymphocytes. However, the study was not able to be reproduced by others and the association between AHH inducibility and lung cancer remained controversial. Since aromatic hydrocarbons are metabolized by several different isozymes of P450, the problem with these studies may well have been due to the lack of specificity of the enzyme assay. The newer technology holds forth the possibility of assessing the relationship between human lung cancer and a specific form of P450. Furthermore, as the technology becomes further refined and available to a wider community of investigators (i.e. the ability to purchase from a vendor an array of specific P450 antibodies) it holds forth the promise of being useful in assessing the induction of specific P450's in response to specific chemical exposures.

REFERENCES

Alvares A.P., Anderson K.E., Conney A.H. and Kappas A.H. (1976) Interactions between nutritional factors and drug biotransformations in man. Proc. Natl. Acad. Sci. USA 73:2501.

Conney A.H., Miller E.C., Miller J.A. (1956) The metabolism of methylated amino azo dyes. V. Evidence for induction of enzyme synthesis in the rat by 3-methyl-cholanthrene. Cancer Res. 16:450.

Conney A.H., Pantuck E.J., Hsiao K-N, Garland W.A., Anderson K.E., Alvares A.P., and Kappas A. (1976) Enhanced phenacetin metabolism in human subjects fed charcoal-broiled beef. Clin. Pharmacol. Ther. 20:633.

Cucinell S.A., Conney A.H., Sansur M., and Burns J.J. (1965) Drug interactions in man. 1. Lowering effect of phenobarbital on plasma levels of bishydroxycoumarin (Dicumarol) and diphenylhydantoin (Dilantin). Clin. Pharmacol. Ther. 6:420.

Dollery C.T., Fraser H.S., Mucklow J.C. and Bulpitt C.J. (1979) Contribution of environmental factors to variability in human drug metabolism. Drug Metab. Rev. 9:207.

Dossing M. (1984) Non-invasive assessment of microsomal enzyme activity in occupational medicine: Present state of knowledge and future perspectives. Int. Arch. Occup. Env. Health 53:205.

Goldberg D.M. (1980) The expanding role of microsomal enzyme induction and its implications for clinical chemistry. Clin. Chem. 26:691.

Guzelian P.S. (1985) Clinical evaluation of liver structure and function in humans exposed to halogenated hydrocarbons. Env. Health Perspec. 60:159.

Hart L.G., Shultice R.W., and Fouts J.R. (1963) Stimulatory effects of chlordane on hepatic microsomal drug metabolism in the rat. Tox. Appl. Pharmacol. 5:371.

Hunter J., Carella M., Maxwell J.D., and Stewart D.A. (1971) Urinary D-glucaric acid excretion as a test for hepatic enzyme induction in man. The Lancet, Mar. 20, 572.

Ideo G., Bellati G., Bellobuono A., and Bissanti L. (1985) Urinary D-glucaric acid excretion in the Seveso area, polluted by tetrachloro-dibenzo-p-dioxin (TCDD): Five years of experience. Env. Health Perspec. 60:151.

Kappas A., Anderson K.E., Conney A.H., and Alvares A.P. (1976) Influence of dietary protein and carbohydrate on antipyrine and theophylline metabolism in man. Clin. Pharmacol. Ther. 20:643.

Kolmodin B., Azarnoff D.L., and Sjoqvist F. (1969) Effect of environmental factors of drug metabolism: Decreased plasma half-life on antipyrine in workers exposed to chlorinated hydrocarbon insecticides. Clin. Pharmacol. Ther. 10:638.

Luoma P.V., Sotaniemi E.A., Pelkonen R.O., Arranto A., and Ehnholm C. (1982) Plasma high density lipoproteins and hepatic microsomal enzyme induction. Eur. J. Clin. Pharmacol. 23:275.

Ohnhaus E.E., and Park B.K. (1979) Measurement of urinary 6-β-hydroxycortisol excretion as an in vivo parameter in the clinical assessment of the microsomal enzyme-inducing capacity of antipyrine, phenobarbitone and rifampicin. Europ. J. Clin. Pharmacol. 15:139.

Ohnhause E.E., Noelpp V.B., and Ramos M.R. (1985) Liver blood flow and enzyme induction in man. Hepato-gastroenterol. 32:61.

Poland A., Smith D., Kuntzman R., Jacobson M., and Conney A.H. (1970) Effect of intensive occupational exposure to DDT on phenylbutazone and cortisol metabolism in human subjects. Clin. Pharmacol. Ther. 11:724.

Saenger P. (1983). 6β-Hydroxycortisol in random urine samples as an indicator of enzyme induction. Clin. Pharmacol. Ther. 34:818.

Strik J.J.T.W.A., Doss M., Schroa G., Robertson L.W., Tiepermann R.V., and Harmsen E.G.M. (1979) Coproporphyrinuria and Chronic Hepatic Porphyria Type A Found in Farm Families from Michigan (USA) Exposed to Polybrominated Biphenyls (PBB). In: Chemical Porphyria in Man, (Strik and Koeman, eds.). Elsevier/North Holland Biomedical Press, New York, pp. 29-53.

Song B-J, Gelboin H.V., and Park S.S. (1985) Monoclonal antibody-directed radioimmunoassay detects cytochrome P450 in human placenta and lymphocytes. Science. 228:490.

Whitfield J.B., Moss D.W., Neale G., Orme M., and Breckenridge A. (1973) Changes in plasma γ-glutamyl transpeptidase activity associated with alterations in drug metabolism in man. Brit. Med. J. 1:316.

13

Experience of testing microsomal enzyme induction in occupational and environmental medine

A. Colombi – Institute of Occupational Health, Clinica L. Devoto, University of Milan, via S. Barnaba 8, 20122 Milano, Italy

INDUCTION AS A TOOL FOR BIOLOGICAL MONITORING

Inductive effects have been documented as a consequence of occupational or environmental exposure to several chemicals but as it can be appreciated from the literature, the number of investigations has been so far limited (Hunter et al., 1972; Conney, 1972; Alvares et al., 1977; Aitio, 1984).
It must be stressed that discovery and early characterization of microsomal induction in the liver occurred more than 30 years ago, as a part of the pioneering work on chemical carcinogenesis by James and Elisabeth Miller and their associates (Brown et al., 1954). Since then a large number of experimental observations have accumulated, but the progresses in understanding and applying enzyme induction in man have been slow because of the lack of suitable methods for monitoring changes in activity of the drug metabolizing enzymes in vivo.

LIMITATIONS OF THE ANALYTICAL METHODS

Theoretically, direct measurement of enzyme activity in bioptic tissue samples before and after exposure to inducers, would represent the most accurate method. However, ethycal and practical limitations do not allow this approach in human studies. Induction in man has to be measured by indirect indices such as changes in the pharmacokinetics of a marker drug or changes in the disposal of endogenous compounds (Goldberg, 1980; Park and Breckenridge, 1981). Non-automatized and time-consuming analytical procedures, complex protocols of measure, and expensive equipments, are the main unfavourable features of the available laboratory tests. The limited number of samples processable, the analytical skilness, and the high technical requirements needed for these tests, give an obvious explanation of the limited surveys up to now performed.

Many improvements have been recently introduced in the analytical performance of several assays.

The development of HPLC techniques for the urinary 6ß-hydroxycortisol measurement instead of the radioimmunoassay (Nakamura & Yakata, 1985), and a simplified method for D-glucaric colorimetric measure (Colombi et al., 1983a) may be considered as a promising start up for a broader practicability of both tests. In the same way has to be regarded the effort to simplify the protocol of antipyrine half-life determination, which has been recently reduced to a single measure performed on saliva sample collected 24h after a self-administered oral dose of the drug (Dossing, 1984).

LACK OF CORRELATION BETWEEN DOSE AND EFFECTS

Among the goals of biological monitoring programs, the most important is the assessment of exposure level based on the relationship between existing the biological indicator and exposure. Therefore, the evaluation of this relationship is the preliminary step for validating the biological indicators. Sometimes the observed absence of correlation does not reflect a real independence from exposure of the measured parameters but is mainly due to limitations in the planning of the study. To establish a relationship between exposure and health effects, two experimental approaches, research or field studies, may be followed.

In research studies, exposure and effect data are generally fully representative and both receive the same care in regard to sampling strategy and analytical performance. This is for example the case of the study of microsomal enzyme induction on volunteers or in carefully controlled patients after administration of drugs. High correlation coefficients have been in fact reported between dose or plasma levels of different drugs (i.e., rifampicin, warfarin, phenitoin and phenobarbital) and indirect indices of enzyme induction (i.e. 6ß-OH-cortisol and D-glucaric acid) (Park and Breckenridge, 1981).

In field studies and, at a lesser extent, in survey programs, the uncertain nature of exposure (referable f.i. to a mixture of different substances) or inadequacy of dose indicators to assess exposure, can affect the investigated relationship. For example, in the study of microsomal enzyme induction in polychlorinated biphenyl exposed workers reported by Maroni et al. (1984), indicators of microsomal enzyme induction (D-glucaric acid and GGT) were both significantly increased when compared with unexposed subjects, but no significant correlation was found between blood total PCB concentration and any of the two parameters. PCBs are complex mixtures of different congeners with different biological activity, and it is not surprising that a coarse measure of exposure does not correlate with a specific biological effect as microsomal induction is. As recent studies indicate, the internal dose assessed through the measurement of the different homologues and isomers in the tissues of exposed subjects, represents a more accurate indicator of the active dose responsible of health effects (Maroni et al., 1984).

Absence of correlation between indicators of exposure and microsomal enzyme induction indices has been also observed with other chemicals. The effect of occupational exposure to DDT on drug and steroid metabolism has been investigated in subjects working in a DDT production plant (Poland et al., 1970). Serum and adipose tissue concentrations of DDT and related substances (p,p'-DDT; p,p'-DDE; o,p'-DDT; p,p'-DDD) were found to be 20 to 30 times higher than those in a control population, and, on a group basis, exposure was associated with lower phenylbutazone serum half life and higher urinary excretion of 6ß-hydroxycortisol. In spite of these clear evidences, no correlation between serum or fat concentrations of DDT or related metabolites and induction indices was found. At that time, the lack of correlation was hypothetically explained more by genetic differences in susceptibility to enzyme induction among different individuals than by inadequacy of internal dose indicator to represent the biologically active dose.

In the light of limited meaning of measurable internal doses, may also be reconsidered the results reported by other authors such as absence of correlation between D-glucaric acid excretion and circulating levels of Aldrin, Dieldrin, p,p'-DDE (Hunter et al., 1972), and D-glucaric acid versus serum p,p'-DDT, p,p'-DDE or Dieldrin in occupationally exposed workers (Morgan et al., 1974).

However, positive results in the correlation between exposure and effects have been obtained with the measure of the excreted dose rather than the serum level. This is the case of the study of microsomal induction in workers manufacturing Endrin (Ottenvanger and Van Sittert, 1979). Endrin is considered a non-persistent organochlorine insecticide which is rapidly metabolized to hydrophilic metabolites, excreted mainly in the faeces and, at a lesser extent, in urine. Urine samples collected after 7-day exposure to endrin, showed a positive significant correlation between urinary levels of anti-12-OH-endrin and D-glucaric acid, suggesting that the measure of the metabolite, rather than the parent compound which was always present in serum below detectable levels, represented an accurate estimate of the biologically active dose.

This study was also completed with a short follow-up on the persistence of measured effects: all subjects were re-examined after a period of 3 days in which no exposure had occurred. Urinary anti-12-OH-endrin concentrations decreased until pre-exposure levels, but D-glucaric acid remained higher when compared to pre-exposure values. Obviously this sampling in contrast to the previous results, did show not correlation between anti-12-OH-endrin and D-glucaric acid. These results underline that strategies of sampling have to be designed carefully. Pharmacokinetic properties of chemicals are critical to allow a representative assessment of exposure and sampling design must also consider the time dependence of the investigated effect. Microsomal induction is generally regarded as an early effect the onset of which requires hours or days depending on dose and pharmacokinetic properties of the inducer. Microsomal enzyme function recovers from xenobiotic-induced changes within a few days after the chemical has been completely eliminated from the body (Dossing, 1984). Therefore, a representative sampling of both exposure and related effect, has to consider the different temporary sequence of the two events.

ABSENCE OF STANDARDIZED SEQUENCES OR BATTERIES OF TESTS

In animals, inducing agents have been shown to influence microsomal enzymes in various ways and to a different extent. The complexity of the phenomenon appears linked to the existence of multiple forms of drug metabolizing enzymes,

which are differently stimulated or inhibited by the various compounds (Connelly and Bridges, 1980).
Therefore to obtain reliable results induction should be evaluated by multiple tests able to respond to the effects of inducing agents on the various enzymes. The limited availability of tests practicable in humans does not allow this approach and the results obtained in human studies always offer indirect and incomplete evidence of the phenomenon. Moreover, the low correlation observed among the tests suggests that they respond differently to different kinds of induction (Hildebrandt et al., 1975; Ohnhaus and Park, 1979; Poland et al., 1970). Consequently we have to use several indicators concomitantly to characterize with best accuracy the effects of exogenous chemicals on enzyme systems.
In the field studies, the choice between sequential or concomitant execution of multiple tests would be of relevant importance for economical reasons. However, the lack of knowledge about the predictive validity of the single positive results for assessing leads to a concomitant use of several tests, thus creating practical limitations.
The knowledge available for the most popular induction test, serum GGT, can emblematically indicate the uncertainty about the biological and clinical meaning of these tests: it is debated whether a rise in GGT activity represents a specific increase in synthesis of this enzyme, an effect of stimulated blood flow in the liver, or an enhanced cellular breakdown. The question, 30 year old, is still open and far from being solved (Penn, 1983).
Many doubts also exist on D-glucaric as a real index of induction since, as observed in animals, increased excretion after phenobarbitone treatment might also be explained by inhibition of glycogen synthesis and increase in carbohydrate flux through the glucoronic acid pathway (Park and Breckenridge, 1981; Goldberg, 1980).

MODIFICATION AND CONFOUNDING FACTORS OF MICROSOMAL ENZYMES ACTIVITY

When studying the relationship between exposure and effect or comparing health effects observed in different groups of exposed subjects, knowledge of the factors other than exposure acting on the investigated effect is required.
Both genetic (endogenous) and environmental (exogenous) factors control the rates and pathways of chemical biotransformation in human beings. These factors have been

shown to be important determinants of interindividual variation in the activity of drug metabolizing enzymes but it is difficult to quantify the relative contribution of genes and environment.

Evidence of <u>gene contribution</u> is available from pharmacokinetic studies on drug metabolism, where differences in sex, age, personal habit and diet have been excluded. In fact, in such studies, a great interindividual variability is observed and a considerable residual variation in the relationship between dose and metabolic rates cannot be explained. Much of this is related to genetic differences in drug metabolism, as had also been shown by the finding of greater interindividual variations in the metabolism of antipyrine, phenylbutazone and nortriptyline in dizygotic than monozygotic twin pairs (Price-Evans, 1977).

Moreover genetic polymorphisms in the oxidative metabolism of many drugs, including antipyrine with at least two phenotypes, have been discovered during recent years and it has been definitely recognized that the pronounced interindividual variation in activity (and inducibility) of microsomal monooxygenases constitute a major source of variability in drug response (Weinshilboum, 1984).

The <u>environmental or exogenous factors</u> influencing metabolism of chemicals are numerous. Obviously, besides use of drugs and exposure to chemicals, ingestion of alcohol, cigarette smoking, and changes in diet may influence human drug metabolism in different ways and at a various extent.

Diet has been recognized as the mayor source of interaction with the environment. Changes in the ratio of protein to carbohydrate or fat and ingestion of charcoal-broiled beef, cabbage and brussels sprouts, and xanthine containing foods are among the more investigated factors that can stimulate (or inhibit) activity of human liver microsomes (Conney et al., 1980). Also factors related to physiological and pathological conditions such as age, pregnancy or liver diseases, may play an important role in modulating the overall activity and the metabolite pattern of liver drug metabolism.

All the above mentioned factors, acting on the biotransformation activity of subjects, may modulate the responde to enzyme-inducing agents, or may act themselves as inducers.

The broad spectrum of their influence on the available indices for measuring enzyme induction is summarized in Table 1. As it can be easily appreciated, all the tests are equally sensitive to different factors but their specificity is very low..

Table 1: Influence of modification and confounding factors on microsomal induction tests.

	Race	Age	Sex	Pregnancy	Alcohol cons.	Smoking habits	Low carbohydrate high protein diet	Coffee - tea	Cabbage, brussels sprouts	Smoked or broiled foods
URINARY D-GLUCARIC ACID		$\neq$	$\neq$	↑	↑	↑			↑	↑
URINARY 6β-OH-CORTISOL						↑				
CLEARANCE OF MARKER DRUGS	$\neq$		$\neq$			↑	↑	↑	↑	↑
γ GT		$\neq$			↑					

$\neq$ = influenced

Although we do not know the extent of the influence of genetic and environmental factors on microsome induction, their presence may act as "confounding or modification factors". For instance they might mask the existence of a dose-effect relationship, or limitate the extent of differences between two investigated groups. Moreover the presence of uncontrollable interfering variables may enlarge so much intra- and inter-individual variability of results, to require more extended investigations to assess significative differences between exposed and reference groups.

In field studies performed to assess exposure to a putative inducer, these variables are naturally occurring in the investigated subjects and many of them cannot be eliminated. The investigators must be aware of the way in which individual variability of response to enzyme induction may affect interpretation of results.

INDUCTION AS SCREENING TEST OF EXPOSURE

In surveillance programs, biological indicators are often used to classify subjects. In general categories which subjects are classified into, are defined according to a dichotomic criterium (f.i. presence or absence of an indicator) or according to ranges of continuous variables (f.i. intensity of exposure or measured effects). The degree of fitness of results in providing correct classification of subjects, is expressed as "validity" of the indicator and may be evaluated by calculating sensitivity and specificity of the indicator in the investigated population.

Also microsomal induction may be used for such a purpose; for instance as a screening tests for evaluating the occurrence or absence of exposure to inducing chemicals.

As an example, the validity of urinary D-glucaric acid as a screening test of induction in patient treated with antiepileptic drugs or in subjects occupationally exposed to PCB has been calculated. Figure 1 shows the results of D-glucaric measurement in urine of epileptic patients treated chronically with phenylidantoin, phenobarbital and multiple therapy of both drugs. The daily assumed dose and the plasmatic levels of the drugs are also reported. The results show a marked influence of antiepileptic drugs on D-glucaric excretion, with values sometimes twenty fold higher than the reference.

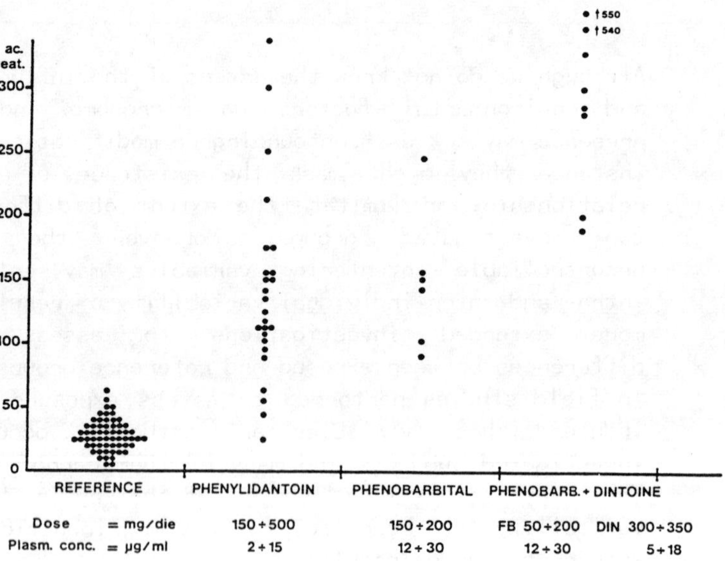

Fig. 1. – Urinary D-glucaric acid in epileptic patients under chronic treatment with anticonvulsant drugs.

In Figure 2 the values are shown as frequency distribution. The discriminating value between effect and no-effect has been set at mean plus 2 SD of the reference values (Colombi, 1983b). This means that values higher than mean plus 2 SD were considered positive. Sensitivity, defined as the ability of

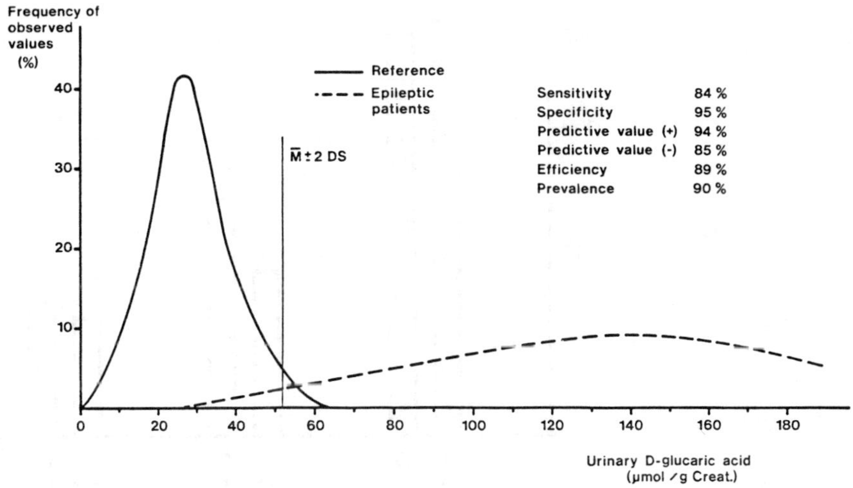

Fig. 2. – D-glucaric as screening test of drug treatment in epileptic patients.

the test to designate as "positive" (with effect) the treated subjects, was equal to 84%. Specificity, defined as the ability of the test to designate as "negative" (without effect) the subjects without treatment, was 95%. The predictive value of a positive result of the test is measured by percentage of positive results that are true positive (positive test associated to drug treatment) on all positive results observed. The high positive predictive value observed of 94% indicated that in epileptic patients an increased excretion of D-glucaric acid has a probability of 94% to be associated with anticonvulsant therapy. This high value is dependent, beside the high sensitivity and specificity of the test, on the high prevalence (high frequency) of the effect "induction" observed in treated epileptic patients.

The test behaves quite differently in subjects occupationally exposed to PCB. The observed mean values obtained in a previous study, were reported in Figure 3 (Maroni et al., 1981). The observed effect on D-glucaric acid excretion was quite moderate when compared with that elicited by anticonvulsant drugs, but on a group basis mean values of excretion in exposed subjects were statistically higher than in controls. The frequency distribution of the values is reported in Figure 4. The number of exposed subjects "positive" to the test was very low (prevalence 12%) and consequently sensitivity and predictive value were also low,

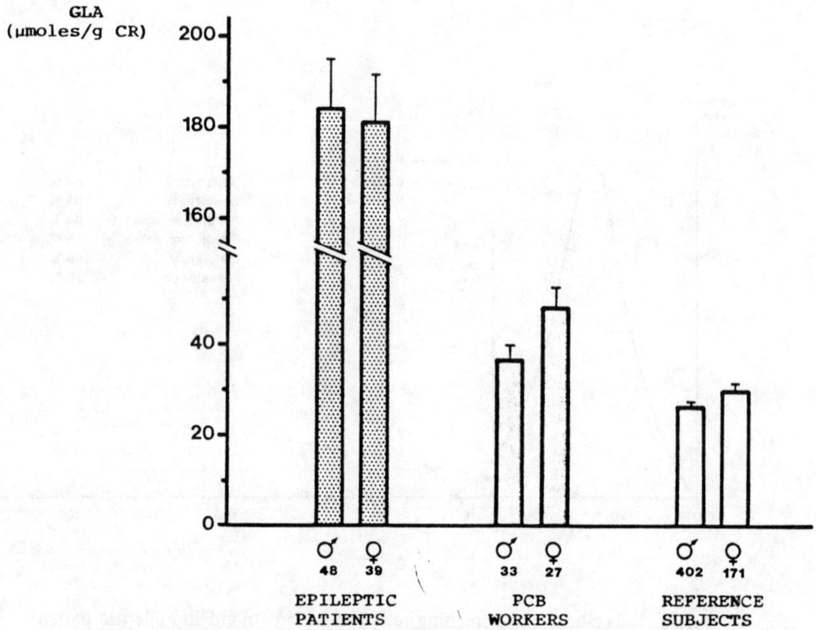

Fig. 3. – Urinary D-glucaric acid excretion in subjects taking antiepileptic medications, in PCB workers and in a reference group.

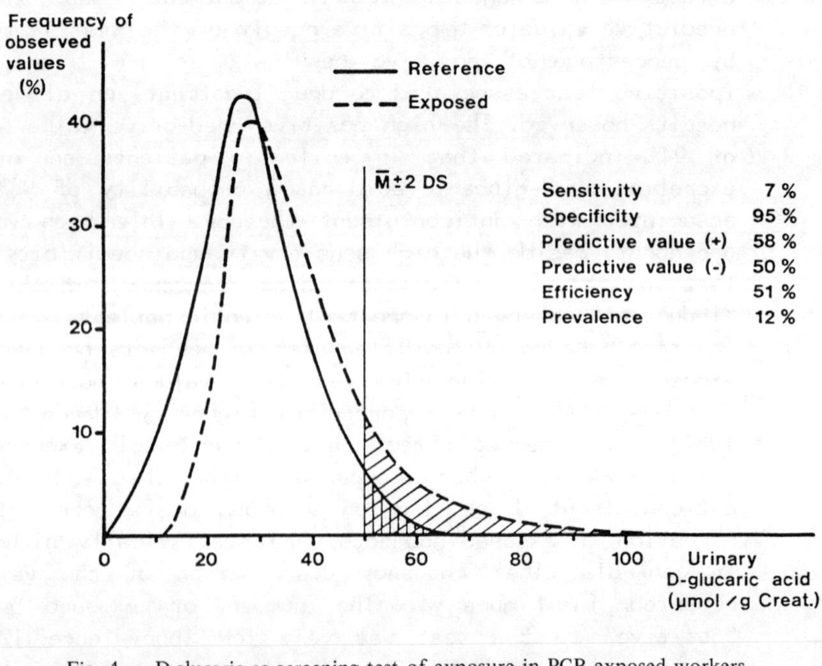

Fig. 4. – D-glucaric as screening test of exposure in PCB exposed workers.

7% and 58% respectively. This means we have only 58 probabilities per cent to classify correctly as exposed, subjects working with PCB and showing altered D-glucaric acid excretion. Many hypotheses may explain this poor sensitivity: first of all, PCB uptake is different among the potentially exposed subjects and only 7% of the subjects had relevant exposure and responded positively to the test; secondly, assuming a similar PCB uptake, the level of exposure is so low that the test is not enough sensitive to reveal it; or thirdly, PCB exposure does not influence D-glucaric excretion at any level of exposure.

The D-glucaric example has been reported to underline the numerous limitations present in using induction as a screening test of exposure. It should be reminded that in order to obtain high values in predictivity of a test, a correct protocol of investigation with measurement in omogenous group of potentially exposed subject and/or studies of a population with high prevalence of the investigated phenomenon are required. These are quite obvious statements but it is well known that negative results tend to be attributed more to poor performance of a test than to wrong experimental design.

As far as microsomal induction is concerned, it means that successful screening investigations can be performed only studying subjects exposed to chemicals with strong evidence of inducing capacity. However, the list of chemicals able to modify microsomal activity it is not as long as it is generally supposed and validation of the different indicators available is still to be performed.

CONCLUSION

Microsomal induction has been increasingly regarded as a sensitive method to evaluate chemical exposure in occupational and environmental medicine after a long experimental phase devoted mainly to pharmacological studies. The aim of such studies was to increase knowledge of how liver reacts to different types of environmental chemicals, and develop methods for monitoring chemical hazards that could supplement or replace the more crude techniques which only detect liver damage after it has became rather severe.

Although there have been several studies on enzyme induction by occupational exposure, many questions on the validity of this approach still remain open. The limited availability of suitable tests has restricted the number of subjects studied and no surveys have been performed to evaluate the predictive

value of the various tests in assessing exposure.
Inadequacies in assessing the relationship between exposure and effects have limited the expansion of microsomal induction as indicator of exposure, and sometimes influenced the correct interpretation of the phenomenon. The limited evidences of inductive effects obtained in studies of occupationally exposed subjects have disappointed the hopes raised by experimental studies.
The discrepancy observed in many cases between experimental and human data may be mainly attributed to the large difference of doses used or involved, but in some cases, inadequacy in evaluating correctly internal doses does not allow to attribuite negative results to absence of exposure or to absence of inductive effects of the chemical for human being.
Regulating mechanisms of differential response to chemical exposure are poorly understood and it is largely accepted that genetic and environmental factors may influence the tests used for induction assessment.
As a future need, the possibility of using induction as early indicator of liver alteration or liver response to chemicals hazard requires a more extensive validation of the tests, with the collection of prospective data. This will allow to elucidate the biological meaning of a single or multiple alteration of different induction tests, and so establish at what level a raised value has prognostic significance for human health.

ACKNOWLEDGEMENTS

The experimental work has been in part supported with the contracts no. 83.02646.56, 84.02332.56 and 85.00581.56 from the National Research Council of Italy (C.N.R.) - Target Project "Preventive Medicine and Rehabilitation"; Subproject "Toxicological Risk", and with a grant of Ministry of Education.

REFERENCES

AITIO A. (1984): Enzyme induction by chemical exposure. In: Biological Monitoring and Surveillance of Workers Exposed to Chemicals. Aitio A., Riihimaki V., Vainio H. eds., Hemisphere Publ. Co., Washington, 309-320.

ALVARES A.P., FISCHBEIN A., ANDERSON K.E., KAPPAS A. (1977): Alterations in drug metabolism in workers exposed to

polychlorinated biphenyls. Clin. Pharmacol. Therap. 22: 140-146.
BROWN R.R., MILLER J.A., MILLER E.C. (1954): The metabolism of methylated aminoazodyes. IV. Dietary factors enhancing demethylation in vitro. J. Biol. Chem. 209: 211-222.
COLOMBI A., MARONI M., ANTONINI C., CASSINA T., GAMBINI A., FOA' V. (1983a): Low-pH method for the enzymatic assay of D-glucaric acid in urine. Clin. Chim. Acta 128: 337-347.
COLOMBI A., MARONI M., ANTONINI C., FAIT A., ZOCCHETTI C., FOA' V. (1983b): Influence of sex, age and smoking habits on the urinary excretion of D-glucaric acid. Clin. Chim. Acta 128: 349-358.
CONNEY A.M., BURNS J.J. (1972): Metabolic interaction among environmental chemicals and drugs. Science 178: 576-585.
CONNEY A.M., BUENING M.K., PANTUCK E.J., PANTUCK C.B., FORTNER J.G., ANDERSON K.E., KAPPAS A. (1980): Regulation of human drug metabolism by dietary factors. In: Environmental Chemicals, Enzyme Function and Human Disease. Ciba Foundation Symposium 76 (new series). Excerpta Medica, Amsterdam, pp. 147-167.
CONNELLY J., BRIDGES J.W. (1980): The distribution and roles of cytochrome P450 in extrahepatic tissues. In: Bridges J.W., Chaseau L.F. eds. Progress in drug metabolism. J. Wiley Chichester 5: 1-111.
DOSSING M. (1984): Non invasive assessment of microsomal enzyme dietary activity in occupational medicine. Present state of knowledge and future perspectives. Int. Arch. Occup. Environ. Health 53: 205-218.
GOLDBERG D.M. (1980): The expanding role of microsomal enzyme induction and its implications for clinical chemistry. Clin. Chem. 26: 691-699.
HILDEBRANDT A.G., ROOTS T., SPECK M., SAALFRANK K., KERFWITZ H. (1975): Evaluation of in vivo parameters of drug metabolizing enzyme activity in man after administration of clemastine, phenobarbital or placebo. Europ. J. Clin. Pharmacol. 8: 327-336.
HUNTER J., MAXWELL J.D., STEWART D.A., WILLIAMS R., ROBINSON J., RICHARDSON A. (1972): Increased hepatic microsomal enzyme activity from occupational exposure to certain organo-chlorine pesticides. Nature 237: 399-401.
MARONI M., COLOMBI A., ANTONINI C., CASSINA T., FOA' V. (1981): D-glucaric acid urinary excretion as a tool for biological monitoring in occupational medicine. In: Organ directed toxicity: chemical indices and mechanisms, Brown S.S. and Davies D.S. eds., Pergamon Press, Oxford, 161-167.

MARONI M., COLOMBI A., FERIOLI A., FOA' V. (1984): Evaluation of porphyrinogenesis and enzyme induction in workers exposed to PCB. Med. Lav. 75: 188-199.
MORGAN D.P., ROAN C.C. (1974): Liver function in workers having high tissue stores of chlorinated hydrocarbon pesticides. Arch. Environ. Health 29: 14-17.
NAKAMURA J., YAKATA M. (1985): Determination of urinary cortisol and 6ß-hydroxycortisol by high performance liquid chromatography. Clin. Chim. Acta 149: 215-224.
OHNHAUS E.E., PARK B.K. (1979): Measurement of urinary 6ß-hydroxycortisol excretion as in vivo parameter in the clinical assessment of the microsomal enzyme inducing capacity of antipyrine, phenobarbitone and rifampicin. Europ. J. Clin. Pharmacol. 15: 139-145.
OTTENVANGER C.F., VAN SITTERT N.J. (1979): Relation between anti-12-hydroxy-endrin excretion and enzyme induction in workers involved in the manufacture of endrin. In: Chemical Porphyria in Man. Strik J.J.T.W.A. and Koeman J.H. eds., Amsterdam, Elsevier North Holland, 123-129.
PARK B.K., BRECKENRIDGE A.M. (1981): Clinical implications of enyme induction and enzyme inhibition. Clin. Pharmacokinetics 6: 1-24.
PENN R., WORTINGTON D.J. (1983): Is serum gamma-glutamyl transferase a misleading test? Brit. J. Ind. Med. 286: 531-535.
POLAND A., SMITH D., KUNTZMAN R., JACOBSON M., CONNEY A.H. (1970): Effect of intensive occupational exposure to DDT on phenylbutazone and cortisol metabolism in human subjects. Clin. Pharmacol. Ther. 11: 724-732.
PRICE-EVANS D.A. (1977): Human pharmacogenetics. In: Drug metabolism from microbe to man. Parke D.V., Smith R.L. eds., Taylor & Francis, London, 369-391.
WEINSHILBOUM R.M. (1984): Human pharmacogenetics. Federation Proceedings 43: 2295-2298.

14

Biochemical indices of lipid peroxidation in occupational and environmental medicine

F. William Sunderman Jr. – Departments of Laboratory Medicine and Pharmacology, University of Connecticut Medical School, 263 Farmington Avenue, Farmington, CT 06032, U.S.A.

ABSTRACT

Free-radical mechanisms of lipid peroxidation are summarized and assays to detect lipid peroxidation products in tissues, body fluids, and breath are discussed (including conjugated dienes in microsomal lipids, lipid hydroperoxides and malondialdehyde precursors in tissue homogenates, lipofuscin pigments in histological sections and tissue extracts, exhalation of volatile hydrocarbons in breath, incorporation of $^{18}O_2$ into tissue lipids, and peroxidative damage to erythrocyte membranes). Evidence that certain toxic metals induce lipid peroxidation in target tissues of experimental animals is cited, including data from the author's laboratory for indices of lipid peroxidation in $NiCl_2$-treated rats. Biochemical methods to assess lipid peroxidation in humans are considered, including a new technique, developed in the author's laboratory, for quantitation of plasma lipoperoxides by HPLC analysis of malondialdehyde precursors. The potential value of such biochemical indices of lipid peroxidation is emphasized for detecting toxicity in persons with occupational and environmental exposures to toxic chemicals.

INTRODUCTION

Lipid peroxidation is a free-radical process whereby certain agents that cause tissue damage (e.g., xenobiotic toxicity, irradiation, aging, nutritional deficiencies) initiate the degradation of polyunsaturated fatty acids in tissues to form hydroperoxy-, keto-, epoxy-, and hydroxy- derivatives, according to an autocatalytic reaction sequence (Figure 1). In step 1, a free-radical ($R^{\bullet}$, a molecule that contains an unpaired electron in an outer orbit) abstracts H^+ from a polyenoic fatty acid (e.g., arachidonic acid) to yield a lipid free-radical, which undergoes resonance shift of double bonds (step 2) to yield a conjugated diene with characteristic spectral absorption at 233 nm. In step 3, the lipid conjugated diene radical reacts with dioxygen to form a lipid peroxy-radical, which abstracts H^+ (step 4) to generate an unstable lipid hydroperoxide. Homolytic decomposition of lipid hydroperoxides (step 5) leads to chain propagation of steps 3 and 4, producing an assortment of alkoxy free-radicals, which may undergo β-scission to yield alkanes and alkenes as metabolic products. More details on mechanisms of lipid peroxidation are presented in recent reviews [1-5].

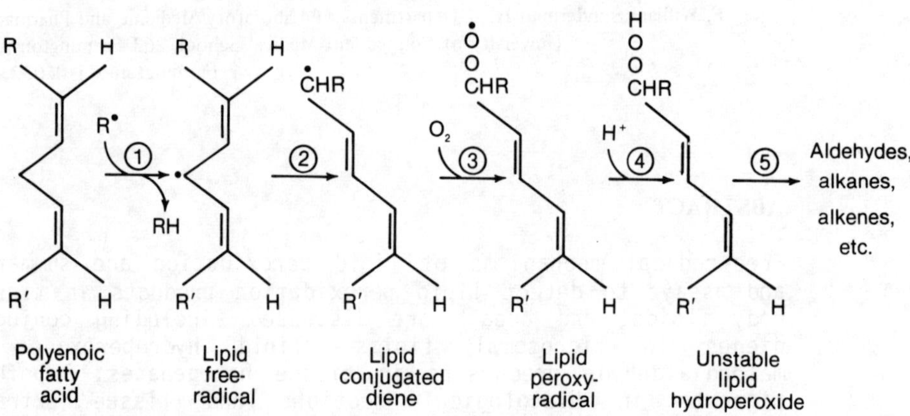

Fig. 1. – Scheme for initiation of lipid peroxidation [1–5].

BIOCHEMICAL INDICES IN ANIMALS

Assays for lipid peroxidation induced in rodents during toxicity trials include (a) measurement of conjugated dienes in lipids extracted from tissue microsomes, based upon spectrophotometry at 233 nm, (b) detection of lipid hydroperoxides in cell membranes by chemiluminescence, (c) quantitation of volatile hydrocarbons in exhaled breath by gas chromatography, (d) identification of lipofuscin pigments in tissues by histochemical reactions or by characteristic fluorescence at 470 nm when excited at 365 nm, (e) spectrophotometric or fluorimetric analysis of lipoperoxides in tissue homogenates by acid hydrolysis to liberate malondialdehyde (MDA), and complexation of MDA with thiobarbituric acid, (f) quantitation of 4-hydroxynonenal and other aldehyde peroxidation products by high-performance liquid chromatography, thin-layer chromatography, or gas chromatography/mass spectrometry, (g) isotope-dilution mass spectrometry of ^{18}O-incorporation into tissue lipids after inhalation of $^{18}O_2$, and (h) detection of erythrocyte membrane damage by timing the filtration of erythrocytes through polycarbonate membranes with 3 μm pores. Further information about such techniques is given in recent articles [6-10].

The various biochemical indices of lipid peroxidation have advantages and drawbacks,- detection of conjugated dienes is specific but imprecise; measurements of exhaled hydrocarbons are non-invasive but subject to interference by products of intestinal bacteria; chemiluminescence techniques are extremely sensitive but lack specificity; quantitation of lipofuscin pigments is sensitive but difficult to standardize; measurement of lipoperoxides by hydrolysis to malondialdehyde and quantitation with thiobarbituric acid is precise and convenient but subject to interferences; analyses of specific aldehydes by HPLC, TLC, or GC-MS are complicated and laborious; determination of ^{18}O-incorporation into lipids is expensive and requires special instrumentation; assay of erythrocyte deformability is simple but non-specific. Slater [1] concluded that the various techniques generally correlate in estimating the relative extent of lipid peroxidation, although the stoichiometries may differ; he cautioned that "it is safest with each particular situation under study to use a variety of methods for cross-checking purposes."

METALS AND LIPID PEROXIDATION

Much attention is currently focused on lipid peroxidation in the pathogenesis of metal toxicity [11], mediated by (a) direct effects of certain metals (e.g., Fe, Cu, Co) on formation of hydroxyl free-radicals from hydrogen peroxide and superoxide via the Fenton and Haber-Weiss reactions [5]; (b) indirect effects of many metals on cellular defenses against peroxidative damage, including depletion of glutathione and inhibition of glutathione peroxidase, superoxide dismutase, or catalase activities [12], and (c) influence of metals on intracellular concentrations of Se, Fe, Cu, and Zn [13]. Recent studies indicate that lipid peroxidation is enhanced in target tissues of rodents exposed to various Cd [13,14], Co [15], Cu [16], Fe [17,18], Hg [19], Ni [20-22], Pb [23], Tl [24], and V [25,26] compounds, based on quantitation of malondialdehyde precursors, measurements of hydrocarbon exhalation and/or assays of conjugated dienes in tissue lipids. The effects of $NiCl_2$ on biochemical indices of lipid peroxidation in rats are illustrated by data in Table 1, derived from studies in the author's laboratory [20-22].

Skepticism about the role of lipid peroxidation in metal-induced toxicity derives primarily from in vitro studies in hepatocyte cultures [27,28], in which antioxidants inhibit lipid peroxidation without blocking cellular efflux of K^+ or lactate dehydrogenase. On the other hand, the role of lipid peroxidation as a molecular mechanism for metal toxicity is supported by findings that metal-induced lipid peroxidation in rodents is enhanced by dietary deficiency of antioxidants, such as selenium and vitamin E, and prevented by administration of vitamin E, selenium, and/or zinc [11].

Table 1: Indices of Lipid Peroxidation in $NiCl_2$–Treated Rats[a].

Assay and units	Samples	$NiCl_2$ dosages (μmol/kg) 0	500	750
Lipoperoxides (nmol MDA/g, dry wt)	Liver	0.51 ± 0.21	1.53 ± 0.54[b]	2.06 ± 0.74[b]
	Kidney	0.39 ± 0.06	0.52 ± 0.05[b]	0.72 ± 0.10[b]
Conjugated dienes (E, 1%, 233 nm)	Hepatic microsomes	0.03 ± 0.02		0.31 ± 0.16[b]
RBC filtration time (sec)	Blood	9.1 ± 1.0		11.3 ± 2.0[b]
Ethene exhalation (nmol/h/kg)	Breath	0.34 ± 0.28	0.71 ± 0.53[b]	0.81 ± 0.31[b]
Ethane exhalation (nmol/h/kg)	Breath	1.21 ± 0.36	1.84 ± 0.83[b]	1.38 ± 0.25

[a] Measurements at 20 - 24 h after sc injection of $NiCl_2$ or vehicle (mean ± SD, 6 - 10 rats/group).
[b] $P < 0.05$ versus vehicle controls.

INDICES OF LIPID PEROXIDATION IN HUMANS

In human subjects, methods for monitoring lipid peroxidation include (a) gas chromatography of ethane, pentane, and other volatile hydrocarbons in exhaled breath [29-31], (b) measurements of lipoperoxides in serum [32-40] or erythrocyte membranes [41,42] by the thiobarbituric acid reaction, and (c) fluorimetric detection of soluble lipofuscin-like substances in serum [8]. Using one or another of these techniques, enhanced lipid peroxidation has been observed in patients with parenteral therapy with lipid emulsion [29], rheumatoid arthritis [31], myocardial infarction [32,39], stroke [32,33], pre-eclampsia [35], alcoholic liver disease [36], diabetic angiopathy [37], burns [38], multiple sclerosis [40], hemodialysis [42], and paraquat poisoning [34].

At present, few clinical laboratories perform biochemical tests for lipid peroxidation, since the methods are complex and unfamiliar. A sensitive, precise, and convenient HPLC technique, recently developed in the author's laboratory, is practical for use in clinical diagnosis. In brief, lipoperoxides in 50 µl samples of EDTA plasma are hydrolyzed (pH 2, 100°C, 60 min) to liberate malondialdehyde (MDA), which is reacted with thiobarbituric acid (TBA) to form the MDA-TBA chromophore. After adjustment to pH 6.8 and removal of precipitated proteins by centrifugation, 20 µl of protein-free extract is injected onto an HPLC column of octadecyl silica gel. The MDA-TBA chromophore is eluted with a mixture (2:3,v/v) of methanol and phosphate buffer (50 mmol/L, pH 6.8) and measured with a spectrophotometric monitor at 532 nm. The retention time of the MDA-TBA chromophore is 4.5 min. In plasma samples from 42 healthy adults, lipoperoxide concentrations averaged 0.39 ± 0.13 µmol MDA/L (range = 0.19 to 0.65). This quantitative technique facilitates the use of plasma lipoperoxide assays as an index of lipid peroxidation in patients exposed to toxic drugs and chemicals.

SUMMARY

Detection of lipid peroxidation induced by free-radical injury of tissue membranes can be achieved by a variety of biochemical techniques. Indices of lipid peroxidation are widely used in toxicity trials in rodents, and, with recent introduction of convenient, non-invasive methods, are rapidly gaining clinical acceptance. Measurements of lipid peroxidation products are useful to detect toxicity in persons with occupational, environmental, or iatrogenic exposures to toxic chemicals.

REFERENCES

1. Slater, S (1984) Free-radical mechanisms in tissue injury. Biochem. J. 222, 1-15.
2. Clavel, JP, Emerit, J, and Thuillier, A (1985) Lipido-peroxydation et radicaux libres. Path. Biol. 33, 61-69.
3. Comporti, M (1985) Lipid peroxidation and cellular damage in toxic liver injury. Lab. Invest. 53, 599-623.
4. Sevanian, A, and Hochstein, P (1985) Mechanisms and consequences of lipid peroxidation in biological systems. Ann. Rev. Nutr. 5, 365-390.
5. Halliwell, B, and Gutteridge, JMC (1985) The importance of free radicals and catalytic metal ions in human diseases. Molec. Aspects Med. 8:89-193.
6. Cadenas, E, Boveris, A, and Chance, B (1984) Low-level chemiluminescence of biological systems. In: Free Radicals in Biology, (WA Pryor, Ed.) Academic Press, New York, Vol. 6, pp 211-242.

7. Frank, H, Hintz, T, and Remmer, H (1980) Volatile hydrocarbons in breath. In: Applied Headspace Gas Chromatography (B Kolb, Ed.) Heyden, London, pp 155-164.
8. Tsuchida, M, Miura, T, Mizutani, K, and Aibara, K (1985) Fluorescent substances in mouse and human sera as a parameter of in vivo lipid peroxidation. Biochim. Biophys. Acta 834, 196-204.
9. Lang, J, Celotto, C, and Esterbauer, H (1985) Quantitative determination of the lipid peroxidation product 4-hydroxy-nonenal by high-performance liquid chromatography. Anal. Biochem. 150, 369-378.
10. Morgan, DL, Dorsey, AF, and Menzel, DB (1985) Erythrocytes from ozone-exposed mice exhibit decreased deformability. Fund. Appl. Toxicol. 5, 137-143.
11. Sunderman, FW Jr (1986) Metals and lipid peroxidation. Acta Pharmacol Toxicol. in press.
12. Younes, M, and Siegers, CP (1984) Interrelation between lipid peroxidation and other hepatotoxic events. Biochem. Pharmacol. 33, 2001-2003.
13. Sugawara, N, and Sugawara, C (1984) Selenium protection against testicular lipid peroxidation from cadmium. J. Appl. Biochem. 6:199-204.
14. Jamail, IS, and Smith, JC (1985) Effects of cadmium on glutathione peroxidase, superoxide dismutase, and lipid peroxidation in rat heart. Toxicol. Appl. Pharmacol. 80, 33-42.
15. Morita, H, Kuno, Y, and Koike, S (1981) Effects of cobalt on superoxide dismutase activity, methemoglobin formation, and lipid peroxide in rabbit erythrocytes. Japan. J. Hyg. 37, 597-600.
16. Dillard, CJ, and Tappel, AL (1984) Lipid peroxidation and copper toxicity in rats. Drug Chem. Toxicol. 7, 477-487.
17. Videla, LA, Fernandez, V, and Valenzuela, A (1985) Effect of ethanol and iron on the hepatic and biliary levels of glutathione and lipid peroxidative indices. Alcohol 2:457-462.
18. Younes, M, Cornelius, S, and Siegers, CP (1986) Ferrous ion supported in vivo lipid peroxidation induced by paracetamol. Res. Comm. Chem. Pathol. Pharmacol. 51, 89-99.
19. Fukino, H, Hirai, M, Hsueh, YM, and Yamane, Y (1984) Effect of zinc pretreatment on mercuric chloride-induced lipid peroxidation in the rat kidney. Toxicol. Appl. Pharmacol. 73, 395-401.
20. Sunderman, FW Jr, Marzouk, AB, Hopfer, SM, Zaharia, O, and Reid, MC (1985) Increased lipid peroxidation in tissues of nickel chloride-treated rats. Ann. Clin. Lab. Sci. 15, 229-236.
21. Donskoy, E, Donskoy, M, Forouhar, F, Gillies, CG, Marzouk, AB, Reid, MC, Zaharia, O, and Sunderman, FW Jr (1986) Hepatic toxicity of nickel chloride in rats. Ann. Clin. Lab. Sci, 16, 108-117.

22. Knight, JA, Hopfer, SM, Reid, MC, Wong, SHY, and Sunderman, FW Jr (1986) Ethene and ethane exhalation in Ni[II]-treated rats, using an improved rebreathing apparatus. Ann. Clin. Lab. Sci., in press.
23. Shafiq-ur-Rehman (1984) Lead-induced regional peroxidation in brain. Toxicol. Lett. 21, 333-337.
24. Brown, DP, Callahan, BG, Cleaves, MA, and Schatz, RA (1985) Thallium induced changes in behavioral patterns: correlation with altered lipid peroxidation and lysosomal enzyme activity in brain regions of male rats. Toxicol. Indust. Health 1, 81-98.
25. Donaldson, J, Hemming, R, and LaBella, F (1985) Vanadium exposure enhances lipid peroxidation in the kidney of rats and mice. Canad. J. Physiol. Pharmacol. 63, 196-199.
26. Elfant, M, and Keen, CL (1985) Vanadium toxicity in adult and developing rats: role of peroxidative damage. Fed. Proc. 44, 497A.
27. Stacey, NH, Cantilena, LR, Jr, and Klaassen, CD (1980) Cadmium toxicity and lipid peroxidation in isolated rat hepatocytes. Toxicol. Appl. Pharmacol. 53, 470-480.
28. Stacey, NH, and Kappus, H (1982) Cellular toxicity and lipid peroxidation in response to mercury. Toxicol. Appl. Pharmacol. 631, 29-35.
29. Wispe, JR, Bell, EF, and Roberts, RJ (1985) Assessment of lipid peroxidation in newborn infants and rabbits by measurements of expired ethane and pentane: Influence of parenteral lipid infusion. Pediat. Res., 19, 374-379.
30. Wade, CR, and Rij, AM Van (1985) In vivo lipid peroxidation in man as measured by the respiratory excretion of ethane, pentane, and other low-molecular-weight hydrocarbons. Anal. Biochem. 150, 1-7.
31. Humad, S, Zarling, EJ, and Skosey, JL (1985) Lipid peroxidation in rheumatoid arthritis: measurement of pentane in breath samples by gas chromatography. Clin. Res. 33, 919A.
32. Satoh, K (1978) Serum lipid peroxide in cerebrovascular disorders determined by a new colorimetric method. Clin. Chim. Acta 90, 37-43.
33. Santos, MT, Valles, J, Aznar, J, and Vilches, J (1980) Determination of plasma malondialdehyde-like material and its clinical application in stroke patients. J. Clin. Pathol. 33, 973-976.
34. Yasaka, S,. Ohya, I, Matsumoto, J, Shiramizu, T, and Sagaguri, Y (1981) Acceleration of lipid peroxidation in human paraquat poisoning. Arch. Intern. Med. 141:1169-1171.
35. Maseki, M, Nishigaki, I, Hagihara, M, Tomoda, Y, and Yagi, K (1981) Lipid peroxide levels and lipid content of serum lipoprotein fractions of pregnant subjects with or without pre-eclampsia. Clin. Chim. Acta 115, 155-161.
36. Suematsu, T, Matsumara, T, Sato, N, Miyamoto, T, Ooka, T, Kamada, T, and Abe, H (1981) Lipid peroxidation in

alcoholic liver disease in humans. Alcoholism: Clin. Exper. Res. 5, 427-431.

37. Nishigaki, I, Hagihara, M, Tsunekawa, H, Maseki, M, and Yagi, K (1981) Lipid peroxide levels of serum lipoprotein fractions of diabetic patients. Biochem. Med. 25, 373-378.
38. Hiramatsu, M, Izawa, Y, Hagihara, M, Nishigaki, I, and Yagi, K (1984) Serum lipid peroxide levels of patients suffering from thermal injury. Burns, 11, 111-116.
39. Aznar, J, Santos, MT, Valles, J, and Sala, J (1983) Serum malondialdehyde-like material in acute myocardial infarction. J. Clin. Pathol. 36, 712-715.
40. Hunter, MIS, Nlemadim, BC, and Davidson, DLW (1985) Lipid peroxidation products and antioxidant proteins in plasma and cerebrospinal fluid from multiple sclerosis patients. Neurochem. Res. 10, 1645-1652.
41. Kobayashi, Y, Yoshimitsu, T, and Usui, T (1983) Evaluation of lipid peroxidation of human erythrocyte hemolysates. J. Immunol. Meth. 64, 17-23.
42. Giardini, O, Taccone-Gallucci, M, Lubrano, R, Ricciardi-Tenore, G, Bandino, D., Silvi, I, Paradisi, C, Mannarino, O, Citti, G, Elli, M, and Casciani, CU (1984) Effects of alpha-tocopherol administration on red blood cell membrane lipid peroxidation in hemodialysis patients. Clin. Nephrol. 21, 174-177.

15

Liver damage and enzyme induction tests among styrene exposed workers

E. Bergamaschi, A. Mutti, M. Ferrari, M. Falzoi, A. Romanelli, G.C. Pasetti*, I. Franchini – Institute of Internal Medicine & Nephrology, Laboratory of Industrial Toxicology &*Department of Infectious Diseases, University of Parma, Italy.

INTRODUCTION

Although styrene toxicity is relatively low, liver damage has been found in animal studies. Some Authors have described histological liver hydropic degeneration, steatosis, congestion [1,2]. Marked elevations of serum enzyme activities, like alanine aminotransferase, aspartate aminotransferase, gamma glutamyl transferase and ornithine carbamoyl transferase have also been reported [1-3].
The same enzyme activities have been shown to be increased among occupationally-exposed workers. However inconsistent results have also been reported [4-6]. Inconsistencies concerning serum bile acids have been found in the available literature [6,7]. So, with regard to the hepatotoxic effects of styrene in exposed workers, opinions vary for at least four reasons:

- differences in selection criteria;
- variability in types and levels of exposure;
- lack of control of confounding variables;
- problems in standardization and comparability of methods.

Moreover, often the results of medical surveillance are used. As a consequence, the above problems can be overcome only with difficulty, because of the lack of any study plan.
Recent research has attracted great attention to

the role of the microsomal enzyme system, suggesting the assessment of microsomal enzyme activity as a possible approach to monitoring occupational hepatotoxicity. Whether styrene can induce microsomal enzymes, is interesting for two reasons:
- first, styrene could be an autoinducer [8];
- second, an induction test could be used as an early indicator, since enzyme induction often precedes toxic effects [9].
The present study was carried out on workers occupationally exposed to styrene in the polyester industry, in order to investigate whether some hepatotoxic or inductive effects could be found and whether exposure - effect / response relationships could be detectable.

SUBJECTS AND METHODS

The exposed group consisted of 126 workers (56 females and 70 males) employed in four factories manufacturing glass-fiber reinforced boats and silos. Their mean age was 39.7 (SD 12.3); the mean duration of exposure to styrene was 9.3 years (SD 5.4). The subjects fulfilled the following selection criteria:
a) prolonged exposure to styrene;
b) no history or clinical signs of hepatic, renal or endocrine diseases;
c) smoking habits of less than 20 cigarettes a day;
d) alcohol intake lower than 80 ml ethanol a day;
e) no regular drug consumption.
The control group consisted of 100 manual workers from the same area. They were healthy workers fulfilling the same selection criteria listed above. Their mean age was 39.3 (SD 8.7) and their distribution according to sex was similar to that of the exposed workers. Each series of samples from exposed and control subjects were examined at the same time.

Biological determinations

Blood. Venous blood samples were drawm before the subjects started work on thursday mornings. Serum aspartate aminotransferase (ASAT), alanine aminotransferase (ALAT), gamma-glutamyl transpeptidase (γ-GT), alkaline phosphatase (AP) and ornithine carbamoyl transferase (OCT) were determined using standard analytical procedures.
Urine. The sum of madelic and phenylglyoxylic acids in the next-morning urine spot samples (i. e. about 15 hours after last exposure) were used as biologi-

cal indicators of internal dose (MPA). The mean value of MPA was 173 (SD 105) mmol/mol of creatinine, corresponding to an 8h TWA concentration of about 25 ppm of styrene in the air.
The urinary excretion of d-glucaric acid (GLC), 6-β-OH-cortisol (6-OH-F) and 17-OH-corticosteroids was also measured by indirect enzyme assay and HPLC, respectively. With the exception of 6-OH-F, which is expressed as the ratio to 17-OH corticosteroids (OHF) the results were related to urinary creatinine.

STATISTICS

All the variables were tested for normality of distribution by using the nonparametric Kolmogorov-Smirnov test at $P > 0.1$. Logarithmic transformation was applied, when appropriate, to obtain a normal distribution. Any significant differences between group means were tested by Student's t test. Comparisons on soubgroups were made by using variance analysis and new Duncan's multiple range test. Pearson's correlation coefficients were used to identify any relationships between variables. Comparisons between prevalences of "abnormal" values were made by the chi-square test.

RESULTS AND DISCUSSION

Styrene exposed-workers showed statistically significant increases in the activity of AP, OCT and -GT when compared with the controls. ALAT, ASAT and serum bile acids did not differ significantly between the two groups. OHF and GLC were also significantly increased (Table 1).

Table 1: Liver function tests among styrene-exposed workers and their matched controls.

TEST	EXPOSED-WORKERS	CONTROL SUBJECTS	p*
	Mean (SD)	Mean (SD)	
AP(μkat/L)	1.78 (0.58)	1.42 (0.57)	.001
OCT(mU/L)	13.65 (3.20)	8.97 (3.61)	.001
SBA(μM/L)	5.50 (2.68)	6.55 (4.53)	n.s.
GLC(mM/M)	4.41 (1.64)	3.47 (1.08)	.001
	GM (GSD)	GM (GSD)	p
ASAT(μkat/L)	0.25 (0.02)	0.25 (0.02)	n.s.
ALAT(μkat/L)	0.25 (0.02)	0.22 (0.02)	n.s.
γ-GT(μkat/L)	0.28 (0.03)	0.20 (0.02)	.001
OHF (%)	5.65 (1.70)	3.00 (1.41)	.001

*Student's t test (on log-transformed values for those variables with log-normal distribution)

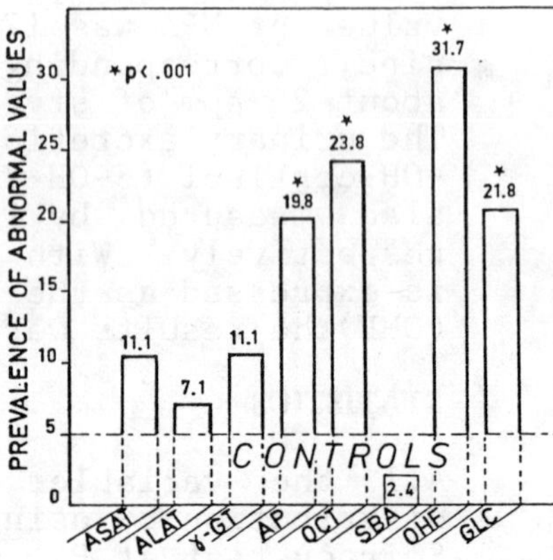

Fig. 1. – Prevalence of values exceeding the 95th percentile of controls for biochemical indicators of liver damage.

Fig 1 shows the prevalence of values exceeding the upper reference limit, i. e. the 95th percentile of control subjects. A high prevalence of abnormal enzyme activities was found among the exposed workers as compared to their controls, the figures ranging from a minimum of 11% for γ-GT to a maximum of 25.8% for OCT. An even higher prevalence of abnormal values among the exposed workers was found for both induction tests, the difference being statistically significant.

The same figures were observed after stratification for styrene exposure, alcohol consumption and microsomal enzyme function.

No significant correlation between duration or intensity of exposure and biochemical indices of liver damage or enzyme induction tests was found, with the exception of OCT and γ-GT, which were significantly correlated with both the intensity and duration of exposure.

The present study might confirm some investigations suggesting that styrene has hepatotoxic and inductive properties [7], since the prevalence of "abnormal" values among the exposed workers was rather high for both some enzyme activities (OCT-AP) and microsomal enzyme induction tests (OHF-GLC).

However, the most widely accepted indicators of liver damage, i. e. serum transaminases, were unchanged. Nor there was any significant increase in serum bile acids, which are presumed to be more sensitive to hepatic injury.

As a result, the observed differences between exposed and control subjects may have weak biological significance, in agreement with the view that occupational exposure to styrene does not cause liver dysfunction. This conclusion would also be suggested by the lack of dose-effect or dose-response relationships for any of the abnormal tests. In spite of an even higher prevalence of abnormal values, enzyme induction tests showed the same trend. Since neither OHF nor GLC were correlated with exposure, the higher prevalence of abnormal values should rely on a presumed higher sensitivity as indicators of effect. Furthermore, the stratification of serum enzyme activities according to the level of both OHF and GLC did not modify the trend already described without any stratification. This suggests that enzyme induction does not interfere with the presumed toxicity of styrene to the liver. Those workers with abnormal induction tests were not protected, nor did they have higher values for liver damage tests. Rather, liver damage and enzyme induction tests showed independent behaviours, abnormal values occurring randomly in all subgroups. The observed effects may thus depend on variables which are not dealt with or on past exposure which is not accounted for in cross-sectional designs. Among these variables, co-exposure to inducing solvents such as acetone or to hepatotoxic chlorinated hydrocarbons may explain our findings. Such exposures in the polyester industry is however limited and can hardly be quantified, being mainly due to cleaning operations not constant over time. Whatever the agent responsible for our findings, it seems to be a weak hepatotoxin, yielding minor changes liver function tests. Their significance to exposed workers' health is difficult to evaluate.

REFERENCES

1) CHAKRABARTI S, BRODEUR J. J. Toxicol. Environ. Health, 1981, 8, 599-607.
2) VAINIO H, JARVISALO J, TASKINEN E. Toxicol. Appl. Pharmacol., 1979, 49, 7-14.
3) PARKKI MG, MARNIEMI J, VAINIO H. Toxicol. Appl. Pharmacol. 1976, 38, 59-70.
4) THIESS AM, FRIEDHEIM M. Scand. J. Work Environ. Health, 1978, 4, suppl. 2, 203-214.
5) HOTZ PA, GUILLEMIN M, LOB M. Scand. J. Work Environ. Health, 1980, 6, 206-215.
6) HARKONEN H, LEHTNIEMI A, AITIO A. Scand. J. Work Environ. Health, 1984, 10, 59-61.
7) EDLING C, TAGESSON C. Brit J Ind Med, 1984, 41, 257-259.
8) LAMBOTTE-VANDERPAER M, NOEL G, ROLLMANN B, MERCIER M, ROBERFROID M. Arch. Toxicol.,1978,1,287.
9) NOTTEN WRF, HENDERSON PT. Int. Arch. Occup. Environ. Health, 1977, 38, 197-207, 209- 220.

Supported by Regione Emilia-Romagna

16

Liver function of workers with pesticides assessed by endogenous parameters and antipyrine test

P.C. Bragt[1], E.J. Brouwer[1], J.H.M. Schellens[2], W.J.A. Meuling[1], A. Poppema[1] and D.D. Breimer[2] – Medical Biological Laboratory TNO[1], P.O. Box 45, 2280 AA Rijswijk, and Center for Bio-pharmaceutical Sciences, University of Leiden[2], P.O. Box 9502, 2300 RA Leiden, The Netherlands.

SUMMARY

The objective of this study was to investigate whether pesticide formulation workers (44 males) mainly working with organophosphorous and chlorinated hydrocarbon pesticides, had symptoms of liver cytotoxicity and/or cholestasis, and to establish whether liver enzyme induction was detectable in the pesticide workers when compared with a control group of dairy workers (29 males). The following serum enzymes were determined: alanine aminotransferase (ALT), aspartate aminotransferase (AST), gammaglutamyltransferase (GT), glutamate dehydrogenase (GLDH) and 5'-nucleotidase (5-ND). Furthermore the clearances of antipyrine and its metabolites were determined in saliva and urine, respectively. The excretion of D-glucaric acid and 6β-hydroxycortisol in the urine was also measured. The results of both the tests on liver cytotoxicity and/or cholestasis, and the tests on liver enzyme induction were not significantly different between the exposed and the control group. It is concluded that there were no detectable effects of pesticides on liver function at the work conditions in the formulation plant.

INTRODUCTION

Many authors have shown that working with pesticides may have an effect on liver function (see Table 1). In general,

Table 1: Exposure to pesticides and effects on liver function.

Exposure kind and level	Effect observed	Reference
Formulation	D-glucaric acid excretion ↑	[1]
DDT formulation (5-52 ppb p,p'-DDE in blood)	No effect on D-glucaric acid	[2]
Aldrin/dieldrin production	Idem	[3]
Endrin production (> 0.13 mg/g creatinine)	D-glucaric acid excretion ↑	[4]
DDT production (573 ± 60 ppb p,p'-DDE in blood)	6β-OHF excretion ↑	[5]
Aldrin/dieldrin production (5-42 ppb in blood)	No effect on 6β-OHF excretion	[6]
Endrin production (< 5 ppb in blood)	6β-OHF excretion ↑	[6]
Malathion production	antipyrine clearance ↑	[7]
Sprayers lindane/DDT (6-10 ppb lindane, 10 ± 9 ppb p,p'-DDE in blood)	antipyrine half-life ↓	[8]
Gardeners and herbicide production workers	antipyrine clearance ↑	[9]

exposure to chlorinated hydrocarbon and some organophosphorous pesticides can lead to proliferation of the hepatic endoplasmatic reticulum, depending on the actually absorbed dose and individual sensitivity. The biological effect of liver enzyme induction, which is also of toxicological importance (e.g. tumor promotion), can be established by noninvasive methods like the excretion of D-glucaric acid [10] and 6β-hydroxycortisol [11] in the urine. Furthermore the half-life and clearance of antipyrine [12] are indicative of the functional liver mass and both parameters are sensitively influenced by inducers and inhibitors of the hepatic drug metabolizing enzyme system (HDMS). The use of antipyrine as a model drug has the advantage that differential stimulation and inhibition of distinct metabolic enzymes of the HDMS can be determined simultaneously without the need of biopsy samples [13].

The availability of these test methods prompted us to perform this study with pesticide workers with reported effects on liver function (Table 1). The purpose was twofold:

- to establish whether effects on liver function could be observed in a modern Dutch pesticide formulation plant;
- to have an impression of the sensitivities of the test methods used.

GROUPS OF INVESTIGATION AND METHODS

Male Dutch workers (n=44) from a pesticide formulation plant represented the "exposed group". During 3 months preceding the study, the following pesticides were formulated:

- herbicides	triazole and triazine compounds urea derivatives chlorophenoxy compounds carbamates organic acids dinitrophenol compounds
- fungicides	dithiocarbamates chloroalkyl thio compounds aromatic hydrocarbons aliphatic and alicyclic compounds organotin compounds heterocyclic nitrogen compounds
- insecticides	organophosphorous compounds chlorinated hydrocarbons carbamates

In view of the changeable batch-wise formulation of pesticides and the lack of information about the relationship between doses of pesticides and the effects on liver function, no attempt was made to establish exposure levels.

Dairy workers (n=29 Dutch males) from a factory at 5 km distance from the formulation plant represented the control group. They were not exposed to organic chemicals. The only compounds were sodium hydroxide, sodium hypochlorite, hydrogen peroxide and nitric acid, mainly intended for cleansing and desinfection purposes.

All workers admitted to the study were in a good state of health and used no drugs. None of them had an anamnesis of liver disease, according to the questionnaire. The characteristics with respect to age, smoking and drinking habits for both groups are presented in Table 2. Only the caffeine index (coffee + 0.6 tea) differed significantly ($P<0.05$, two-tailed Wilcoxon test).

Table 2: Characteristics of the exposed group and the control group.

	Controls	Exposed
Age	40.2 (21-59)	35.7 (19-59)
Smoking (sig./day)	9.8 (0-32.5)	14.2 (0-41)
Alcohol (gl./w)	11.5 (0-38)	12.4 (0-50)
Coffee + 0.6 tea (cups/day)	9.1 (3-21)	7.9 (0-25)*

* $p < 0.05$ (two tailed Wilcoxon test)

The following tests were performed:

- alanine aminotransferase (ALT), aspartate aminotransferase (AST), glutamate dehydrogenase (GLDH) (Boehringer test kit), alkaline phosphatase (AP), gamma-glutamyl transferase (GT) (Abbott test kit) and 5'-nucleotidase (5-ND) (Sigma Diagnostics test kit) in serum from a venous blood sample;
- D-glucaric acid (DGA) in a 24 hours urine collection, according to a modification of the method of Colombi et al [14,15];
- 6β-hydroxycortisol (6β-OHF) according to a newly developed HPLC method [16] and 17-hydroxycorticosteroids (17-OHCS) (bioMerieux test kit) in a 24 hours urine collection;
- antipyrine (AP) in saliva obtained with cotton plugs 8, 24 and 32 hours after the ingestion of 250 mg antipyrine (Brocacef) dissolved in 50 ml of tap water, and the antipyrine metabolites 3-hydroxymethyl antipyrine (HMA), 4-hydroxy antipyrine (OHA) and norantipyrine (NORA) in a 24 hours and a subsequent 8 hours urine collection, according to Teunissen et al [17].
- The clearances of antipyrine and its metabolites were calculated from the area under the elimination curve, after extrapolation to $t=\infty$. The bioavailability was estimated at 100% and the volume of distribution at 40 litres. The saliva clearance of antipyrine was also calculated with the one-sample test proposed by Døssing et al [18] for the respective collection times.
The coefficient of variation of all methods was less than 8%.

RESULTS

Serum enzymes

The results are presented in Table 3. Only the GLDH activity differed significantly between the exposed and the control group (two-tailed Welch or Student t-test).

Table 3: Serum enzyme activities in pesticide workers and dairy workers.

Enzyme	Activity (U/L serum) Controls (n=29)	Exposed (n=44)
ALT	7.3 (3.4-22.4)	6.3 (1.7-21.6)
AST	6.7 ± 0.84	6.8 ± 0.39
AP	68.8 ± 3.41	66.4 ± 1.53
GLDH	2.3 (1.1-6.0)	1.9 (0.9-4.3)*
GT	18.8 (9-49)	18.6 (9-51)
5-ND	4.2 ± 0.24	4.6 ± 0.20

* $P < 0.05$ (two-tailed Student test)

Test on liver enzyme induction

Neither the DGA and the 6β-OHF excretion rate (whether or not corrected for total adrenal activity through the 17-OHCS excretion rate), nor the antipyrine clearance and half-life from salive showed a significant difference between the exposed and the control group (two-tailed Welch or Student t-test, Table 4). The same was true for the clearances of HMA, OHA, HMA and residual AP in the urine (Figure 1).

Table 4: Enzyme induction parameters in pesticide workers and dairy workers.

Test	Controls (n=25-29)[1]	Exposed (n=37-44)
D-glucaric acid (µmoles/24 h)	42.3 (24.5-70.9)	42.7 (20.0-90.7)
6β-hydroxycortisol (µg/24 h)	242.9 ± 12.4	244.2 ± 11.1
6β-OHF/17-OHCS	0.026 ± 0.010	0.029 ± 0.015
Antipyrine $T_{\frac{1}{2}}$ (h)	10.8 (6.2-20.3)	10.3 (6.0-18.5)
Antipyrine clearance (ml/min)	58.0 (26-80)	55.8 (23-103)

[1] Some samplings were not performed correctly by the workers. This lead to a reduction in the number (n) of tests.

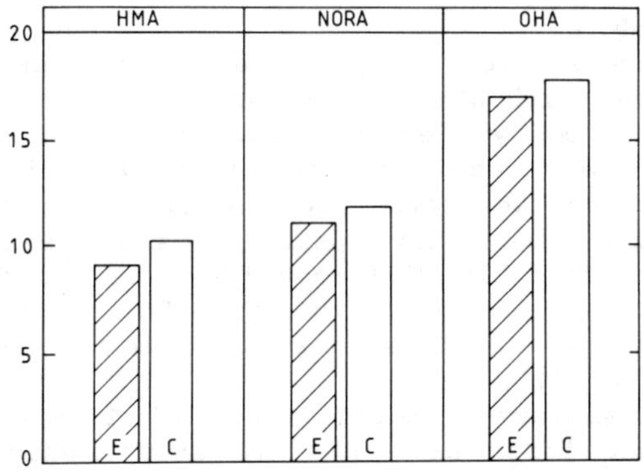

Fig. 1. – Clearances for production of HMA, NORA and OH in the exposed (E) and control (C) group.

CORRELATION BETWEEN TWO METHODS FOR THE ESTIMATION OF AP CLEARANCE

The best correlation between the one-sample test and the area-under-the-curve method was observed 24 hours after the ingestion of AP (Figure 2). The correlation coefficient was 0.9617 (95% confidence limits 0.9401 and 0.9762). After 8 hours the correlation coefficient was 0.9139 (CL 0.8670 and 0.9461) and after 32 hours it was 0.9352 (CL 0.8994 and 0.9596). The 8 and 32 hours data are not presented graphically.

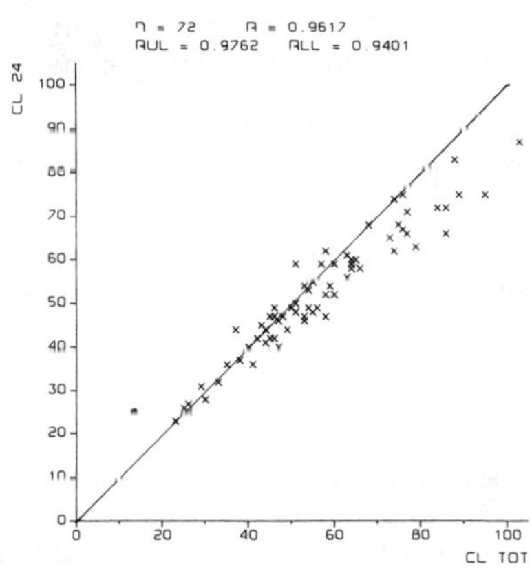

Fig. 2. – Correlation between a multi-sample and a one-sample method for the calculation of the antipyrine clearance from saliva. Solid line is 45° line.

DISCUSSION

Some studies [5,9] indicate that changes in HDMS activity may occur without evidence of (sub-)clinical liver pathology as reflected by elevated serum enzyme values. Changes in HDMS activity are considered to be an early biological effect of especially chlorinated hydrocarbon pesticides. In the present study the serum enzyme activities in pesticide workers were not significantly different from the activities in a control group, with the exception of GLDH, which was lower in the exposed group. This has to our opinion no biological significance.

Liver enzyme induction could not be established with the different tests used in this study, when pesticide workers were compared with the control group. There are two possible explanations:

- the exposure was too low to result in HDMS activity changes;
- the pesticide workers were exposed to both inhibiting and inducing compounds at the same degree.

Alvares [19] mentions the inhibitory properties of organophosphorous compounds on the HDMS (mainly insecticides) whereas chlorinated hydrocarbon pesticides (herbicides, fungicides and insecticides) have inducing properties [4-6,8]. When the pesticide workers in this study were divided into 3 groups, namely workers with powdered insecticides (n=8), workers with liquid insecticides and herbicides (n=13) and service personnel (n=16), no difference between the tests of HDMS function could be established (data not shown). Thus, it is assumed that the low level of exposure is responsible for the lack of effects on HDMS activity, rather than the pattern of exposure.

Finally, we have been able to confirm the findings of Døssing et al [18] that the one-sample clearance test of antipyrine from saliva is convenient for use in industrial populations. The optimal sampling time appeared to be 24 hours after dosing of antipyrine.

ACKNOWLEDGEMENT

This work was supported by a grant from the Directorate General of Labour of the Dutch Ministry of Social Affairs and Employment.

REFERENCES

[1] Seutter-Berlage, F., M.A.P. Wagenaars-Zegers, J.M.T. Hoog Antink and H.A.T.M. Custers. Urinary thioether en D-glucaric acid excretion after industrial exposure to pesticides. In: Chemical Porphyria in Man; ed. by J.J.T.W.A. Strik and J.H. Koeman. Amsterdam, Biomedical Press, Elsevier/North-Holland, 1979, p. 131-135.

[2] Morgan, P.D. and C.C. Roan. Liver function in workers having high tissue stores of chlorinated hydrocarbon pesticides. Arch. Environ. Hlth. 29 (1974) 14-17.

[3] Hunter, J., J.P. Maxwell, D.A. Stewart and R. Williams. Increased hepatic microsomal enzyme activity from occupational exposure to certain organochlorine pesticides. Nature 237 (1972) 399-401.

[4] Ottevanger, C.F. and N.J. van Sittert. Relation between anti-12-hydroxy-endrin excretion and enzyme induction in workers exposed to endrin. In: Chemical Porphyria in Man ed. by J.J.T.W.A. Strik and J.H. Koeman, Amsterdam, Elsevier/North-Holland Biomed. Press, 1979, p. 123-129.

[5] Poland, A., D. Smith, R. Kuntzman, M. Jacobson and A.H. Conney. Effect of intensive occupational exposure to DDT on phenylbutazone en cortisol metabolism in human subjects. Clin. Pharmac. Therap. 11 (1970) 724-732.

[6] Jager, K.W. Aldrin, dieldrin, endrin and telodrin; an epidemiological and toxicological study of long-term occupational exposure. Amsterdam, Elsevier Publishing Company, 1970.

[7] Uppal, R., P.R. Sharma and R.R. Chaudhury. Effect of pesticide exposure on human microsomal enzyme induction. Human Toxicol. 1 (1982) 155-158.

[8] Kolmodin-Hedman, B. Changes in drug metabolism en lipoproteins in workers occupationally exposed to DTT and lindane. Arh. Hig. Rada. 24 (1973) 289-296.

[9] Døssing, M. Changes in hepatic microsomal enzyme function in workers exposed to mixtures of chemicals. Clin. Pharmacol. Ther. 32 (1982) 340-346.

[10] Sotaniemi, E.A., F. Medzihradsky and G. Eliasson. Glucaric acid as an indicator of use of enzyme-inducing drugs. Clin. Pharmac. Ther. 15 (1973) 417-423.

[11] Ohnhaus, E.E. and B.K. Park. Measurement of urinary 6β-hydroxycortisol excretion as an in vivo parameter in the clinical assessment of the microsomal enzyme inducing capacity of antipyrine, phenobarbitone and rifampicin. Eur. J. Clin. Pharmacol. 15 (1979) 139-145.

[12] Vesell, E.S. The antipyrine test in clinical pharmcology: conceptions and misconceptions. Clin. Pharmacol. Ther. 26 (1979) 275-286.

[13] Danhof, M., D.P. Krom and D.D. Breimer. Studies on the different metabolic pathways of antipyrine in rats: Influence of phenobarbital en 3-methylcholanthrene treatment. Xenobiotica 9 (1979) 695-702.

[14] Colombi A., M. Maroni, C. Antonini, T. Cassina, A. Gambini and V. Foà. Low-pH method for the enzymatic assay of D-glucaric acid in urine. Clin Chim Acta 128 (1983) 337-347.

[15] Meuling, W.J.A. and J.J. van Hemmen. An improved semi-automated enzymatic determination of D-glucaric acid in urine (this book).

[16] Bragt P.C. and A. Poppema, unpublished.

[17] Teunissen, M.W.E., J.E. Meerburg-Van der Torren, N.P.E. Vermeulen and D.D. Breimer. Automated HPLC-determination of antipyrine and its main metabolites in plasma, saliva and urine, including 4,4'-dihydroxyantipyrine. J. Chromatogr. 278 (1983) 367-378.

[18] Døssing, M., H. Enghuse Poulsen, P.B. Andreasen and N. Tygstrup. A simple method for determination of anti-pyrine clearance. Clin. Pharmacol. Ther. 32 (1982) 392-396.

[19] Alvares, A.P. Interaction between environmental chemicals and drug biotransformation in man. Research Review. Clin. Pharmacokin. 3 (1978) 463-477.

17

Assessment of early hepatotoxicity during exposure to solvent mixtures

G. Franco, R. Fonte, G. Tempini, F. Candura – Dipartimento di medicina preventiva, occupazionale e di comunità dell' Università di Paviá – Fondazione Clinica del lavoro di Paviá

ABSTRACT

To assess the potential hepatotoxicity of solvent mixtures, we determined serum bile acid (SBA) concentrations as compared with conventional liver function tests in a selected group of workers occupationally exposed to a mixture of organic solvents (mostly toluene, xylene, acetone, n-butylacetate, n-butanol, ethylacetate) and in a reference group. The results demonstrate the higher sensitivity in detecting liver dysfunction achieved with the SBA test as compared with conventional hepatic function tests. As increased SBA concentrations are considered to reflect an impairment of anion transport across the liver, higher SBA levels in the group of workers exposed to organic solvents might be explained as a slight and early liver failure sign.

INTRODUCTION

Liver injury has long been associated with occupational exposure to a wide variety of chemicals. In particular, several chemicals used as solvents are generally cosidered potentially hepatotoxic in man (Døssing and Skinhøj, 1985). Recent studies suggest that occupational exposure to organic solvents at levels ranging from about half to slightly more than the TLV proposed by the ACGIH (1984) do not cause any increase in liver enzyme activities (Waldron et al.,1982, Kurppa et al.,1983, Lundberg et al.,1985, Tagesson, 1985). However, other studies report abnormal liver function tests in chemical workers exposed to a variety of solvents (Sotaniemi et al.,1982, Fishbein et al.,1983, Tagesson, 1985).

Despite these controversial findings, the failure to find liver impairment may signify either that solvent exposure was insuffi= cient to cause injury or dysfunction or that the parameters used were not sufficiently sensitive to detect any hepatic change. Conventional liver function tests can in fact be rather insensi= tive and are unable to rule out the existence of subclinical dis= ease. However, a further possibility has been put forward: the use of serum bile acids (SBA) as indicators of hepatocytic func= tion (Editorial,1982). Because of their higher sensitivity and specificity for chemical liver injury than for non-chemical hepa= tic disease (Liss et al.,1985), SBA levels have been assessed as indicators of effects in occupational medicine. To study liver function during exposure to solvent mixtures, we determined SBA concentrations as compared with conventional liver function tests using a group of workers occupationally exposed to a mix= ture of organic solvents and in a reference group.

MATERIALS AND METHODS

A group of workers exposed to organic solvents were selected from among 55 workers at a chemical factory producing fillers varnishes using the following criteria: (i) exposure to solvent mixtures for over 2 years, (ii) daily ethanol consumption less than 50 g, (iii) no history of hepatic disease, (iv) no drug in= take in the previous 3 months. Subjects admitted to the study satisfied all the above criteria. The exposed group was thus restricted to 30 male subjects, whose mean age was 36 years (age range 22-56), and mean exposure duration was 10 years (range 2-26). Table 1 shows the exposure levels in the breathing zone measured by personal sampling in the last 6 years. The workers were exposed to between 6 and 9 solvents, mostly toluene,xylene, acetone, n-butylacetate, n-butanol, ethylacetate. All figures

Table 1: Exposure levels in the last 6 years.

Solvent	Exposure level (range mg/m^3)	Median exposure (mg/m^3)	TLV-TWA (mg/m^3)
Acetone	5 - 1448	130	1780
n-butylacetate	18 - 1683	154	710
n-butanol	8 - 96	25	150
n-hexane	289 - 699	494	180
Ethylacetate	24 - 739	165	1400
Methylene chloride	44 - 642	268	350
Methylisobutylketone	25 - 124	109	205
Styrene	3 - 390	117	215
Toluene	3 - 311	65	375
Xylene	8 - 658	123	435

are expressed as 8-hrs Time Weighted Average and compared with the TLV-TWA adopted by ACGIH. Controls (mean age 35 years, age range 22-53) were chosen from among unskilled blue collar workers using the above criteria, but excluding, of course, exposure to solvents and other known hepatotoxic chemicals. A 10 ml blood sample was taken from fasting control and currently exposed sub=jects to assess SBA concentrations and activities of alanine aminotransferase (ALT,EC 2.6.1.2), aspartate aminotransferase (AST,EC 2.6.1.1), and γglutamyl transferase (GGT,EC 2.3.2.2). The serum was separated within 20 mins and kept at -20°C until analysis was carried out. ALT,AST and GGT determinations were carried out by conventional methods. SBA levels were measured by an enzymatic method (Enzabile, Nyegaard). Reference values for liver enzyme activities were those commonly used by our labora=tory. Having found that the distribution of SBA levels in the reference group was Gaussian, the upper limit (=5.6 µmol/l) was calculated as a mean value of the reference group (=2.8 µmol/l) plus 2 s.d. (1 s.d.=1.4). Student's t-test and chi-square test were used for the statistical assessment.

RESULTS AND DISCUSSION

The mean levels of liver enzyme activities in the exposed and control groups were similar (ALT: 0.31 ± 0.23 vs 0.27 ± 0.21; AST: 0.27 ± 0.13 vs 0.28 ± 0.15; GGT: 0.40 ± 0.24 vs 0.45 ± 0.33) and no stat=istical differences were found between the 2 groups (Fig.1). Mean SBA levels increased in the exposed group (8.0 ± 6.0vs 2.8 ± 1.4)

Fig. 1. – Distribution of serum liver enzyme activities in the group of exposed to solvent mixture (●) and in controls (○). Dotted lines indicate the upper limits of normal.

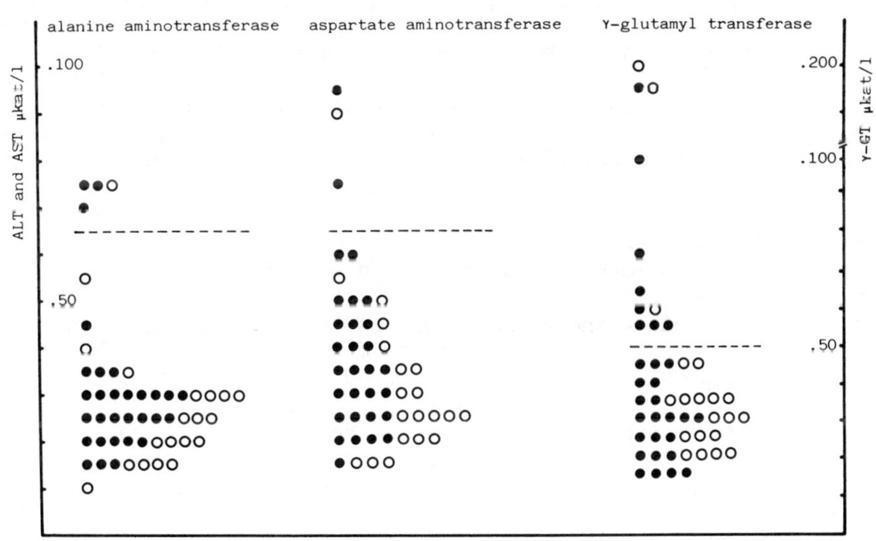

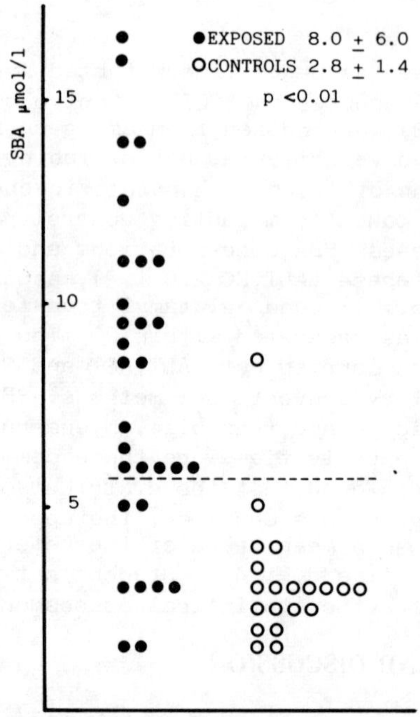

Fig. 2. – Distribution, mean level and standard deviation of serum bile acids (SBA) concentrations in the group of workers exposed to solvent mixture (●) and in controls (○). Dotted line indicates the upper limit (x+2s.d.) of the controls.

and the difference as compared with the controls was significant ($p < 0.01$) (Fig.2). Biological monitoring of workers occupationally exposed to chemicals and especially to organic solvents is based both on the assessment of the absorbed dose and indices of effect. Nevertheless, existing methods for detecting early signs of liver impairment or dysfunction seem to be rather insensitive. However, increased SBA levels were observed in styrene exposed workers (Vihko et al.,1984, Edling and Tagesson, 1984), in solvent mixtures exposed workers (Vihko et al.,1984) and in vinyl chloride exposed workers (Vihko et al.,1984, Liss et al.,1985), even in the absence of increased liver enzyme activities, whereas another study did not confirm these observations in styrene exposed workers (Härkonen et al.,1984).

This study shows that mean SBA levels were higher and differ significantly within the group of subjects occupationally exposed to solvent mixtures as compared with the control group. In 22 out of 30 exposed workers (73%) SBA levels were higher than 5.6 μmol/l (the cut-off value) as compared with 1 out of 20 controls (5%). Because there is no evidence that cases or controls had abnormal alcohol intake, or other possible causes of liver injury, the results are interpreted as a consequence of exposure to the chemicals.

CONCLUSIONS

The observation of higher SBA levels in the group of workers cur= rently exposed to organic solvents might be explained as a change in hepatocyte function. As increased SBA concentrations are con= sidered as reflecting an impairment of anion transport across the liver (Kaplowitz et al.,1973), this study may suggest that -in the absence of any other factor affecting the liver- an increased SBA levels were due to exposure to organic solvents, either suggest= ing hepatocyte injury or simple competition for liver uptake of solvents or their metabolites. Hence, we suggest that the dis= crepancies with other reports might be caused by the lesser sen= sitivity of serum liver activities used for detecting hepatocytic function or even by the different degree, type and duration of exposure to chemicals.

Conventional liver function tests seem to be rather insensitive for early liver disease detection. In fact their measurement does not rule out the existence of subclinical diseases. However, knowledge of exposure, the use of sensitive and specific biologi= cal tests, and careful clinical evaluation make it possible to obtain important information about the assessment of hepatotoxic risks (Fishbein et al.,1983). Among the tests to be used, SBA de= termination in biological monitoring of workers exposed to poten= tially hepatotoxic chemicals might be proposed.

REFERENCES

ACGIH. TLVs Threshold limit values for chemical substances in the work environment adopted by ACGIH for 1984-85.(1984) Cincinnati.

DØSSING M, SKINHØJ P. Int Arch Occup Environ Health 1985;56:1-21.

EDITORIAL. Lancet 1982;2:1136-1138.

EDLING C, TAGESSON C. Br J Ind Med 1984;41:257-259.

FISHBEIN A, ROSS RR, LERMAN Y. Lancet 1983;1:129.

HÄRKONEN H, LEHTUIEMI A, AITIO A. Scand J Work Environ Health 1984;10:59-61.

KAPLOWITZ N, KOK E, JAVITT NB. JAMA 1973;225:292-293.

KURPPA K, TOLA S, HERNBERG S, TOLONEN M. Lancet 1983;1:129.

LISS GM, GREENBERG RA, TAMBURRO CH. Am J Med 1985;78:68-75.

LUNDBERG I, HAKANSSON M. Br J Ind Med 1985;42:596-600.

SOTANIEMI EA, SUTINEN SE, SUTINEN SI, ARRANTO AJ, PELKONEN RO. Acta Med Scand 1982;212:207-215.

TAGESSON C. Scand J Work Environ Health 1985;11(suppl 1):101-103.

VIHKO R, VIHKO P, MAENTAUSTA O, PAKARINEN A, JANNE O, YRJANEHEIK= KI E. In: Aitio A, Rihimaki V, Vainio H (Eds) Biological monitor= ing and surveillance of workers exposed to chemicals. Hemisphere Publishing Corporation.(1984) Washington.

WALDRON HA, CHERRY N, VENABLES H. Lancet 1982;2:1276.

18

Serum collagen peptides as markers of hepatic damage

M.C. Cantaluppi, G. Annoni, M.F Donato, P. Lampertico, B. Khlat, N. Dioguardi, M. Colombo – Clinica Medica III, Università di Milano

ABSTRACT

Serum type III procollagen peptide (PIIIP), a marker of type III collagen metabolism, was high during hepatic fibroplasia. In view of the link between fibrosis and progressive hepatic disease, PIIIP serum levels might be useful for monitoring the course of chronic liver disease. Normal levels of PIIIP range from 4 to 14 ng/ml. PIIIP in patients with chronic disease was significantly elevated in 81% of those with chronic active hepatitis and active cirrhosis. It was normal in all with alcoholic steatosis, in 95% of those with chronic persistent hepatitis and 63% of those with idiopathic hemochromatosis. In 76 patients with acute viral hepatitis, who were followed for two years, sPIIIP was significantly elevated during the acute phase was normal by the 6th month in all patients who recovered, but remained elevated in those who developed chronic active hepatitis. sPIIIP levels were correlated with serum ALT and the histologic parameters of hepatic inflammation. This indicated that sPIIIP may help differentiate cases with progressive liver disease and fibrosis from those with inactive fibrosis or benign inflammation.

INTRODUCTION

It has been suggested that serum procollagen type III peptide levels may serve as a marker for fibroplasia during lung, bone, bone-marrow or liver diseases (1-3). Serum levels of PIIIP mainly reflect release of the aminoterminal portion from cells of organs in which there is biotransformation of type III collagen (4). High levels of sPIIIP have been previously reported in adult patients with progressive liver diseases (alcoholic hepatitis, chronic active hepatitis and active cirrhosis), in close correlation with histological evidence of hepatic fibrosis (5-6). To determine the clinical value of sPIIIP as an index of hepatic fibrogenesis, we elected to study its serum levels in patients with liver diseases of different etiology. Results of these studies were correlated with major laboratory parameters of hepatic damage and with liver histology.

MATERIALS AND METHODS

401 untreated adult patients with acute and chronic liver diseases were studied. The epidemiological and clinical data for the patients are shown in Table 1. Acute viral hepatitis (AVH) was defined according to internationally agreed criteria (7). Chronic hepatitis and cirrhosis were diagnosed on the basis of clinical, biochemical and morphological criteria (8). Activity of the disease was scored by the histological activity index of Knodell (HAI) (9). 40 healthy individuals served as controls. Blood samples were taken from patients with AVH at the ALT peak and after 2 and 4 weeks and 3, 6, 12, 18 and 24 months. Sera aliquots were kept frozen at -20°C until assayed. sPIIIP was measured with a commercial radioimmunoassay kit (Behring) that contains an 125-I labelled bovine peptide and a specific antibody produced in rabbits.

RESULTS

In 40 healthy controls,sPIIIP levels were 4-14 (mean: 9.5)ng/ml. It was significantly elevated in all patients with AVH (range: 12.5-153 ng/ml). sPIIIP remained elevated on the average in the 76 patients studied prospectively for 24 months (range: 8.4 22.2 ng/ml) and in 4 who developed chronic active hepatitis. sPIIIP returned to normal values in 59 patients who recovered and in 7 who developed chronic persistent hepatitis.

81% of the patients with chronic active hepatitis and active cirrhosis had significantly high PIIIP (range 14.6-41.6, mean 18, ng/ml). Conversely, all patients with alcoholic steatosis, 95% of those with chronic persistent hepatitis and 63% of those with idiopathic hemochromatosis had normal sPIIIP levels (Table 1). In patients with AVH, sPIIIP levels were correlated with ALT, AST, and bilirubin (r= 0.74 - 0.73 - 0.70; $p < 0.01$), and with albumin in alcoholic patients (r= 0.51; $p < 0.05$). In chronic active hepatitis sPIIIP was also correlated with the

the histological parameters for hepatic inflammation ($r= 0.46$, $p < 0.01$), and in patients with cirrhosis, with those for fibrosis ($r= 0.30$, $p < 0.05$).

Table 1: Serum PIIIP levels in patients with miscellaneous liver disease.

Diagnosis	Pts (No)	Males	Mean age (yrs)	Mean sPIIIP (ng/ml)
AVH	126	71	40	35.7±16.3
CPH	35	22	37	9.7± 2.7
CAH	57	40	38	18.9± 7.3
AC	40	29	57	19.5± 6.6
PBC	15	0	51	28.1± 8.3
HC	15	9	63	32.4±17.1
S	11	11	46	8.5± 2.2
SF	38	34	44	11.5± 4
C	34	24	50	25.1±18.9
F.IHC	9	6	29	11.9± 4.6
C.IHC	21	20	48	13.9± 6.1
Total	401			

AVH = acute viral hepatitis; CPH = chronic persistent hepatitis; CAH = chronic active hepatitis; AC = active cirrhosis; PBC = primary biliary cirrhosis; HC = hepatic cancer; S-SF-C = alcoholic steatosis, steatofibrosis, cirrhosis; F.IHC = fibrosis + idiopathic hemochromatosis; C.IHC = cirrhosis + idiopathic hemochromatosis.

DISCUSSION

The prospective study of selected patients with AVH has partially confirmed previous data (10). The majority of the patients with unresolved hepatitis and high serum PIIIP have chronic active hepatitis or cirrhosis, while unresolved hepatitis patients with normal serum PIIIP had lesser inflammatory changes in the liver. This indicates that procollagen peptides may be of help for predicting progressive liver disease. On the assumption that synthesis, pool size and degradation of collagen are altered during liver disease, we had expected all patients with progressive liver disease to have high sPIIIP. However, in the cross-sectional study of patients with chronic liver disease, some of those with chronic active hepatitis or cirrhosis had normal levels of sPIIIP. A discrepancy between PIIIP and liver histology might be due to individual differences in the turnover of hepatic type III collagen (11) or to different degree of associated inflammation. Since during AVH sPIIIP is correlated with ALT and bilirubin and at the time of onset of hepatitis serum PIIIP was twice as high as in the phase of chronic active hepatitis, there appears to be an association between serum procollagen levels and the degree of hepatic inflammation. A similar trend for sPIIIP to be higher in patients with more severe inflammatory lesions in the liver had been demonstrated in patients with chronic hepatitis, as well as in patients with alcoholic liver disease. However, in addition to hepatic necroinflammation, such other factors as hepatocyte regeneration, de-novo synthesis of the reticular framework and enhanced leakage of peptide during intracellular degradation of the procollagen molecule might contribute to raising serum levels of PIIIP (12). Whatever the mechanism for increased procollagen in AVH may be, the persistently high serum PIIIP levels in patients who later developed chronic active hepatitis indicate that this marker of collagen metabolism may be useful for monitoring the outcome of AVH. The availability of a serum marker that predicts progressive disease in the liver might help us to select better those patients who should be evaluated histologically.

REFERENCES

1) Low R.B., Cutroneo K.R., Davis G.S. et al. 1983,Lab.Investig. 48 (6): 755-59.
2) Simon L.S.,Krane S.M., Wortman P.D. et al. 1984. J. of Clin. Endocrin. and Metabolism 58: 110-20.
3) Hochweiss S., Fruchtman S., Hahn E.G. et al. 1983. Am. J. of Hematol. 15: 343-51.
4) Fessler J.H., Fessler L.I. 1978. Ann. Review Biochem.47:129.
5) Rohde H.,Vargas L.,Hahn E.G.et al.1979. Eur. J. Clin.Invest. 9: 451-59.
6) Colombo M.,Annoni G.,Donato M.F. et al. 1983. Am.J. Clin. Pathol. 80 (4): 499-502.
7) Leevy C.M., Popper H., Sherlock S. 1976. Chicago: Year Book Medical Publishers Inc.: 9-21.
8) De Groote J.,Desmet V.J.,Gediqk P; et al. 1968. Lancet ii: 626-8.
9) Knodell T.W., Wollman J. 1981. Hepatology 1(5): 431.
10) Colombo M.,Annoni G., Donàto M.F. et al. 1985. Hepatology 5: 475-9.
11) Bienkowski R.S., Cowan M.J., McDonald J.A. et al. 1978. J. Biol. Chem. 253: 4356-63.
12) Fleischmayer R., Timpl R., Tuderman L. et al. 1981. Proc. Natl. Acad.Sci. USA 78: 7360-4.

19

Serum type III procollagen peptide : detection of pulmonary fibrosis and its application in occupational lung diseases

I. Okazaki, K. Miura, Y. Kobayashi, H. Kondo – Department of Preventive Medicine and Public Health, School of Medicine, Keio University

ABSTRACT

The serum levels of type III procollagen peptide were measured in 68 healthy controls, 24 cancer patients of which 13 had the complication of pulmonary fibrosis after cancer therapy, and 24 patients with silicosis. The cases showing high values revealed interstitial pneumonitis followed by rapid progressive pulmonary fibrosis caused by cancer therapy. In these cases the elevated serum levels decreased to normal levels after 6 months or one year, although advanced pulmonary fibrosis persisted. There was no correlation between the serum levels and the degree of pulmonary fibrosis in either group of patients. These results indicate that this assay may be useful for detecting individuals with abnormal constituents proceeding to rapid progressive pulmonary fibrosis among workers exposed to hazardous gas or dust.

INTRODUCTION

Since Rohde et al. (1979) developed an assay method for serum type III procollagen peptide (P-III-P) and reported its application in chronic liver diseases, the assay has been applied not only in chronic liver diseases (Raedsch et al. 1982; Niemela et al., 1983; Colombo et al. 1983; Maruyama et al., 1984), but also in pulmonary fibrosis (Low et al., 1983; Kirk et al., 1983; Okazaki et al., 1983; Anttinen et al., 1986; Begin et al., 1986). Low et al. (1983) and Begin et al. (1986) demonstrated increased bronchoalveolar lavage levels of P-III-P in pulmonary fibrosis, i.e., increased neosynthesis of type III collagen in pulmonary fibrosis. However, in occupational medicine, bronchoalveolar lavage is not suitable for surveys to detect pulmonary fibrosis among healthy workers. We investigated the limit of usefulness of determining serum P-III-P level in pulmonary fibrosis, especially for suitable application in occupational medicine.

P–III–P AS FIBROGENIC MARKER

Collagen-producing cells such as fibroblasts, endothelial cells and muscle cells can synthesize procollagen within the cells. Once procollagen is secreted extracellularly, both N-terminal and C-terminal peptides of procollagen are cleaved by endopeptidases to form collagen molecules which cross-link to form collagen fibrils, as shown in Figure 1. Therefore, the released peptides of procollagen may reflect the degree of biosynthesis of collagen and increased petide levels in serum may suggest the presence of inflammation or the beginning of fibrosis.

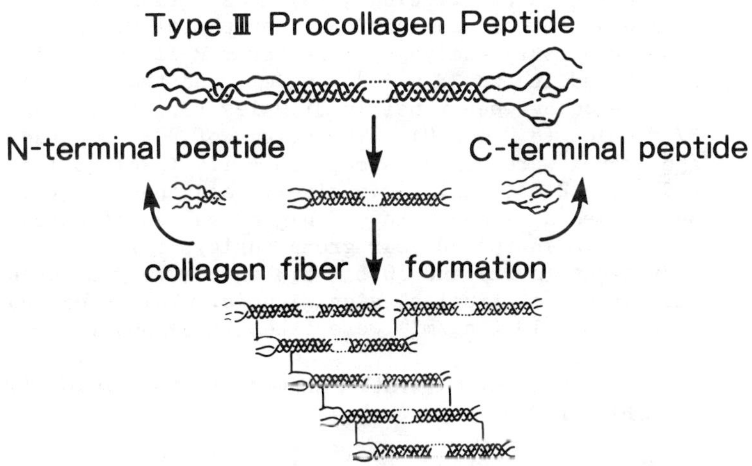

Fig. 1. – Procollagen peptide and formation of collagen fibrils.

Pulmonary fibrosis is characterized morphologically by disorganization of the normal alveolar structures, altered cellular population and disordered interstitial collagen. Recent research on collagen polymorphism in the lung has revealed that type I collagen is located in the interstitium of alveolar septa and type III collagen in the septa and the perivascular wall (Bradley et al., 1974; Seyer et al., 1976; Madri et al., 1980). The reaction of connective tissue to injury in wound healing or the fibroproliferative process is characterized first by neosynthesis of type V collagen and type IV collagen, followed by neosynthesis of type III collagen, after which type I collagen fibers are prominently deposited (Gay et al., 1983).

ASSAY METHOD FOR SERUM P–III–P

The serum P-III-P levels were measured using a radioimmunoassay according to the method of Rohde et al. (1979). Commercially available research kits including the specific antibody against bovine type III procollagen peptide, unlabeled bovine peptide (Col 1-3) standard and ^{125}I-labeled peptide (Col 1-3) were purchased through Hoechst (Tokyo). Reproducibility of results obtained with different test kits was achieved with an error of less than 5%. Statistical calculations are based on Student's t-test. Data are expressed as mean + SD.

SERUM PEPTIDE LEVELS IN 68 HEALTHY ADULTS

Sixty-eight healthy subjects met the following criteria: (a) no abnormal physical findings, (b) no abnormal laboratory tests (urine and stool examinations, peripheral blood cell counts and blood chemistry analysis). The serum P-III-P levels in 68 healthy adults was 8.6 + 2.4 ng/ml, and there was no statistical difference between those in 31 males (9.1 + 2.6) and those in 37 females (8.2 + 2.0). The values of 20-, 30- and 40-year-old age groups were almost same, 7.6 + 1.5 (n=6), 8.1 + 2.8 (n=12) and 8.1 + 2.7 (n=15) respectively, and those over 50 years of age showed a tendency toward higher values of serum peptide (8.6 + 1.8 in the 50-year group (n=14), 8.7 + 1.7 in the 60-year group (n=12) and 10.8 + 2.3 in the 70-year group (n=9)), but without statistical significance. Values exceeding the mean + 2SD (13.4 ng/ml) were taken to be abnormal.

DETECTION OF PULMONARY FIBROSIS AS A COMPLICATION OF CANCER THERAPY

Before cancer therapy the serum P-III-P levels in 12 cases with lung cancer were 11.0 + 4.3 ng/ml and those in three other patients with breast, colon and uterine cancers were within normal range. It has been reported that serum P-III-P levels in cancer patients are not high if liver involvement is not observed (Bolarin et al., 1982). Therefore, cancer cell infiltration did not elevate serum levels. Twenty-four patients (17 with lung cancer, 2 with breast cancer, 1 with thymoma, 1 with esophageal cancer, 1 with colon cancer, 1 with uterine cancer and 1 with malignant lymphoma) were treated with radiation or anti-cancer drugs. After cancer therapy 13 patients with complicating pulmonary fibrosis revealed 24.3 + 8.0 in serum (range of 9.3 to 42.4), while the remaining 11 cases without complication of pulmonary fibrosis showed lower values of 16.1

+ 2.5 (range: 12.4 - 19.4) with statistical significance (p 0.01). However, the serum levels of the peptide were not associated with the degree of pulmonary fibrosis revealed by chest x-ray, but with the grade of progression in clinical course. That is, all three cases exceeding 30 ng/ml suffered from the rapid progressive type of interstitial pulmonary fibrosis after cancer therapy. A 41-year-old man with lung cancer underwent surgery. The serum P-III-P level was 14.2 ng/ml before surgery. After the operation he was treated with several kinds of anti-cancer drugs including bleomycin-derivative peplomycin, adriamycin and cis-platinum. The patient had the complication of pulmonary fibrosis revealed by the chest x-ray findings showing some infiltration and a fibrous shadow in the right upper field as well as an abnormal linear shadow on the left side. The serum level of the peptide increased gradually, as shown in Figure 2.

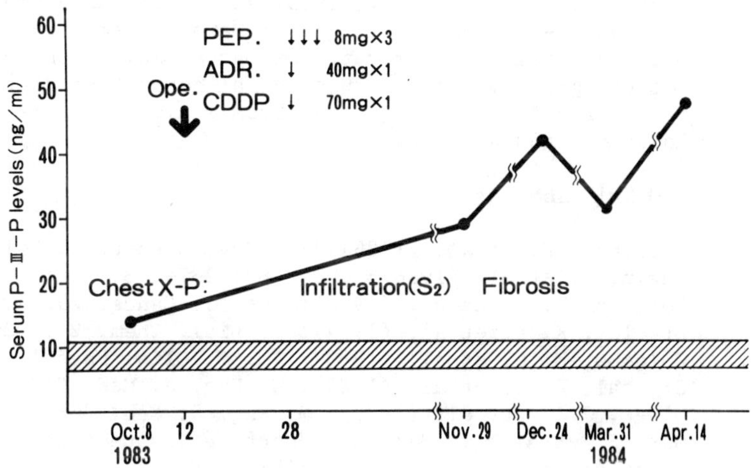

Fig. 2. – Serum P–III–P levels related to treatment with anti-cancer drugs in a 41-year-old male with squamous cell cancer. The serum value increased gradually after some infiltration and fibrous shadow in the right upper field (S_2) as well as an abnormal linear shadow on the left side appeared on chest x-rays. PEP: peplomycin, ADR: adriamycin, CDDP: cis-platinum.

APPLICATION OF SERUM P–III–P ASSAY TO OCCUPATIONAL MEDICINE

Twenty-four silicosis patients with or without pulmonary tuberculosis revealed slightly higher values of serum P-III-P level of 12.6 + 6.5 compared with the values in 68 healthy adults, with statistical significance (p 0.01). The presence of pulmonary tuberculosis in silicosis patients increased the tendency to higher serum levels of the peptide (13.7 + 8.5 (n=12) vs. 11.5 + 2.5 (n=12)), but without statistical significance. There were 7 cases showing abnormal values exceeding 13.4 ng/ml, 3 without pulmonary tuberculosis and 4 with. The serum P-III-P levels in silicosis patients were not correlated with degree of pulmonary fibrosis revealed by chest x-ray findings, and lower than those in cancer patients with complication of rapid interstitial pulmonary fibrosis after radiation therapy or treatment with anti-cancer drugs. The difference between the serum values in both groups with pulmonary fibrosis may be due to a different clinical progress and a different mechanism of fibrogenesis (that is, differences in site of early lesions (Seaton et al., 1984), population of inflammatory cells (Crystal et al., 1984), immunological reactions (Crystal et al., 1984), ability to synthesize collagen by fibroblasts (Phan et al., 1985) and collagenase production by macrophages (Gadek et al., 1979; Marom et al., 1980). In occupational medicine, however, this assay may be useful for detecting individuals with abnormal constituents proceeding to rapid progressive pulmonary fibrosis among workers exposed to hazardous gas or dust.

REFERENCES

Anttinen, H. et al. (1986): Am. Rev. Respir. Dis. 133: 88-93.
Begin, R. et al. (1986): Chest 89: 237-243.
Bolarin, D. M. et al. (1982): Int. J. Cancer 29: 401-405.
Bradley, K. H. et al. (1974): J. Biol. Chem. 249: 2674-2683.
Colombo, M. G. et al. (1983): Am. J. Clin. Pathol. 80: 499-502.
Crystal, R. G. et al. (1984): N. Eng. J. Med. 310: 154-166.
Gadek, J. E. et al. (1979): N. Eng. J. Med. 301: 737-742.
Gay, S. et al. (1983): In Connective Tissue Diseases. Wagner, et al. ed. Williams and Wilkins, Baltimore/London. 120-128.
Kirk, J. M. E. et al.(1983): Thorax 38: 712-713.
Low, R. B. et al. (1983): Lab. Invest. 15: 69-74.
Madri, J. A. and Furthmayer, H. (1980): Human Path. 11:353-366.
Marom, Z. V. I. et al. (1980): Am. Rev. Respir. Dis. 121: 859-867.
Maruyama, K. et al. (1984): Acta Hepatol. Jpn. 25: 24-32.
Niemela, O. L. et al. (1983): Gastroenterol. 85: 254-259.
Okazaki, I. et al. (1983): J. UOEH 5: 461-467.
Phan, S. H. et al. (1985): J. Clin. Invest. 76: 241-247.
Raedsch, R. et al. (1982): Z. Gastroenterol. 20: 738-743.
Rohde, H. et al. (1979): Eur. J. Clin. Invest. 9: 451-459.
Seaton, A. (1984): In Occupational Lung Diseases. Morgan, W. K. C. and Seaton, A. eds. W. B. Saunders Co., Philadelphia, 250-294 and 323-376.
Seyer, J. M. et al. (1976): J. Clin. Invest. 57: 1498-1507.

20

The lipid risk factor of coronary heart disease in human exposure to carbon disulphide

Teresa Wrońska-Nofer, Wojciech Laurman – Department of Biochemistry, Institute of Occupational Medicine, Lodz, Poland

ABSTRACT

The serum cholesterol binding reserve (SCBR) as well as the level of total cholesterol, LDL- and HDL-cholesterol were evaluated in men occupationally exposed to CS2 and in unexposed controls. Two exposed to CS2 groups consisted of workers of average age 48 and 29 years, respectively. The concentration of CS2 in the air was 40-116 mg/m3. Control groups matched the exposed ones. The obtained data provided evidence that exposure to CS2 reduced serum capacity to bind cholesterol. The decrease in SCBR value in exposed men was significant and markedly greater than the observed alterations in the level of total-, LDL- and HDL-cholesterol.

INTRODUCTION

The experimental studies provided evidence that atherosclerotic effect of CS2 is secondary to the disturbances induced by this chemical in cholesterol metabolism in the liver /1,2,3/. Hyperlipemia and dyslipoproteinemia /3,4/ resulting from metabolic disturbances as well as disturbed by CS2 metabolism of lipids directly in the arterial wall /5,6/ contribute to the development of atherosclerotic process in animals under the experimental conditions /3,7/. As lipids and lipoproteins play a key role in the development of atherosclerotic process /8,9/ the measurement of serum total cholesterol, LDL-cholesterol and HDL-cholesterol has been applied widely as a indicator of risk for coronary heart disease (CHD) in population.Nevertheless, in a number of studies some other parameters e.g. apoproteins, the capacity of serum lipoprotein to bind cholesterol /10,11/ were used to evaluate the risk of CHD.

In the present study, serum cholesterol binding reserve, which define the capacity of serum lipoprotein to bind cholesterol, has been measured in men occupationally exposed to CS_2. The value of this parameter has been compared with the alterations produced by CS_2 in the total cholesterol, HDL- and LDL-cholesterol, factors accepted as indicators of the risk of atherogenesis.

MATERIALS AND METHODS

86 male patients in two groups aged 21-35 and 36-65 years, employed in viscose rayon plant for 1-13 and 5-35 years, respectively, were examined in this study. The value of CS_2 concentration at work environment during last 8 years was 40-116 mg/m3.

As the controls 80 male subjects in two groups aged 21-35 years and 36-63 years were randomly selected among individuals referred to the Outpatient Clinic and to the Clinic of Occupational Disease at the Institute of Occupational Medicine in Lodz. The patients of control groups were matched the exposed ones. Serum cholesterol, lipoproteins cholesterol and SCBR were determined in individuals of groups intoxicated with CS_2 and controls simultaneously. LDL and HDL fractions were separated by sequential ultracentrifugation of serum as described by Hatch and Less /12/. The measurements of serum cholesterol and lipoproteins cholesterol concentrations were performed enzymatically, using the cholesterol oxidase/4-amino-phenazone phenol method (Boehringer Mannheim monotest cholesterol). Serum cholesterol binding capacity was determined according to Borresen and Berg /13/. The significance of differences between the groups intoxicated with CS_2 and controls was calculated by the Student t-test.

RESULTS AND DISCUSSION

The number of studies provided evidence that people suffering from coronary atherosclerotic heart diseases had increased serum concentration of total cholesterol, triglicerides, LDL- and VLDL-choleste-

rol, whereas HDL-cholesterol concentration in these patients was lower than in the controls /8,9/.
Our study indicated that the same type of alterations was attributed to the occupational exposure to CS2 at concentration 40-116 mg/m3. The mean level of total cholesterol and LDL-cholesterol in subjects exposed to CS2 (older group) was higher than the mean level in those unexposed. The average differences for total- and LDL-cholesterol amounted to 19 mg/dl and 44 mg/dl, respectively. In contrast with the LDL-cholesterol concentration of HDL-cholesterol was slightly lower, but significant (Table 1). These differences were found to be more significant in older group of men working for a longer period in exposure to CS2 than those in men of younger group with shorter duration of exposure (Table 1).

Table 1: The content of total-, LDL and HDL-cholesterol in serum of men chronically exposed to CS2.

	Age (years)	Cholesterol (mg/dl) Total	LDL	HDL
A-group of younger men (exposure time av.8 years)				
Control (29)	28,5±4,3	172±23	114±25,1	45±10,1
CS2 (34)	29,2±4,8	190±35	148±26xx	44±11,1
B-group of older men (exposure time av.21 years)				
Control (51)	48,9±6,7	190±25	126±16,7	40±11,8
CS2 (52)	47,8±6,9	209±38xx	170±27xx	36±5,1x

The mean values ± S.D. are given; number of men is given in brackets. x,xx - significantly different from control group at p < 0,005 and 0,001.

The role of lipoproteins (LDL and HDL) in atherosclerosis development is associated with the function of both fractions in cholesterol transport. Impaired binding of cholesterol with HDL and decreased transport of HDL-bound cholesterol from the peripheral tissue back to the liver may influence the development of the atherosclerotic process. Clinical examinations of man with myocardial infarction provide evidence that man with this disease has reduced capacity to bind cholesterol /10,11/. The results of this study indicate that chronic occupational intoxication with CS2 also reduces the serum capacity to bind cholesterol. The value of SCBR in young men exposed to CS2 was four times lower than that of the controls (Table 2). The reduced value of SCBR was negatively correlated with alterations produced by CS2 in total- and LDL-cholesterol level. Among the individuals exposed to CS2 reduction of SCBR value was accompanied with the elevation of total- and LDL-cholesterol concentration in the serum, whereas in unexposed subjects opposite trend was observed.

Table 2: Serum cholesterol binding reserve (SCBR) in men chronically exposed to CS2.

Group	Age (years)	Exposure time (years)	SCBR (mg/dl)
Control (22)	37,7±8,7 (24–56)	-	134,0±39,1
CS2 (23)	35,9±7 (24–52)	3-23	44,7±21,9x

The mean values ± S.D. are given; number of men is given in brackets. x - significantly different from control group ($p < 0,001$).

Our preliminary study has not provided explanation for the obtained data. It may be assumed that the low SCBR value resulted from CS2 induced lowering of the level of serum HDL - the particle which has the highest capacity to bind cholesterol. However, quantitatively inconsiderable decrease in HDL under the influence of CS2 (only in older men) seems to shake this hypothesis.
It may be supposed that CS2 due to lipolytic property may interact with lipoproteins and, through this mechanism, it affects their structure and reduces binding capacity. To elucidate this question and to collect more information on the significance of SCBR as a CHD risk indicator in human exposure to CS2, further studies have been undertaken.

Grant from WHO, Geneva, No. GL/RES/OCH/020/PB/84

REFERENCES

1. Wrońska-Nofer T.: Studies on in vivo incorporation of 14C-acetate and 14C-mevalonate into cholesterol in the liver of rat intoxicated with carbon disulphide. Int. Arch. Arbeitsmed. 34, 221-229, 1975.

2. Wrońska-Nofer T.: Effect of carbon disulphide intoxication on fecal excretion of end products of cholesterol metabolism. Int. Arch. Occup. Environ. Health 40, 261-265, 1977.
3. Wrońska-Nofer T.: Some disorders in metabolism of cholesterol and their contribution to the development of experimental atherosclerosis in rats exposed to carbon disulphide. Med. Pracy 30, 121-134, 1979 (in Polish).
4. Wrońska-Nofer T., Laurman W., Szendzikowski S., Kołakowski J., Węgrowski J.: Lipids alterations in serum and arterial wall of rabbits chronically exposed to carbon disulphide. Med. Pracy 31, 311-318, 1980 (in Polish).
5. Wrońska-Nofer T.: Aortic phospholipids synthesis in experimental carbon disulphide intoxication in rats. Int. Arch. Occup. Environ. Health 36, 229-234, 1976.
6. Wrońska-Nofer T., Parke M.: Influence of carbon disulphide on metabolic processes in the aorta wall: Study of the rate of cholesterol synthesis and the rate of influx of 14C-cholesterol from serum into the aorta wall. Int. Arch. Occup. Environ. Health 42, 63-68, 1978.
7. Wrońska-Nofer T., Szendzikowski S., Laurman W.: The effect of carbon disulphide and atherogenic diet on the development of atherosclerotic changes in rabbits. Atherosclerosis 31, 33-39, 1978.
8. Comai K., Feldman D.L., Goldstein A.L.,Hamilton J.G.: Atherosclerosis: An overview. Drug Dev. Res. 6, 113-125, 1985.
9. Levy R.I.: Cholesterol, lipoproteins, apoproteins, and heart disease: Present status and future prospects. Clin. Chem. 27, 653-662, 1981.
10. Hsia S.L., Chao Y.S., Hennekens C.H., Reader W.B.: Decreased serum cholesterol-binding reserve in premature myocardial infarction. Lancet ii, 1000-1004, 1975.
11. Hsia S.L., Briese F., Hoffman I.: Cholesterol -binding reserve and myocardial infarction. Lancet i, 799, 1976.
12. Hatch F.T., Less R.S.: Practical methods for plasma lipoprotein analysis. Adv. Lipid Res. 6, 2, 1968.
13. Borresen A.L., Berg K.: Serum reserve cholesterol binding capacity: The relative importance of different lipoprotein classes. Artery 9, 96-119, 1981.

21

Occupational exposure indicators to tetrachloroethylene in a dry-cleaning shop

Lj. Skender, V. Karačić, D. Prpić-Majić, S. Kežić – Institute for Medical Research and Occupational Health, Zagreb

ABSTRACT

In order to estimate the intensity of exposure to tetrachloroethylene as well as to determine the best indicator of exposure, 19 subjects employed in a dry-cleaning shop were studied. In each subject the concentration of tetrachloroethylene in blood and trichloroethanol and trichloroacetic acid in blood and urine were determined by gas chromatography. Venous blood and urine samples were collected on Monday morning before the start of work and on Thursday after the working day. In addition to characteristic indicators of tetrachloroethylene exposure the activity of a sulfhydryl enzyme δ-aminolevulinic acid dehydratase was determined in all samples of blood on Monday and Thursday and serum alanine aminotransferase, aspartate aminotransferase and γ-glutamyltransferase were estimated in serum samples of blood taken on Thursday only.

INTRODUCTION

Tetrachloroethylene (PERC) is extensively used world-wide as a solvent and therefore the permanent monitoring of workers exposed to PERC is highly recommended. However, before developing a biological monitoring programme the toxicokinetics of the solvent must be known. PERC is initially biotransformed by microsomal mono-oxygenase to an epoxide which is considered to be responsible for biological actions of PERC. The epoxide rearranges spontaneously by intramolecular migration of a chlorine atom to trichloroacetyl chloride, which hydrolyses to trichloroacetic acid. Human capacity to metabolise PERC is limited and the compound is mainly excreted unchanged in exhaled air; less than 3% is excreted in urine as trichloroacetic acid (TCA) (Fernandez et al. 1976). In addition to TCA, Ikeda et al. (1972) detected trichloroethanol (TCE) in the urine of workers exposed to PERC. It seems that humans are less sensitive to the acute toxic damage of PERC because of lower PERC metabolism for humans in comparison with rats and mice (Reichert 1983). There is still a lack of epidemiological findings on PERC activity and it is not easy to make any conclusion regarding PERC health risk. The objectives of the study were:

- To estimate the intensity of PERC exposure.
- To determine the best indicator of PERC exposure.
- To check the possible adverse effect of PERC on the liver.

SUBJECTS AND METHODS

The examined workers are employed in a dry-cleaning shop and their characteristics are presented in Table 1.

Table 1: Characteristics of the Examined Group.

Female	15
Male	4
Age (years)	37.3 ($\bar{X}$), Range: 19-53
Duration of exposure (years)	16.6 ($\bar{X}$), Range: 2-30
Consumption of alcohol	4 - moderate
	15 - nil

Venous blood and urine samples were collected on Monday morning before the start of work (after three days off) and on Thursday, 30 min. after the working day. The concentrations of PERC in blood and TCE and TCA in blood and urine were determined by gas chromatography (Kežić et al. 1985). A sample of PERC used in the investigated dry-cleaning shop was also analysed. In samples of blood on Monday and Thursday erythrocyte δ-aminolevulinic acid dehydratase (ALA-D) was determined (Berlin et al. 1974). In serum samples of blood taken on Thursday alanine aminotransferase (ALT), aspartate aminotransferase (AST) and γ-glutamyltransferase (γ-GT) were estimated. The results of PERC, TCA and ALA-D in blood and TCE and TCA in urine are presented as median (M) and range (R) values. The significance of the differences in parameters examined were tested by median test. The correlation analyses were done by the standard statistical method.

RESULTS

The PERC used in the investigated dry-cleaning shop was found to be very pure, less than 0.5% of trichloroethylene (TRI) was found. In all blood samples collected on Monday and Thursday PERC and TCA were present in measurable amounts, while TCE was detectable in only four subjects. Therefore only PERC and TCA results were presented (Table 2). In urine both TCE and TCA were present in measurable amounts (Table 3). TRI was not detected in any of the blood samples.

Table 2: Median (M) and Range (R) Values of PERC and TCA in Blood (PERC–B and TCA–B) Taken on Monday (I) Before Work and on Thursday (II) After Work.

Statistical parameter	PERC-B µmol/L		TCA-B µmol/L	
	I	II	I	II
M	2.96	8.44	6.49	8.57
R	1.63-31.78	4.76-79.97	1.65-17.08	1.71-20.93
N	19	19	19	19

The median PERC value on Thursday was about three times higher than on Monday, while the increase in TCA was only 32%. A very good correlation was found ($r=0.964$; $p<0.001$) between the PERC concentrations in blood taken on Monday and Thursday.

Table 3: Median (M) and Range (R) Values of TCE and TCA in Urine (TEC–U and TCA–U) Collected on Monday (I) Before Work and on Thursday (II) After Work.

Statistical parameter	TCE-U µmol/mmol creatinine		TCA-U µmol/mmol creatinine	
	I	II	I	II
M	0.03	0.13	2.26	2.34
R	0-1.71	0-0.95	0.47-15.76	0.81-10.81
N	19	19	19	19

It is evident that the median value of TCE-U on Thursday was about four times higher than on Monday although TCE-U on Monday was found in a trace concentration in 10 samples and on Thursday in 4 samples. Contrary to these findings there was no difference at all between the TCA concentrations.

In erythrocyte ALA-D activity there was no significant difference ($p>0.10$) between blood samples collected on Monday and Thursday and all values were within a normal range i.e. 26.1 U/L Ercs. Concerning the activities of the serum enzymes investigated increased activities of AST were found in two subjects and of ALT and γ-GT in seven subjects. An outstanding increase in all three enzymes studied (AST=55 U/L; ALT=96 U/L; γ-GT=120 U/L) was found only in a subject with a markedly high concentration of PERC-B ($PERC_I$=31.78 µmol/L; $PERC_{II}$=79.97 µmol/L) in comparison with other workers. The same worker has been working in the dry-cleaning shop for 30 years.

Concerning the differences of PERC-B, TCA-B, TCE-U and TCA-U between Monday and Thursday the only significant difference was found in PERC-B concentrations ($\chi^2=5.156$; $p<0.05$).

The relationship between examined parameters is presented in Table 4.

Table 4: The Relationship Between Examined Parameters.

Examined parameters	r	p
$PERC_I$-B:TCA_I-B	0.643	< 0.01
$PERC_I$-B:TCE_I-U	0.096	> 0.10
$PERC_I$-B:TCA_I-U	0.966	< 0.001
$PERC_{II}$-B:TCA_{II}-B	0.638	< 0.01
$PERC_{II}$-B:TCE_{II}-U	0.752	< 0.001
$PERC_{II}$-B:TCA_{II}-U	0.721	< 0.01

The significant ($p<0.01$) and highly significant ($p<0.001$) correlations were found between all examined parameters except Monday samples of PERC-B and TCE-U.

DISCUSSION

Of the PERC-B concentrations determined on Monday after three days off in seven workers PERC-B was found to be higher than 6.03 µmol/L, the "biological tolerant value" for blood sampling 16 hrs. after the end of exposure recommended by Deutsche Forschungsgemeinschaft (1984). This finding is in agreement with the long biological half-life of PERC of 144 hrs. (Ikeda 1977). Compared with the results of Lauwerys et al. (1983) it indicates a PERC level exposure more than 50 ppm. The ratio between the lowest and the highest blood PERC concentrations was 19.5 on Monday and 16.8 on Thursday respectively. These results together with the highly significant correlation ($r=0.964$; $p<0.001$) between PERC-B concentrations (I:II) indicate a stable chronic exposure to PERC.

A lack of significant correlation ($r=0.096$; $p>0.10$) between $PERC_I$-B and TCE_I-U in contrast to significant correlation ($r=0.752$; $p<0.001$) between $PERC_{II}$-B and TCE_{II}-U could be explained by the short biological half-life of TCE (10-15 hrs). Since the biological half-life of TCA is much longer (70-100 hrs) a good correlation between PERC and TCA in blood as well as in the urine samples on both Monday and Thursday was found.

The fact that there was no significant change in TCA in blood and urine during the working week could be due to the continuous exposure to PERC and to the long biological half-life of TCA.

Considering the presented results i.e. nonsignificant change in TCA-B, TCA-U and TCE-U during working week in spite of good correlation between examined parameters, the determination of neither TCE nor TCA can be used for estimation of PERC exposure. Therefore it may be concluded that PERC determination in blood is the only biological parameter which can be recommended as an indicator for occupational exposure to PERC.

In rats exposed to TRI Koizumi et al. (1984) found an inhibition of ALA-D in blood. There is a difference in the kinetics and rates of metabolic processes between TRI and PERC, although both biotransform over an epoxide which could bind to sulfhydryl groups. The asymmetry increases the reactivity and electrophilicity of the chlorinated ethylenes and their corresponding epoxides. The asymmetrically substituted TRI is much more mutagenic than the symmetrically substituted PERC (Anders and Jakobson 1985). Therefore, no inhibition of a sulfhydryl enzyme ALA-D in the blood of workers exposed to PERC could be explained by the difference in TRI and PERC metabolism as well as of lower PERC metabolism for humans in comparison with rats and mice.

The findings of increased activities of AST, ALT and -GT indicate the adverse effect of PERC on the liver. Non-uniform data on PERC hepatogenicity require further study.

ACKNOWLEDGEMENT

The authors wish to thank Mrs. Antonija Keršanc who performed ALA-D analyses.

REFERENCES

Anders MW, Jakobson I (1985) Scand.J.Work Environ.Health 11, Suppl. 1, 23-32.
Berlin A, Schaller KH (1974) Z.Klin.Chem.Klin.Biochem. 12, 389-390.
Deutsche Forschungsgemeinschaft (1984) Report No. XX Commission for the investigation of health hazards of chemical compounds in the work area, Verlag Chemie, Weinheim, 1984.
Fernandez J, Guberan E, Caperos J (1976) Am.Ind.Hyg.Assoc. 37, 143-150.
Ikeda M, Obtsuji H, Imamura T, Komoike Y (1972) Br.J.Ind.Med. 29, 328-333.
Ikeda M (1977) Environ.Health Perspect. 21, 239-245.
Kežić S, Skender Lj, Karačić V, Prpić-Majić D (1985) Yugoslav. Physiol.Pharmacol.Acta 21, Suppl. 3, 167-168.
Koizumi A, Fujita H, Sadamoto T, Yamamoto M, Kumai M, Ikeda M (1984) Toxicology 30, 93-102.
Lauwerys R, Herbrand J, Buchet JP, Bernard A, Gaussin J (1983) Int.Arch.Occup.Environ.Health 52, 69-77.
Reichert D (1983) Mutation Res. 123, 411-429.

22

PCB congeners in adipose tissue lipid and serum of past and present transformer repair workers and a comparison group

A. Fait, E.A. Emmett, E. Grossman, E.D. Pellizzari – Institute of Occupational Medicine, University of Milan, Italy. Center for Occupational and Environmental Health and Department of Biostatistics, The Johns Hopkins University, Baltimore, Maryland, U.S.A. and Research Triangle Institute, Research Triangle Park, N.C., U.S.A.

SUMMARY

Serum and adipose tissue lipid concentrations of individual PCB congeners were determined from 35 transformer repair workers currently exposed to PCBs, 17 previously exposed and 56 non-exposed workers, using Fused-silica capillary gas chromatography with electron capture detector. Peaks identity was confirmed by mass spectrometry. 89 PCB peaks were identified and chemically defined. The pattern of individual PCB congeners was similar in exposed and non-exposed subjects. The pattern of PCB peaks in our exposed workers was different from that in Yusho patients and in other studied PCB exposed populations, suggesting different toxic potentials from PCBs in different types of exposure.

INTRODUCTION

Commercial Polychlorobiphenyls (PCBs) are a mixture of congeners with different patterns and degrees of chlorination. The assessment of PCB exposure in humans has usually been performed through measurement of the total serum or adipose PCB levels. This index may not be predictive of the biological effect of a PCB mixture: experimental studies have in fact demonstrated that metabolism and toxicity of PCB congeners differ greatly depending upon the number and position of the chlorine atoms on the biphenyl rings (Safe, et al. 1985).

Further, the relative disposition of various congeners in serum and adipose tissue is not well understood so that factors which might govern the choice of the tissue for measurement are not clear.

METHODS

As part of a larger clinical-epidemiological study on transformer repair workers in Washington, D.C. area (Emmett, et al in press) we quantitated specific PCB congeners in serum and adipose tissue lipid from individuals with present and past exposure to Aroclor 1260 and from a nonexposed comparison group.

Subjects were 35 transformer repair workers currently exposed to Aroclor 1260, 17 previously exposed and a comparison group of 56 nonexposed workers from the same employer. All participants were male. Past exposed workers had been away from exposure to PCBs for an average of 4.5 years. Samples were analyzed by Fused silica capillary gas chromatography with electron capture detector (FSCGC) and verification of identity and quantity of each PCB congener was performed by FSCGC with negative ion chemical ionization mass spectrometry, using pure individual standards (Pellizzari, et al. in press). Chromatographic peaks were grouped in homolog categories according with the number of chlorine substituents in the biphenyl rings. About 50% of all observations on congeners concentrations were unquantifiable, since they fell below the limit of detection of the analytical method: statistical techniques for left and interval censored data were applied to utilize incomplete data provided from chemical analysis (Self and Grossman, in publication).

RESULTS AND DISCUSSION

89 PCB peaks were identified and chemically defined. Total serum and adipose PCB_s concentrations for the three exposure groups are reported in Table 1.

In all groups PCBs were much more concentrated in adipose tissue, as expected. The median total concentration in currently exposed was 43.7 ppb in serum and 3180 ppb in adipose tissue: both values were higher than in comparison subjects. On the other hand, total PCB levels in past exposed workers were not different from those in comparison group: since high exposure levels had been reported in the past, these findings suggest elimination over time.

71 peaks out of the total of 89 which were identified represented congeners which could be assigned to a single homolog group, although in 24 instances the peak was constituted by 2 or 3 different congeners. 18 peaks were unassignable to a

single homolog group since the peak represented 2 or more congeners with different degrees of chlorination. In both serum (Table 2) and adipose tissue (Table 3) hexa-CB and hepta-CB congeners predominated in all groups, while nona-CB and lower chlorinated PCBs were present in relatively low concentrations.

In both the transformer repair workers and the comparison group the same seven PCB peaks predominated both in serum (Table 4) and adipose tissue (Table 5).

All the seven peaks reported presented significantly higher values among currently exposed than comparison subjects. The qualitative pattern of major PCBs congeners peaks was rather similar in both occupationally exposed workers and comparison subjects, suggesting exposure to PCB mixture of similar composition in both workers exposed to Aroclor 1260 and the environmentally exposed population of the Washington, D.C. area.

The source of the PCB exposure appears to influence the pattern of PCB_s found in tissues and should be taken into account in comparing results from different populations. Tab. 6 shows different patterns of major serum PCB congeners identified in different exposed populations: transformer repair workers exposed to Aroclor 1260, capacitor workers exposed mainly to Aroclor 1016 (Wolff et al, 1982) and Yu-Cheng patients accidentally exposed to rice oil contaminated with Kanechlor (Masuda et al, 1985).

Because the toxicity of PCB congeners varies widely it seems higly likely that the different composition of PCB isomers in different groups indicates different toxic potentials. Thus it appears insufficient to compare PCB_s exposed populations merely by tissue concentrations of total PCB_s.

REFERENCES

Emmett, E.A., Maroni., Schmith, J.M., Levin, B.K., Jeffreys, J., Studies of transformer repair workers exposed to PCBs: (1) study, design, PCBs concentrations, questionnaire and clinical examination results. (Jour.Occup.Med., in press)

Masuda, Y., Kuroki, H., Haraguchi, K. and Nagayama, J.: PCB and PCDF congeners in the blood and tissues of Yusho and Yu-Cheng patients. (Environm. Health Persp. 59:53-58, 1985)

Pellizzari, E.D., Mosely, M.A., Cooper, S.D.: Recent advances in the analysis of Polychlorinated Biphenyls in environmental and biological media (Jour.Chromatog., in press)

Safe, S., Bandiera, S., Sawyer, T., Robertson, L., Safe, L., Parkinson, A., Thomas, P.E., Ryan, D.E., Reik, L.M., Levin, W., Denomme, M.A., Fujita, T.: "PCBs structure-functions relationships and mechanisms of action" (Environ. Health Persp. 60:47-56, 1985)

Self, S.G. & Grossman, E.A.: Linear rank test for interval censored data with application to PCB levels in adipose tissue of transformer repair workers (in press)

Wolff, M.S., Thornton, J., Fischbein, A., Lilis, R., Selikoff, I.J.: Disposition of polychlorinated biphenyl congeners in occupationally exposed persons (Toxicol and Appl. Pharmacol. 62:294-306, 1982)

Table 1: Total serum and adipose PCB_s concentration (PPB) by exposure category.

	EXPOSURE CATEGORIES			TEST FOR DIFFERENCES*		
	CURRENT EXPOSED	PAST EXPOSED	COMPARISON GROUP	CURRENT VS. COMPARISON	PAST VS. COMPARISON	CURRENT VS. PAST
SERUM						
N	35	17	56			
MEDIAN	43.7	30.0	16.1	S	NS	NS
(25%-75% QUARTILES)	(20.1-78.9)	(5.8-60.7)	(8.2-24.4)			
ADIPOSE						
N	33	15	52			
MEDIAN	3180	821	888	S	NS	S
(25%-75% QUARTILES)	(1360-5870)	(708-1870)	(516-1190)			

* LINEAR RANK TEST of SELF AND GROSSMAN
S = SIGNIFICANT ($P < 0.0167$)
NS = NOT SIGNIFICANT

TABLE 2
CONCENTRATIONS (PPB) OF SERUM PCB HOMOLOG GROUPS IN THE THREE EXPOSURE CATEGORIES, EXPRESSED AS MEDIAN VALUE

	EXPOSURE CATEGORIES			TEST FOR DIFFERENCES*		
HOMOLOG GROUP	CURRENT EXPOSED (N=35)	PAST EXPOSED (N=17)	COMPARISON GROUP (N=56)	CURRENT VS COMPARISON	PAST VS COMPARISON	CURRENT VS PAST
MONO-CB	0.32	0.30	0.26	S	NS	NS
DI-CB	0.56	0.92	0.78	NS	S	NS
TRI-CB	1.48	0.71	0.45	S	NS	NS
TETRA-CB	0.92	0.99	0.61	NS	NS	NS
PENTA-CB	2.00	1.66	0.83	S	NS	NS
HEXA-CB	8.39	4.95	2.71	S	NS	NS
HEPTA-CB	5.75	3.70	1.58	S	S	NS
OCTA-CB	2.69	1.45	0.78	S	S	NS
NONA-CB	0.05	0.22	0.13	NS	NS	NS
MULTIPLE CHLORINATION**	4.12	8.22	3.18	NS	NS	NS

**SUM OF THE PEAKS UNASSIGNABLE TO A SINGLE HOMOLOG GROUP BECAUSE OF CONGENERS OF MULTIPLE CHLORINATION

* LINEAR RANK TEST OF SELF & GROSSMAN
S = SIGNIFICANT ($P < 0.0167$)
NS = NOT SIGNIFICANT

Table 3: Concentrations (PPB) of adipose PCB homolog groups in the three exposed exposure categories, expressed as median value.

HOMOLOG GROUP	EXPOSURE CATEGORIES: CURRENT EXPOSED (N =33)	PAST EXPOSED (N =15)	COMPARISON GROUP (N =52)	TEST FOR DIFFERENCES*: CURRENT VS COMPARISON	PAST VS COMPARISON	CURRENT VS PAST
MONO-CB	1.46	1.17	1.24	NS	NS	NS
DI-CB	4.78	10.80	5.46	NS	NS	NS
TRI-CB	15.10	16.40	14.50	NS	NS	NS
TETRA-CB	77.70	31.40	38.30	S	NS	NS
PENTA-CB	284.00	65.40	66.80	S	NS	S
HEXA-CB	813.00	217.00	240.00	S	NS	S
HEPTA-CB	661.00	215.00	198.00	S	NS	S
OCTA-CB	292.00	93.80	104.00	S	NS	S
NONA-CB	22.50	13.20	15.90	NS	NS	NS
MULTIPLE CHLORINATION**	525.00	217.00	126.00	S	NS	S

**SUM OF THE PEAKS UNASSIGNABLE TO A SINGLE HOMOLOG GROUP BECAUSE OF MULTIPLE CHLORINATION

* LINEAR RANK TEST OF SELF & GROSSMAN
S = SIGNIFICANT ($P < 0.0167$)
NS = NOT SIGNIFICANT

Table 4: Major PCB congeners identified in serum in the exposed and comparison groups. Concentrations expressed as median value (PPB).

		EXPOSURE CATEGORY		TEST FOR DIFFERENCES*
PEAK NUMBER	STRUCTURAL ASSIGNMENT	CURRENT EXPOSED (N = 35)	COMPARISON GROUP (N = 56)	CURRENT VS COMPARISON
62	2,3,5,6,3',4'5'/ 2,3,4,5,2',4',5' (hepta-CB)	3.40	0.80	S
49	2,3,4,2',4',5'(hexa-CB)	2.00	0.78	S
54	2,4,6,3',4',5'/ 2,4,5,2',4',5'/ 2,3,4,5,2',5' (hexa-CB)	1.90	0.73	S
59	2,3,4,5,2',3',4' (hepta-CB)	1.60	0.48	S
40	2,4,5,3',4' 3,4,5,2',3' (penta-CB)	1.00	0.52	S
69	2,3,4,5,2',3',5',6'/ 2,3,4,5,6,2',3',5' (Octa-CB)	1.20	0.39	S
87	2,3,5,6,2',4',5'/ 2,3,4,2',3',4'/ 2,3,4,5,2',4',6' (mixed)	1.00	0.33	S

*Generalized Wilcoxon Test S = Significant ($P < 0.0167$) NS = Not Significant

Table 5: Major PCB congeners identified in adipose in the exposed and compari comparison groups. Concentrations expressed as median value (PPB).

PEAK NUMBER	STRUCTURAL ASSIGNMENT	EXPOSURE CATEGORY: CURRENT EXPOSED (N = 33)	EXPOSURE CATEGORY: COMPARISON GROUP (N = 52)	TEST FOR DIFFERENCES*: CURRENT VS COMPARISON
62	2,3,5,6,3',4'5'/ 2,3,4,5,2',4',5' (hepta-CB)	350	110	S
49	2,3,4,2',4',5'(hexa-CB)	270	84	S
54	2,4,6,3',4',5'/ 2,4,5,2',4',5'/ 2,3,4,5,2',5' (hexa-CB)	210	73	S
59	2,3,4,5,2',3',4' (hepta-CB)	190	60	S
40	2,4,5,3',4' 3,4,5,2',3' (penta-CB)	144	41	S
69	2,3,4,5,2',3',5',6'/ 2,3,4,5,6,2',3',5' (octa-CB)	120	43	S
87	2,3,5,6,2',4',5'/ 2,3,4,2',3',4'/ 2,3,4,5,2',4',6' (mixed)	120	23	S

* Generalized Wilcoxon Test S = Significant ($P < 0.0167$) NS = Not Significant

Table 6: The seven major serum PCB congener peaks in various PCBs exposed populations studied by different authors.

Transformer repair workers* (Fait - 1986)			Capacitor Workers** (Wolff - 1982)			Yu-Cheng Patients*** (Masuda - 1985)		
Peak Rank	Peak No.	Structure	Peak Rank	Peak No.	Structure	Peak Rank	Peak No.	Structure
-	19	3,4,2'/2,4,4' TRI-CB	(1)	1	2,4,4' TRI-CB	-	-	-
-	25	2,4,5,4' TETRA-CB	(2)	6	2,4,5,4' TETRA-CB	-	-	-
-	31	2,4,3'4'/3,4,5,3' TETRA-CB	(4)	8	2,4,3'4' TETRA-CB	-	-	-
(1)	62	2,3,5,6,3'4'5'/2,3,4,5,2'4'5' HEPTA-CB	-	33	2,3,4,5,2'4'5' HEPTA-CB	(5)	6	2,3,4,5,2'4'5' HEPTA-CB
(2)	49	2,3,4,2'4'5' HEXA-CB	(7)	25	2,3,4,2'4'5' HEXA-CB	(1)	4	2,3,4,2'4'5' HEXA-CB
(3)	54	2,4,6,3'4'5'/2,4,5,2'4'5'/ 2,3,4,5,2'5'HEXACB	(6)	22	2,4,5,2'4'5' HEXA-CB	(3)	3	2,4,5,2'4'5' HEXA-CB
(4)	59	2,3,4,5,2'3'4' HEPTA-CB	-	35	2,3,4,5,2'3'4' HEPTA-CB	(7)	7	2,3,4,5,2'3'4' HEPTA-CB
(5)	40	2,4,5,3'4'/3,4,5,2'3' PENTA-CB	(3)	20	2,4,5,3'4' PENTA-CB	(2)	1	2,4,5,3'4' PENTA-CB
(6)	69	2,3,4,5,2'3'5'6'/2,3,4,5,6,2',3',5' OCTA-CB	-	-		-	-	-
(7)	87	2,3,5,6,2'4'5'/2,3,4,2'3'4'/2,3,4,5,2'4'6'mixed-CB		29	2,3,4,2'3'4' HEXA-CB	-	-	-
				26	2,3,5,6,2'4'5' HEPTA-CB	-	-	-
-	56	2,3,4,3'4'5'/2,3,4,5,3'4' HEXA-CB	-	32	2,3,4,5,3'4' HEXA-CB	(4)	5	2,3,4,5,3'4' HEXA-CB
-	84	2,3,5,2'4'5'/2,3,4,3'4'/2,3,4,6,3'5' MIXED	-	23	2,3,4,3'4' PENTA-CB	(6)	2	2,3,4,3'4' PENTA-CB

* AROCLOR 1260 exposure

** AROCLOR 1016, 1242, 1254 exposure

*** KANECHLOR 400, 500 exposure

23

An improved semi-automated enzymatic determination of D-glucaric acid in urine

W.J.A. Meuling and J.J. van Hemmen – Department of Occupational Toxicology, Medical Biological Laboratory T.N.O., P.O. Box 45, 2280 AA Rijswijk (The Netherlands)

SUMMARY

The method we propose here, represents an optimized determination of D-glucaric acid by means of the β-glucuronidase inhibition test. The modifications we propose are stabilisation of the D-glucaric acid/1,4-Glucarolactone equilibrium after boiling by rapid cooling in ice use of a linear calibration curve and use of an automated assay. The coefficient of variation for the proposed method is; within-run 1.7% and inter-day 6.4%. Up to 120 samples per day can be analysed easily by one technician.

INTRODUCTION

Many drugs, environmental chemicals and food-additives can induce hepatic microsomal enzymes [1,2,3]. The urinary excretion of D-glucaric acid (GLA) is used by several authors as an index for this phenomenon [4,5]. Since 1963, when Marsh [6] first established GLA as a normal constituent of human urine, several methods, colorimetric [7], chromatographic [8,9,10] and enzymatic [11,12,13] have been developed. Most of these methods, however, may lack in accuracy and reproducibility, since relative large discrepancies in reference values have been reported even when the same method was used [14,15,16]. Among these methods, the most frequently used are the enzymatic assays based on the original method of Marsh [6], although it has been modified by several investigators [11,13,14,17,18]. The method described here is based on the method of Colombi et al. [13]. In order to improve the level of precision of the method some modifications are described. These modifications, make it possible to estimate GLA in urine with good reproducibility.

MATERIALS AND METHODS

Chemicals

Phenolphtalein-β-D-glucuronide,β-glucuronidase (Type L II: from Limpets, Patella Vulgata) approximately 900 U/ml and D-saccharic acid (calcium salt) were purchased from Sigma Chemical Co., St. Louis, MO, USA. All other chemicals used were of analytical grade.

Urine treatment

For the determination of GLA in a 1 ml urine sample, 4 ml of a sodium acetate buffer (0.5 M, pH 3.8) was added. A set of standard solutions of GLA was treated similarly. The tubes were capped, heated for 40 min. at 100°C, immediately afterwards cooled and kept in melting ice.

Assay

The enzymatic assay with β-glucuronidase using phenolphtalein-β-D-glucuronide as a substrate was carried out within 2 hours after boiling. Using a 1:11 dilution of treated urine or standard solutions with the sodium acetate buffer, the inhibition of β-glucuronidase was established. The assay was performed at 37°C using an ABA-100 bichromatic analyser (Abbott Laboratories, Pasadena, CAL, USA). After 30 minutes the reaction was stopped by adding (1:1 v/v) glycine buffer (2.0 M, pH 11.5). The intensity of the colour of the liberated phenolphtalein was measured at 550 nm.

RESULTS AND DISCUSSION

In the original method of Marsh [6] the determination of GLA is based on the principle that by boiling aqueous solutions of GLA at pH 2, GLA is partial converted to 1,4-glucarolactone (1,4 GL) a strong inhibitor of the enzyme β-glucuronidase. Crucial in this method is the equilibrium between GLA and 1,4 GL which is dependent on pH and temperature. Differences in procedures led to non-consistent results and a number of modifications of the original method in the literature [11,13-18]. Recently Colombi and co-workers [13] studied these factors extensively, which resulted in a method in which boiling of aqueous solutions of GLA and the enzymatic assay were carried out at the same pH. Although this method yielded good results, we observed that it is important that time and temperature after boiling should be controlled, otherwise erroneous results were found. Better results were obtained when the boiled solutions of GLA were kept in ice for up to several hours, as depicted in the figure.

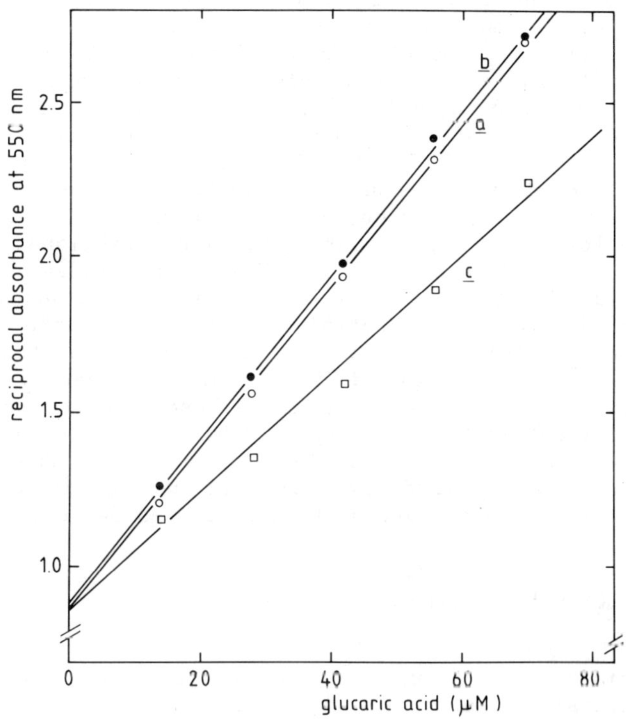

Fig. 1. – Results obtained with standard solutions of GLA treated as described in Materials and methods. Linear curves were obtained by plotting the reciprocal absorbances versus concentrations of GLA. (a) : The solutions were kept in ice for 7 minutes after boiling; (b) : ibid for 120 minutes; (c) : The solutions were kept at room temperature for 120 minutes. Each point represents the average of a duplicate determination.

It is concluded therefore, that at room temperature the equilibrium is not 'frozen', e.g. some lactone may be reconverted to GLA. Although different investigators studied this equilibrium, none of them dealt with stabilisation after boiling. Also none of the methods reported so far, has described any standardisation of the time between boiling and performing the assay. In contrast to Colombi et al. [13] our study used linear plots comparable to those of Dixon [19]. A very good linear relationship was observed between the reciprocal values of the optical densities and the concentrations of GLA used. Linearity was observed with concentrations of GLA up to 150 µM. Higher concentrations require dilution of the sample prior to analysis. To investigate the precision of the method, a pooled urine sample obtained from healthy men was analysed independently several times on one day. The within-run precision was determined, resulting in a coefficient of variation of 1.7% (n=6). The inter-day precision was determined, by analysis of the pooled urine sample on 8 days, resulting in a coefficient of variation of 6.4%. All these results were obtained by duplicate determinations. In urine samples collected over 24 hours from 29 healthy non-smoking men, not taking any medication, in age ranging from 19-59 years, the following average amount of GLA was established:

43,6 ± 10,9 µmol/24 h (Mean ± SD), range : 24,5 - 70,9 µmol/24 h. When the amounts were related to the creatinine content of the urine sample, the following average amount was calculated:

2,77 ± 0,13 µM/M creatinine (Mean ± SD) range: 1,9 - 4,8 µM/M creatinine. These results were in good agreement with values reported by other authors [11,16,18]. When the method is carried out manually, about 40 samples can be analysed daily. With our semi-automated version we were able to analyse up to 120 samples per day.

ACKNOWLEDGEMENT

This work was supported by a grant from the Directorate General of Labour of the Dutch Ministry of Social Affairs and Employment.

REFERENCES

1 Aarts, E.M. (1965) Biochem. Pharmacol. 14,359-363.

2 Davidson, D.C, McIntosh, W.B., and Ford, J.A. (1974) Clin. Sci. Mol. Med. 47, 279-283.

3 Latham, A.N. (1974) J. Pharm. Pharmacol. 26,284-286.

4 Siest, G., Batt, A.M., and Galteau, M.M. (1977) Ann. Biol. Clin. 35 , 425-432.

5 Hunter, J., Maxwell, J.D., Carrella, M., and Stewart, D.A. (1971) Lancet i, 572-575.

6 Marsh, C.A. (1963) Biochem. J. 86, 77-86.

7 Ishidate, M., Matsui, M., and Okada, M. (1965) Anal. Biochem. 11, 176-189

8 Warrander, A., Waring, R.H. (1978) Xenobiotica 8, 605-609.

9 Laakso, E.I., Tokola, R.A., and Hirvisalo, E.L. (1983) J. Chromatogr. 278, 406-411.

10 Walters, D.G., Lake, B.G., Bayley, D., and Cottrell, R.C. (1983) J. Chromatogr. 276, 163-168.

11 Simmons, C.J., Davis, M., Dordoni, B., and Williams, R. (1974) Clin. Chim. Acta 51,47-51.

12 Marsh, C.A. (1985) Anal. Biochem. 145, 266-272.

13 Colombi, A., Maroni, M., Antonini, C., Cassina, T., Gambini, A., and Foà, V. (1983) Clin. Chim. Acta 128, 337-347.

14 March, J., Turner, W.J., Shanley, J., and Field, J. (1974) Clin. Chem. 20, 1155-1158.

15 Latham, A.N. (1975) J. Pharm. Pharmacol. 27, 612-615.

16 Colombi, A., Maroni, M., Antonini, C., Fait, A., Zocchetti, C., and Foà, V. (1983) Clin. Chim. Acta 128, 349-358.

17 Jung,K., Scholz, D., Schreiber, G. (1981) Clin. Chem. 27, 422-426.

18. Fiedler, K., Schröter, E., Cramer, H. (1980) Eur. J. Clin. Pharmacol. 18, 429-432

19. Dixon, M. (1953) Biochem. J. 53, 170-171

24

Health surveillance of a potential TCCD-exposed industrial population in Seveso: pattern of some liver-related biochemical indicators

G. Assennato – Istituto di Medicina del Lavoro dell'Università, Bari
P. Cannatelli – Servizio di Medicina del Lavoro Ospedale, Desio
I. Ghezzi – I.S.T., Genova

On July 10 1976 a runaway reaction during the production of trichlorophenol in the ICMESA plant at Seveso led to the discharge of 2, 3, 7, 8 - tetrachlorodibenzo-p-dioxin into the open air. Technical details of the accident have been thoroughly discussed in many papers [1][2].

The work-force at the ICMESA plant is considered to be a group at high risk; it is assumed that they were chronically exposed to a mixture of chemicals containing low TCDD-levels as well as acutely exposed at the time of the accident and subsequently, until the plant was closed.

A comprehensive health surveillance program was established by the lombardy region in order to detect preclinical toxic effects, in 1981 a cross-sectional study was carried out.

The clinical and pathological findings in the literature suggest that a major site of the toxic action of TCDD is the liver [3]. Therefore, the general objectives of our study was to evaluate whether 5 years from the accident there was any TCDD-related hepatotoxic effect in the ICMESA group as compared to the reference groups and to analyze time trends of several biochemical indicators of hepatotoxicity in the ICMESA group.

METHODS

STUDY DESIGN: The cross-sectional study was carried out in 1981 and was based on the comparison between the ICMESA group and two reference groups consisting of workers of two plants, not considered to be exposed to hepatotoxic substances.
The following data were obtained for each subject:

A) Data from an interview schedule with items related to alcohol intake, smoking status, drug consumption, occupational history etc.
B) Physical examination.
C) Laboratory measures, consisting of several biochemical indicators (serum AST,

ALT, alkaline phosphatase, gamma-glutamyltranspeptidase, total cholesterol, triglycerides, urinary porphyrins, urinary A.L.A.).
The clinical chemistry activity was performed at the laboratory of clinical pathology, DESIO hospital.
DATA ANALYSIS: One-way and two ways anova and ancova were used for the comparison of the continuous variables. Chi-square test for independent samples was used to compare the distribution of categorical variables.

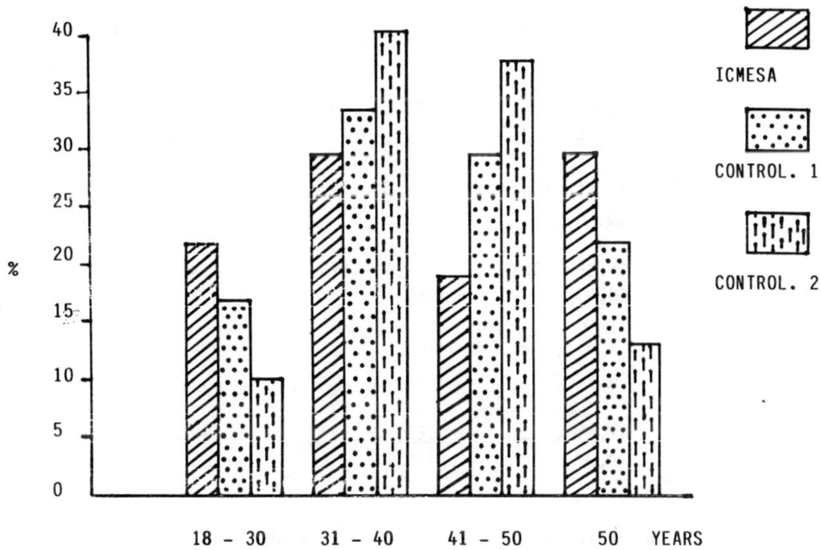

Fig. 1. — Age distribution of the study populations.

RESULTS

CROSS-SECTIONAL STUDY: The response rate was fairly high (89% in the ICMESA group, 100% in the control groups). In Fig.1 the age distribution is shown for the study groups; the ICMESA age distribution being similar to the 1st control group. There are some differences for the lenght of employment being shorter in the ICMESA group, due to the high turnover rate before the accident. No cloracne case was observed in the ICMESA group. The laboratory data are shown in table 1, 2. no significant between groups difference was detected, except for the mean values of urinary porphyrins and alkaline phosphatase stratifying the data by alcohol intake, the mean value of urinary porphyrins is significantly higher in the ICMESA group only for the subgroup of heavy drinkers.
LONGITUDINAL STUDY: The response rate of the ICMESA group (196 subjects) during the follow-up period was quite unsatisfactory: only 77 workers were always present at the annual examinations. As far as the biochemical indicators are concerned, a significant decrease was observed over the first three years.
SGOT and SGPT values showed an increasing trend from 1976 to 1978, the values being almost unchanged, subsequently. Mean serum cholesterol showed an increasing trend from 1976 to 1978 eventually returning to the 1976 level.

Table 1: ICMESA cross-sectional study: mean values, sample size (n), standard deviation and statistical significance for selected biochemical indicators in the study groups.

	ICMESA GROUP	CONTROL GROUP (1)	CONTROL GROUP (2)	P
AST	17,93 (158)	17,10 (176)	16,50 (101)	N.S.
	8,26	5,94	4,89	
ALT	24,61 (158)	22,16 (176)	20,88 (101)	N.S.
	20,71	8,98	8,69	
ALP	30,49 (158)	27,48 (177)	28,45 (101)	.01
	10,78	7,13	7,74	
GGT	34,74 (158)	34,07 (177)	26,91 (101)	N.S.
	40,45	28,71	25,59	
U-Porf	94,54 (152)	79,44 (173)	83,11 (100)	.05
	60,47	50,69	52,49	
Chol	191,82 (158)	190,49 (177)	203,39 (101)	.05
	39,84	43,43	43,94	
Trig	142,45 (158)	161,42 (177)	136,29 (100)	N.S.
	119,73	162,44	85,64	

P .001 BETWEEN DRINKERS IN THE ICMESA GROUP.

P .001 BETWEEN STUDY GROUPS FOR "HEAVY DRINKERS".

P = N.S. IN THE OTHERS SUBGROUPS AND BETWEEN GROUPS.

Table 2: ICMESA cross-sectional study: mean values, sample size (n), standard deviation and statistical significance by drinking habit in the study population for urinary porfiryns.

	ICMESA GROUP	CONTROL GROUP (1)	CONTROL GROUP (2)
NON DRINKERS	70,13 (15)	77,92 (12)	72 (5)
	47,57	54,60	33,11
LIGHT DRINKERS	77,33 (27)	71 (25)	78,24 (17)
0 - 25 cc/day	47,32	50,55	37,44
MODERATE DRINKERS	73,36 (25)	85,43 (37)	70,44 (16)
26 - 50 cc/day	32,83	47,75	42,24
HEAVY DRINKERS	102,16 (82)	77,89 (97)	82,80 (61)
50 cc/day	52,28	45,96	37,39

Table 3: ICMESA follow-up study (entire cohort): mean values, sample size (n), standard deviation for selected biochemical indicators, over time.

	1976	1977	1978	1979	1980	1981
AST	14,94 (147)	14,44 (145)	18,74 (124)	17,90 (116)	18,20 (105)	17,93 (158)
	8,22	8,40	19,48	8,39	7,31	8,26
ALT	12,87 (147)	18,01 (145)	26,57 (124)	23,03 (116)	24,73 (105)	24,61 (158)
	10,56	10,05	45,02	13,88	15,09	20,71
ALP	35,51 (147)	33,06 (145)	29,29 (124)	31,46 (116)	30,10 (105)	30,49 (158)
	14,37	13,98	10,60	11,99	11,42	10,78
GGT	30,70 (147)	29,29 (144)	32,93 (123)	32,42 (116)	32,09 (105)	34,74 (158)
	33,62	33,64	30,46	28,15	29,97	40,45
U-Ala	2,29 (75)	3,44 (119)	2,87 (122)	3,32 (113)	3,13 (105)	3,45 (153)
	1,30	1,77	1,27	1,44	1,35	1,54
Chol	201,04 (133)	215,22 (143)	215,26 (124)	218,44 (115)	197,44 (105)	191,82 (158)
	43,86	44,71	45,42	43,37	37,93	39,84
Trig	139,98 (131)	115 (143)	135,59 (123)	139,15 (114)	135,18 (105)	142,46 (158)
	75,02	63,76	63,91	67,48	81,37	119,73

CONCLUSIONS

TCDD exposure in the ICMESA group was not so heavy as previously suggested. No case of cloracne was detected, however, at the cross-sectional study (1981) there was a significantly higher mean value of urinary porphyrins and alkalin phosphatase, compared to the reference groups. The higher mean value in the ICMESA group was significant only in the heavy drinkers, so suggesting an interaction between chemical exposure and alcohol intake. The longitudinal analisis of data from the surveillance program shows a decreasing trend for the mean value of alkalin phosphatase which is striking considering that, in 1981, in spite of the decreasing trend it was still significantly higher in the ICMESA group compared to both reference groups. The analisys of cross-sectional and longitudinal data shows that some changes for indicators considered to be markers of TCDD activity (alkaline phosphatase and urinary porphyrins) did occur in the ICMESA working population, whether or not such association is actually causal, it is very hard to establish, given the mixture of other chemicals such population was exposed to in the past.

REFERENCES

[1] BISANTI L. ET AL. (1980). ACTA MORPH. ACAD. HUNG., 28, 134-157
[2] BONACCORSI A., FANELLI R., TOGNOLI G., (1978): AMBIO, 7, 234-239
[3] OLSON J.R. ET AL. (1980): TOXICOL. APPL. PHARMACOL., 56, 78-85
[4] PAZDEROVA-VEJLUPKOVE ET AL. (1980): PRACOV. LEK., 32, 204-209

REPORT OF SESSION II

Prepared by R.J. Rubin, F.W. Sunderman, and A. Colombi

Markers of enzyme induction are subtle indicators of pre-toxic (or adaptive) responses to exposure to a variety of chemicals including a number of widely used drugs. These adaptive responses are readily reversible and by themselves are not indicators of toxicity. Although there are a number of different methods for assessing enzyme induction in humans the most widely used and validated methods are: (1) urinary excretion of D-glucaric acid; (2) urinary excretion of 6β-OH cortisol which when normalized for urinary cortisol excretion allows collection of spot samples; and (3) antipyrine clearance or t 1/2 as assessed by blood or saliva sampling. Any of these methods, although lacking in specificity, can be used to assess enzyme induction in humans.

In assessing enzyme induction one must be aware of a large number of confounding factors which have to be controlled for in any protocol design. These factors include, but are not limited to, genetics, diet, smoking, and alcohol consumption.

Additional needs in this field include further study of the relative sensitivity of the various methods to a broader spectrum of xenobiotics. Although there are some data that the urinary excretion of 6β-OH cortisol is the most sensitive marker for several different drug inducers, there is a paucity of data for determining relative sensitivity to a number of environmental xenobiotics. Most likely, however, one will not be able to develop a hierarchy of relative sensitivities that will fit all exposures since the assays are either too general (glucaric acid excretion) or conversely too specific (6β-OH cortisol excretion and antipyrine clearance or t 1/2). Perhaps the resolution to this problem will come with the further development of monoclonal antibodies for assaying specific cytochrome P450s by radioimmunoassay.

Other assays that require additional validation include the urinary excretion of porphyrins as markers of induction of the heme pathway and carbon monoxide breath analysis as a marker of increased heme turnover.

A major question that arises with evidence of enzyme induction is "what is the health implication?". Some argue that any alteration from normal in a biological system is evidence of toxicity. Others argue that changes such as enzyme induction are merely adaptive responses which because they enhance the excretion of the inducer from the body are beneficial to the exposed individual. Bridging these two positions is evidence, mostly from animal studies but also confirmed in humans, that induction by one chemical can greatly alter the toxicity of a second agent. A major body of evidence exists that induction by one chemical can greatly alter the response to chemical carcinogens (either increase or decrease carcinogenicity). Thus, although the inductive response may be viewed simply as an adaptation, the toxic response to other xenobiotics must be kept in mind. There is a strong need for long-term follow-up of the health status (particularly for cancer) for individuals with evidence of enzyme induction. Finally, there is a major need for correlation of internal dose of a xenobiotic with inductive response. Such data would allow a greater understanding of conflicting data that has been collected for population exposures as well as for individual differences in inductive response.

With regard to in vivo assessment of actual toxic effects to the liver, there are available a variety of classical serum enzyme tests (ALT, AST, γ-GT, etc.). It should be appreciated that elevations in any of these enzyme assays represent leakage of cytosolic enzymes from the liver cell and thus reflect ongoing damage. There appears to be little to offer in terms of advantage of one assay system over another. In fact, a battery of assays is usually needed. Thus, there is a need for development and validation of sensitive and specific assays such as serum bile acids as markers of liver anion transport. For example, evidence was presented that workers exposed to styrene had no significant changes in serum bile acids while there were marked changes in a variety of other non-specific enzyme assays. On the other hand, workers exposed to a number of organic solvents exhibited significant changes only in serum bile acids, suggesting a specific impairment of this functional activity.

Another area in need of considerable development is that of markers of oxidative stress. Considerable animal data suggest that oxidative stress to a cell represents a major mechanism of toxicity. A number of assays that have been widely used in these animal experiments need to be evaluated and validated in humans. Two methods that were presented for consideration were (1) assay of malondialdehyde (MDA) or other lipoperoxides excreted in urine; and (2) ethane exhalation.

Finally, evidence was presented for the use of serum type III procollagen peptides (PIIIP) as markers of chronic hepatic damage or interstitial pulmonary fibrosis. While taking advantage of the most modern knowledge in collagen synthesis and its role in chronic fibrosis, this assay will usually reflect massive, ongoing toxicologic response. The need here is for sensitive and specific assays for detection of early alterations in collagen synthesis at a time when the process is still readily reversible. Further developments in collagen biochemistry and pathophysiology hold forth the promise of such developments.

Part III

Biochemical indices of liver toxicity: porphyrin metabolism

Coordination:
G. Secchi (Milano, Italy)
A. Farulla (Roma, Italy)

25

Porphyrin metabolism as a target of exogenous chemicals

G.H. Elder, A.J. Urquhart – Department of Medical Biochemistry, University of Wales College of Medicine, Cardiff CF4 4XN, U.K.

ABSTRACT

Different mechanisms by which exogenous chemicals may affect haem biosynthesis are described. Compounds that inhibit enzymes and/or accelerate haem turnover in liver produce specific alterations in porphyrin metabolism and excretion that may serve as indicators of toxic effects in humans.

INTRODUCTION

It has been known for many years that exogenous chemicals may disturb porphyrin metabolism in humans and other mammals. For some of these compounds, particularly those causing experimental porphyrias (Sweeney, 1985), relationships between specific perturbations of haem biosynthesis and patterns of porphyrin accumulation in tissues and excreta have been delineated. Such patterns can be used to predict the sites of action of exogenous chemicals within the pathway of haem biosynthesis. Conversely, when a chemical is known from animal studies to have a specific effect on haem biosynthesis, tests for the predicted abnormality of porphyrin metabolism provide a potential method for assessing exposure in humans.

HAEM BIOSYNTHESIS AND PORPHYRIN METABOLISM

Adult humans synthesise about 7 µmol haem/kg body weight/day, of which around 80% is produced in erythroid tissue. Most of the rest is synthesised in the liver, where about half the haem produced is used for the formation of haemoproteins of the cytochrome P450 series.

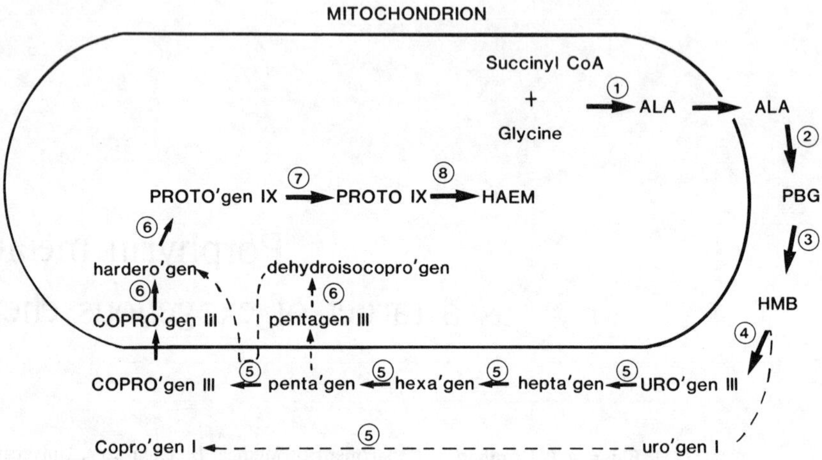

Fig. 1. – The pathway of haem biosynthesis. ALA, 5–aminolaevulinate; PBG, porphobilinogen; HMB, hydroxymethylbilane; uro'gen, copro'gen, etc. uroporphyrinogen, coproporphyrinogen, etc; proto, protoporphyrin. Enzymes are ALA-synthase (1), PBG synthase (2), PBG-deaminase (3), uroporphyrinogen III synthase (4), uroporphyrinogen decarboxylase (5), coproporphyrinogen oxidase (6), protoporphyrinogen oxidase (7), and ferrochelatase (8). Normally, less than 1% of HMB is converted to uroporphyrinogen I and the alternative route via dehydroiso-coproporphyrinogen is not quantitatively significant.

The pathway of haem biosynthesis (Figure 1) is identical in all tissues, although the relative activities of the individual enzymes are different in erythroid tissue and liver (Kappas et al, 1983). The porphyrinogen intermediates of the pathway are unstable and readily oxidised to the corresponding porphyrins, which are not metabolized further. In normal circumstances, these unstable intermediates are present at low concentrations and are turned over rapidly thus largely preventing their conversion to porphyrins.

In the liver the supply of haem for haemoprotein assembly is closely controlled by the activity of ALA synthase, the rate-limiting enzyme of the pathway, which is under negative feedback regulation by haem which is present in an regulatory haem pool (Kappas, 1983). Depletion of the regulatory haem pool, for example by haemoprotein assembly following stimulation of apocytochrome P450 synthesis by an exogenous chemical, leads to a secondary induction of ALA synthase that persists until the haem pool is replenished. Control mechanisms in erythroid and other tissues are different: ready inducibility of ALA synthase by a wide range of exogenous and endogenous compounds appears to be an unique feature of hepatic haem synthesis.

In normal circumstances, at least 85% of the ALA produced is metabolized to haem. The rest is either excreted unchanged in the urine or converted to intermediates which are excreted (Table 1). Urine and bile contain both porphyrinogens and porphyrins derived from them by autoxidation in tissues before excretion.

Table 1: Porphyrin concentrations in normal subjects.

	Total porphyrin	Main components
Liver	0-3 nmol/g	Proto
RBC	40-1700 nmol/litre	Zinc-proto
Plasma	5-20 nmol/litre	Proto, uro
Urine	30-320 nmol/litre	Copro (80%), uro (15%)
Faeces	<290 nmol/g dry wt	Proto (73%), copro (24%)

Urine also contains ALA (0-50 μmol/l) and PBG (0-9 μmol/l)
Figures for total porphyrin include porphyrinogens.

The immediate fate of intermediates that are lost from the pathway, either by diffusion or oxidation, appears to be determined by cell permeability and the availability of binding sites within the cell. Thus coproporphyrin(ogen) and protoporphyrinogen diffuse from hepatocytes even at low intracellular concentrations whereas uroporphyrin(ogen) and protoporphyrin IX tend to be retained within the cell. Once outside the cell, the route of excretion is largely determined by solubility properties. The more hydrophilic and more highly carboxylated intermediates appear predominantly in the urine while the most hydrophobic compound, protoporphyrin IX, is excreted in the bile (Table 1) (Elder, 1980).

Alterations in the pattern of excretion and accumulation in tissues of porphyrins and their precursors can be produced by a number of mechanisms. Exogenous chemicals may disturb porphyrin metabolism by any of these mechanisms, either separately or in combination (Table 2). When a chemical inhibits one of the enzymes that convert ALA to haem, negative feed-back regulation by haem enables compensatory changes to occur in liver which tend to restore the rate of hepatic haem synthesis to normal. ALA synthase activity increases and leads to an increase in production of the substrate of the defective enzyme. The relationship between decreased enzyme activity and abnormalities of porphyrin excretion is very variable in humans. These inter-individual differences in the degree of the compensatory response have not been explained.

Increased activity of ALA synthase, either due to primary stimulation by a chemical or secondary to a chemically -induced increase in haem turnover may also lead to over production of haem precursors. Chemicals that both inhibit enzymes and independently stimulate ALA synthase are particularly effective in causing porphyrin overproduction (Smith & De Matteis, 1980). Exogenous chemicals may also influence the route of excretion. Compounds that cause cholestasis produce coproporphyrinuria by diverting coproporphyrin I from the bile to the urine. Thus chemically-induced coproporphyrinuria does not necessarily indicate a primary effect on haem synthesis.

Table 2: Exogenous chemicals and porphyrin metabolism.

Effect on	Overproduction/excretion of
A. Haem biosynthesis	
1. Enzyme inhibition	Substrate
2. ALA overproduction	
(i) 1°	ALA
(ii) 2° to increased haem turnover	copro, proto
B. Excretion	
1. Cholestasis	copro I (urine)

EXOGENOUS CHEMICALS AND PORPHYRIN METABOLISM

The chick embryo hepatocyte culture has been widely used as an experimental model for investigating the effects of drugs and other chemicals on haem biosynthesis (Marks, 1985). The large number of compounds that produce porphyrin accumulation in this system can be divided into two groups. The first and larger group contains compounds that in general do not produce detectable abnormalities of porphyrin metabolism in mammals or normal humans. Many chemicals in this group induce cytochrome P450. In mammals, induction of cytochrome P450, although accompanied by some increase in ALA formation (Table 2), is not usually associated with measurable increases in porphyrin concentrations in tissues or in urine. However there is some evidence that the coproporphyrinuria that accompanies prolonged exposure to some chemicals, may be related to sustained induction of cytochrome P450 (Strik & Koeman, 1979).

The second group contains chemicals that have specific effects on haem biosynthesis, some of which also induce cytochrome P450 synthesis. Compounds in this group predictably disturb porphyrin metabolism in mammals, including humans. They fall into four main categories: alcohols, heavy metals, suicide substrates of cytochrome P450 and polyhalogenated aromatic hydrocarbons (PHAH). All, apart from heavy metals, selectively affect hepatic haem synthesis. The toxic actions of alcohol, lead and other heavy metals on haem biosynthesis have been reviewed (McColl & Goldberg, 1980; Marks, 1985) and will not be considered here.

SUICIDE SUBSTRATES OF CYTOCHROME P450

2-isopropyl-4-pentenamide (AIA) and 3,5-bis(carbethoxy)-2,4,6-trimethyl-1,4-dihydropyridine (DDC) have long been known to cause experimental hepatic porphyrias in rodents that are characterized by sustained induction of ALA synthase and massive accumulation of porphyrin in the liver (Sweeney, 1985). Over the past few years it has become clear that both these compounds are representatives of a large group of chemicals that perturb hepatic haem synthesis to varying degrees because they are oxidized by cytochrome P450 to reactive species which alkylate the haem prosthetic group of the cytochrome with the eventual formation of N-alkylporphyrins.

Chemicals that act as suicide substrates in this way include a variety of heterocyclic compounds (e.g. DDC, sydnones, griseofulvin, 1-aminobenzotriazole) and compounds containing terminal olefine and acetylene groups (e.g. AIA, norethisterone, allyl barbiturates, vinyl chloride) (Marks, 1985; Ortiz de Montellano et al, 1986). Their effect on haem biosynthesis in the liver depends on the nature of the N-alkylporphyrin product and on the extent to which the suicide inactivation reaction leaves an apocytochrome that can be reconstituted with haem from the regulatory pool.
DDC exemplifies a small group of compounds (e.g. some DDC analogues, sydnones, griseofulvin) that promote the formation of N-alkylprotoporphyrins containing small alkyl substituents (e.g. methyl, ethyl, vinyl) that act as potent inhibitors of ferrochelatase (De Matteis et al, 1980; Ortiz de Montellano et al, 1986). Inhibition of this enzyme leads to marked accumulation in the liver, and increased biliary excretion, of its substrate, protoporphyrin IX. In this type of suicide inactivation reaction, alkyl groups are transferred from the chemical to the pyrrole nitrogen atoms of the cytochrome P450 haem, thus displacing iron and producing green-coloured, red-fluorescent N-alkylporphyrins.

In reactions with AIA and other chemicals with terminal olefinic or acetylinic groups the whole compound appears to be transferred to the pyrrole nitrogen atoms. The bulky N-alkylprotoporphyrins that result from this reaction are unable to enter the active centre of ferrochelatase and do not act as inhibitors. N-alkylprotoporphyrins produced by transfer of large subsituents (e.g. the benzyne moiety of 1-aminobenzotriazole) are similarly non-inhibitory (Ortiz de Montellano et al, 1986). AIA selectively depletes the haem of cytochrome P450 leaving an apocytochrome that can be reconstituted by haem (Correia et al, 1981). AIA therefore markedly increases hepatic haem turnover leading to a secondary increase in ALA-synthase activity with massive overproduction of coproporphyrin and protoporphyrin (Table 2). On the other hand, 1-aminobenzotriazole depletes cytochrome P450 by a mechanism that appears to prevent the formation of a reconstitutable apocytochrome. As a result, it has very little effect on haem biosynthesis (Ortiz de Montellano et al, 1986). Other suicide substrates, including those that produce ihibitors of ferrochelatase, appear to vary in their capacity to stimulate haem turnover by this mechanism. In addition to the compounds described above, some pesticides containing diazinon analogues, which have been implicated in the pathogenesis of human porphyria, may produce N-alkylporphyrins (Nichol et al, 1983).

Little is known about the metabolism of N-alkylporphyrins. They readily permeate hepatocytes in vivo and in culture. However the possibility that they may be detectable in plasma or in faeces does not appear to have been explored.

POLYHALOGENATED AROMATIC HYDROCARBONS

In the late 1950s large numbers of cases of cutaneous porphyria were observed in S.E. Turkey among individuals who had eaten seed wheat that had been treated with the fungicide, hexachlorobenzene (HCB). Animal experiments confirmed that HCB caused a chronic hepatic porphyria and subsequently showed that this action was shared by some other PHAHs: 2,3,7,8-tetrachlorodibenzo-(p)-dioxin (TCDD), hexabromobenzene and certain polychlorinated- and polybrominated-biphenyls (Smith & Francis, 1980; Sassa et al, 1986). All these compounds also induce cytochrome P450 species in the liver.

Susceptibility to PHAH-induced porphyria varies markedly between individuals, strains and species (Elder, 1978). In rodents there is some correlation between susceptibility and inducibility of arylhydrocarbon hydroxylase and other enzymes controlled by the Ah locus (Greig et al, 1984) but other factors are also important. The development of porphyria is prevented by iron deficiency (Sweeney et al 1979) and enhanced by iron-overload (Smith & Francis, 1983) or, in rats, by oestrogens (Rizzardini & Smith, 1982).

Other halogenated hydrocarbons, for example the lower chlorinated benzenes, do not appear to cause porphyria (Carlson, 1977). The small increase in liver and urinary porphyrin concentrations that they provoke at high doses is probably related to sustained induction of cytochrome P450.

PHAH cause porphyria by decreasing the activity of uroporphyrinogen decarboxylase in the liver (Table 3).

In animals receiving HCB mixed with their diet, there is a delay of two weeks or more before the enzyme activity starts to decrease progressively (Elder, 1978), which may not be observed after a single large dose of either HCB or TCDD (Smith & Francis, 1986). The decrease in enzyme activity is accompanied by accumulation of uroporphyrin and heptacarboxylic porphyrin in the liver (Table 3) and by excretion of these and other porphyrins derived from substrates and intermediates of the uroporphyrinogen decarboxylase reaction (Elder, 1978). The porphyrin excretion pattern produced by these changes may be sufficiently specific to act as an indicator of uroporphyrinogen decarboxylase inhibition even when the total amounts of porphyrin excreted are within normal limits. Porphyria persists for at least 14 weeks after a single oral dose of HCB (Smith & Francis, 1986), and continues for several weeks after withdrawal of HCB from the diet and elimination of more than 95% of HCB from the liver (Koss et al, 1983).

In humans, a partial deficiency of hepatic uroporphyrinogen decarboxylase activity is the underlying biochemical abnormality in porphyria cutanea tarda (PCT). In some patients with this condition, which is the commonest of the human porphyrias, the enzyme defect is inherited in an autosomal dominant fashion (familial or type II PCT) but in the majority there is no evidence for inheritance (sporadic or type I PCT). This latter type of PCT appears to be indistinguishable from PHAH-induced porphyria (Elder, 1978) except that there is no history of exposure to halogenated chemicals.

Table 3: Effect of HCB on haem biosynthesis. Male C57/BL6 mice were fed an HCB (0.02% w/w) diet following an i.p. injection of iron dextran (12.5 mg iron). Uroporphyrinogen decarboxylase (UROD) activities in liver (A) (nmol/min/g wet wt.) and after uptake by solid-phase anti-UROD antibody (B) (pmol/min/0.1) are shown (means for 4 animals). *P 0.05 vs controls.

Days on diet	URO D		Liver porphyrin (pmol/g)
	A	B	
4	6.5	1.15	29
8	7.4	1.03	49
14	9.2	1.08	81
21	*7.4	*0.92	148
28	*4.8	*0.76	890
35	*1.5	*0.30	17550

In addition to the many cases reported following HCB ingestion in Turkey, overt PCT or subclinical PCT (possession of the characteristic porphyrin excretion pattern without symptoms) has also been reported following exposure to TCDD. Poland & Knutson, 1982) and polycholorinated biphenyls (Chang et al, 1980). In individuals exposed to PHAHs, the significance of minor changes in porphyrin excretion may be difficult to assess because of lack of suitable control studies while, for isolated cases of PCT, coincidental occurence of sporadic PCT, unrelated to PHAH exposure, is often hard to exclude.

In spite of considerable investigation, the mechanism whereby PHAHs decrease uroporphyrinogen decarboxylase activity in the liver remains unknown. The progressive fall in uroporphyrinogen decarboxylase activity is accompanied by a corresponding accumulation of material that competes with active enzyme for binding to a specific solid-phase antibody to uroporphyrinogen decarboxylase (Table 3). This material has been provisionally identified as catalytically-inactive immunoreactive enzyme in experiments that suggest that the PHAH-induced fall in enzyme activity is unaccompanied by a decrease in enzyme protein (Elder & Sheppard, 1982). PHAHs therefore appear to inactivate the active centre(s) of uroporphyrinogen decarboxlase by a process that is sufficiently localised to avoid alteration of major epitopes. Possible mechanisms for selective inactivation of the active centre(s), perhaps by modification of essential thiol groups, that have been suggested include direct inhibition by PHAHs (Kawanishi et al 1983), reaction with a PHAH metabolite (Debets et al, 1980), inhibition by a specific inhibitor produced in the liver in response to PHAHs (Cantoni et al, 1984) and oxidative inactivation by some form of reactive oxygen species (Ferioli et al, 1984). The last mechanism is attractive because it might explain the synergistic action of iron in this type of porphyria and because uroporphyrinogen decarboxylase can be inactivated by oxidation in vitro without loss of immunoreactivity (Elder et al, 1986); but at present it is not clear how this postulated oxidative reaction is directed specifically at the active centre of uroporphyrinogen decarboxylase.

REFERENCES

Cantoni, L., Fiume, D.D., Rizzardini, M., Ruggieri, R.(1984). In vitro inhibitory effect on porphyrinogen carboxylyase of liver extracts from TCDD treated mice. Toxicol. Lett., 20, 211.

Carlson, G.P. (1979). Brominated benzene induction of hepatic porphyria, Experimentia, 35(4) 513.

Correia, M.A., Farrell, G.C., Olson, S. et al (1981). Cytochrome P450 heme moiety: the specific target in drug-induced heme alkylation. J. Biol Chem., 256: 5466.

Debets, F.M.H. Hamers, W.J., Strik, J.J. (1981). Biotransformation and porphyrinogenic action of

hexachlorobenzene and its metabolites in a primary liver cell culture. Toxicology, 19, 185.
De Matteis, F., Gibbs, A.H. (1980). Drug-induced conversion of liver haem into modified porphyrins. Evidence for two classes of products. Biochem. J., 187: 285.
Elder, G.H. (1978). Porphyria caused by hexachlorobenzene and other polyhalogenated aromatic hydrocarbons, in Handbook of Experimental Pharmacology, Vol. 44, De Matteis, F. and Aldridge, W.N., Eds., Springer-Verlag, Berlin, 157.
Elder, G.H., Roberts, A.G., Urquhart, A.J. (1986). Acquired uroporphyrinogen decarboxylase defects: molecular mechanisms, in Porphyrins and Porphyrias, Nordmann, Y. (ed.) John Libby-Inserm, Paris, 147.
Ferioli, A., Harvey, C., De Matteis, F. (1984) Drug-induced accumulation of uroporphyrin in chick hepatocyte cultures. Biochem. J., 244: 769.
Kappas, A., Sassa, S. and Anderson, E.E. (1982) The porphyrias, in The Metabolic Basis of Inherited Disease, Stanbury, J.B., Wyngaarden, J.B. and Fredrickson, D.S., Eds. McGraw-Hill, New York, 1301.
Kawanishi, S., Seki, Y. and Sano, S. (1983) Uroporphyrinogen decarboxylase. J. Biol. Chem. 258, 4285.
Koss, G., Seubert, S., Seubert, A. (1983). Studies on the toxicology of hexachlorobenzene. Arch. Toxicol., 53: 13.
Marks, G.S. (1985). Exposure to toxic agents: the hem biosynthetic pathway and hemoproteins as indicator. CRC Clinical Reviews in Toxicology, 15,: 151.
McColl, K.E.L., Goldberg, A. (1980). Abnormal porphyrin metabolism in disease other than porphyria. Clinics in Haematology, 9, 427.
Nichol, A.W., et al. (1983). The site of inhibition of porphyrin biosynthesis by an isomer of diazinon in rats. Biochem. Pharmacol. 32, 2653.
Ortiz de Montellano, P.R., Costa, A.K., Grab, E.P. et al. (1986). Cytochrome P450 destruction and ferrochelatase inhibition. In Porphyrins and Porphyrias, Nordmann, Y. (ed.). John Libby-Inserm, Paris, p. 109.
Poland, A., Knutson, J.C. (1982).2,3,7,8-Tetrachloro-dibenzo-p-dioxin and related halogenated aromatic hydrocarbons: examination of the mechanism of toxicity. Annu. Rev. Pharmacol. Toxicol. 22, 517.
Rizzardini, M., Smith, A.G. (1982). Sex differences in the metabolism of hexachlorobenzene by rats and the development of porphyria in females. Biochem. Pharmacol 312, 3543.
Sassa, S., Sugita, O., Ohnuma, N. et al (1986). Studies of the influence of chloro-substituent sites and conformational energy in polychlorinated biphenyls on uroporphyrin formation in chick embryo liver cell cultures. Biochem. J. 235: 291.
Smith, A.G., De Matteis, F. (1980). Drugs and the hepatic porphyrias. Clinics in Haematology, 9: 371.
Smith, A.G., Francis, J.E. (1980). Relative abilities on a molar basis of hexachloro-, hexafluoro- and

hexabromobenzene to decrease liver uroporphyrinogen decarboxylase activity and cause porphyria in female rats. Res. Commun. Chem. Pathol. Pharmcol. 28, 377.

Smith, A.G., Francis, J.E. (1983). Synergism of iron and hexachlorobenzene inhibits hepatic uroporphyrinogen decarboxylase in inbred mice. Biochem. J., 214: 909.

Smith, A.G., Francis, J.E. (1986). Induction of porphyria in inbred mouse strains by polyhalogenated aromatics. In Porphyrins and Porphyria, Nordmann, Y. (ed.)., John Libby-Inserm, Paris, p.127.

Strik, J.J.T.W.A., Koeman, J.H. (1979). Chemical Porphyria in Man. Strik, J.J.T.W.A. and Koeman, J.H. Eds. Elsevier, Amsterdam.

Sweeney, G.D., Jones, K.G., Cole, F.M. (1979). Iron deficiency prevents liver toxicity of 2,3,7,8-tetra-chlorodibenzo-(p)-dixoin (TCDD). Science, 204: 332.

Sweeney, G.D. (1985) Experimental porphyria. Clinics in Dermatology, 3: 125.

26

Chronic hepatic porphyria induced by chemicals: the example of dioxin

M.O. Doss, A.M. Colombi – Clinical Biochemistry at the Faculty of Medicine of the Philipp University, D-3550 Marburg, Federal Republic of Germany, and Instituto Medico Milanese, Milano, Italy

ABSTRACT

Exposure to dioxin triggered a clinically chronic hepatic porphyria (porphyria cutanea tarda) in family members with hereditary uroporphyrinogen decarboxylase deficiency. The remission occured after chloroquin therapy. A hereditary disposition seems to be necessary for biochemical and clinical expression of chronic hepatic porphyria after a unique dioxin exposure. Non-susceptible persons develop after exposure only a symptomatic coproporphyrinuria without clinical relevance.

INTRODUCTION AND DEFINITION

Chronic hepatic porphyria (CHP) is the most frequent porphyria. CHP manifests itself as porphyria cutanea tarda (PCT) by characteristic skin symptoms and is generally associated with chronic liver disease. Red-cell uroporphyrinogen decarboxylase (UD) activity is decreased in the 'hereditary' type whereas, in the 'sporadic' type, it is normal. However, hepatic UD is lowered in all CHP/PCT patients (Table 1).

We prefer this term 'chronic hepatic porphyria' because the central organ of this porphyria disease process is the liver. The denotation 'chronic hepatic porphyria' was based on three facts: a) The centre of this disturbance is the liver; b) the porphyria develops slowly, shows from the beginning a chronic course, which passes various characteristic subclinical phases, biochemically defined, and separable from each other (Table 2) (5,7); c) another reason for referring to the chronic hepatic disturbance in porphyria metabolism which leads to 'porphyria cutanea tarda' as 'chronic hepatic porphyria' is the possibility of experimental imitation of the chronic hepatic metabolic disturbance. Whereas so far a porphyria of the acute type could only be partially imitated, namely by inducing δ-aminolevulinic acid synthase with allylisopropylacetamide, it is possible, to a greater extent, to imitate a CHP in its biochemical characteristics by hexachlorobenzene, polyhalogenated biphenyls and dioxin.

Table 1: Factors responsible for expression of chronic hepatic porphyria.

A	G e n e r a l Concordance of hepatic lesion and porphyrin accumulation due to inherited or toxic uroporphyrinogen decarboxylase deficiency leading to increased urinary porphyrin excretion with dominance of uro- and heptacarboxyporphyrin.
B	G e n e t i c (endogenic) Hereditary uroporphyrinogen decarboxylase deficiency in liver and blood cells. Uroporphyrinogen decarboxylase deficiency only in hepatic tissue (?).
C	E x o g e n i c (toxogenetic) Alcohol Drugs - Estrogens, hormonal oral contraceptives. Toxins - Hexachlorobenzene, dioxin, polyhalogenated biphenyls, vinyl chloride.

Table 2: Differentiation of chronic hepatic porphyrias into clinically latent phases (CHP type A, B and C) and the manifest state (CHP type D; i.e. porphyria cutanea tarda).

CHP phase	Urinary porphyrins (μmol/24 h)	Constellation (%) Uro	Hepta	Copro
A	0.3 - 0.8	< 40	< 10	> 40
B	0.5 - 1.3	> 40	< 10	< 40
C	0.8 - 2.4	> 50	> 20	> 10
D	> 2.5	> 50	> 20	< 10
Normal	< 0.2	< 30	< 5	> 60

MOLECULAR PATHOGENESIS

Pathogenesis of CHP including its clinically manifest phase, PCT, is complex (Table 1). This CHP disease process was classified in two different types depending on the activity of UD in erythrocyte (7,10,19,20,25,37). In the hereditary or 'familial' type subnormal erythrocyte UD activity follows an autosomal dominant trait. In the 'sporadic', 'non-hereditary' or 'toxic' type erythrocyte UD activity is normal. Liver UD activity is decreased in both types. A disturbance of hepatic UD is a general premise for the development of CHP. It remains to be determined, to what extent hereditary factors (12,18) effect all cases of the 'sporadic' form with its restriction of diminished UD to the liver (9). Alcohol and estrogens are important exogenous factors (7,28). We know, that not only in animals but also in humans an exclusively toxic CHP can be evoked by polyhalogenated aromatic compounds (33), especially by hexachlorobenzene (26). It is caused by a decreased UD activity in the hepatic tissue, but not in erythrocytes (8,32,36).

2,3,7,8-Tetrachlorodibenzo-p-dioxin (TCDD), a contaminant of various environmental chemicals, is the most potent porphyrinogenic agent known (13). It induces hepatic δ-aminolevulinic acid synthase (22) and inhibits hepatic UD (27,30,31,38). In the liver a 2000- to 4000-fold increase of uro- and heptacarboxyporphyrins was observed in the mouse after oral administration of TCDD over 4 weeks (13). An accumulation of these porphyrins is characteristic for CHP in man and animals. In contrast to chronic dioxin exposure, a single oral dose of TCDD did not produce porphyria (15).

After an accident in a chemical plant near Seveso (Italy) (4) two persons, exposed to low levels of TCDD, developed CHP type D or porphyria cutanea tarda, respectively; they belong to a family of 66 persons partially living in the TCDD-contaminated area (34). Evidence for a genetic predisposition is presented in this study by determination of erythrocyte UD (7) in affected and non-affected family members.

CLINICAL BIOCHEMICAL STUDIES

Two members of a family living in Seveso developed PCT after exposure to dioxin caused by an accident in a plant. Erythrocyte UD activity was diminished in 1978 and 1982 to the same level of ~50% reflecting a genetic abnormality (Table 3). Their father and one sister also showed an UD deficiency. Furthermore, a reduced UD could be detected in 5 relatives of the father. In 1977, the propositus (son, born 1953, Tables 3 and 4) developed a PCT (CHP type D) with a total porphyrin excretion of ~2.5 mg (uroporphyrin ~ 70%). A year later, after remission of clinical symptoms under chloroquine therapy, the porphyria process was in the CHP type C phase (34), retreating to a subclinical type A without clinical relevance in 1982 (Table 4). Porphyrin precursor excretion (δ-aminolevulinic acid and porphobilinogen) was normal in all cases.

Table 3: Decrease of red cell uroporphyrinogen decarboxylase activity in members of an Italian family exposed of dioxin from Seveso.

Subjects		UD activity: % of controls[a] 1978	1982	PCT-symptoms[b] 1977/78	1982
1. Father	*1925	63	62	-	-
2. Mother	*1928	87	95	-	-
3. Son	*1953	60	50	+	-
4. Daughter	*1957	63	49	+	-
5. Daughter	*1961	57	65	-	-

[a] Uroporphyrinogen decarboxylase (UD) activity of controls: 19.8 ± 3.6 ($\bar{x}$ ± s) µmol coproporphyrin/l erythrocytes · h (n = 96).

[b] Cutaneous symptoms of porphyria cutanea tarda.

Table 4: Urinary porphyrin excretion in family members according to Table 1 in 1982.

Subjects	Porphyrins (nmol/24 h)[a] Uro	Hepta	Hexa	Penta	Copro	CHP type
1	968	241	28	14	116	C
2	13	3	1	3	63	-
3	32	13	3	6	107	A
4	111	30	5	10	119	A/B
5	11	1	1	3	87	-
Upper norm	30	4	3	5	120	

[a] Urinary porphyrin precursors (δ-aminolevulinic acid and porphobilinogen) as well as tri- and dicarboxyporphyrin excretion were normal.

The father and the second daughter (Table 3) as well as the other relatives (34) with subnormal UD (one brother of subject 1 and his children) did not develop PCT, only a subclinical CHP type A (Table 5). Reinvestigations initiated in 1982 by subject 5 showed a normal porphyrin excretion (Table 4). Prolongation of porphyrinuria in the father was influenced by alcohol. The course of CHP severity, progression and remission is outlined in Table 5.

Subjects 3 and 4 (Table 3) were treated with chloroquine over a period of one year, followed by an improvement of clinical symptoms and a significant decrease of porphyrin excretion over CHP type C down to type A (Table 4). Histologic examination of liver biopsies of the propositus (subject 3) showed signs of 'chronic hepatitis'.

Table 5: Course of chronic hepatic porphyria (porphyria cutanea tarda) in TCDD-exposed family members from 1977 to 1982.

Subjects	Chronic hepatic porphyria type 1977		1978		1982	UD activity
1	A	→	D	→	C	↓
2	-	→	A	→	-	normal
3	D	→	C	→	A	↓
4	D	→	C	→	B/A	↓
5	-	→	A	→	-	↓

The family's father (subject 1) did not develop a PCT after the Seveso accident; total urinary porphyrin excretion was determined at 0.3 nmol/24 h; his PCT was triggered by alcohol in 1978 (alcohol consumption > 0.5 l wine/day). Under continuous alcohol influence the disease process did not remit to phase A as in cases 3 and 4, but rather remained in the decompensated latent phase of a CHP type C (Table 5). The mother of the family showed normal UD activity (Table 3); she demonstrated only a transitory elevation of porphyrin excretion in the sense of an incipient CHP type A as in sporadic, not family-bound cases.

13 (22%) of 60 persons from Seveso (not belonging to the above family) exposed to TCDD had a secondary coproporphyrinuria. 5 of these 13 people showed a transition constellation to CHP type A (5) with a slight increase of uro- and heptacarboxyporphyrin and coproporphyrin dominance (urinary samples from 1977). In 3 of 5 'transition cases' porphyrinuria was more severe in 1980 than in 1977 whereas pathologic porphyrinuria had returned to the normal range in the other 10 persons.

COMMENTS

A deficient UD activity in red cells was recognized as a marker for the inherited type of CHP (7,10,19,37). A mutation in 'familial' PCT by immunoreactive UD has been established (11,25), which supports our results on different types of PCT (6). An increase of uroporphyrin and heptacarboxyporphyrin in urine and liver is impossible without an inhibition of UD in the liver, so that a toxic effect of various organochemicals (33) on hepatic UD appears as the most plausible cause for a CHP as well as for the transition of coproporphyrinuria to CHP (5). Thus, reduced hepatic UD activity is the specific defect in CHP, which must not necessarily be 'intrinsic' or inherited in all cases (12), because a purely toxic pathogenesis of CHP can be related to e.g. hexachlorobenzene (8,29,32,36) and dioxin (2,13,15) in animals, and was observed in man after an environmental hexachlorobenzene (26), polyhalogenated biphenyl (35) and dioxin intoxication (3,24). TCDD inhibits hepatic UD (16,27,30,31) analogous to hexachlorobenzene and polychlorobiphenyls (17). As shown in experimental hexachlorobenzene-induced CHP the activity of UD in red cells (and bone marrow) is not disturbed, whereas the hepatic enzyme activity of UD drops to 5% of the initial value after 7 weeks of intoxication (36). Chemicals such as hexachlorobenzene, polychlorinated biphenyls, TCDD and chlorinated phenols are capable of evoking CHP in the absence of an inherited enzyme defect concerning UD (1). These findings may support the concept of an inherited and a merely acquired form of CHP including PCT (Table 1).

A PCT after dioxin exposure due to a genetic disposition could only be found in two family members by determination of UD deficiency. 60 examined persons showed a secondary coproporphyrinuria only in 22% of all cases, with a transition constellation to CHP type A in 5 persons. None of the 'sporadic' cases developed PCT. A unique acute exposure and TCDD intake is evidently not sufficient for the development of a clinical CHP. Our clinical observation that a unique acute TCDD exposure will not evoke a porphyria in people without a genetic susceptibility has an experimental explanation: TCDD-porphyria studies in the rat by acute versus chronic exposure to TCDD demonstrated quite clearly that porphyria is only induced by the chronic and not by the unique TCDD administration (15). Urinary porphyrins and their porphyrin pattern represent a highly sensitive parameter for the different degrees of intoxication with TCDD (2): Chronic administration of different TCDD amounts to rats proved that coproporphyrinuria develops first depending on the amount and duration of the intoxication, which will eventually result in a CHP with an increase of heptacarboxyporphyrin and uroporphyrin excretion after several months. The continuous transition to a CHP, reproduced in animal experiments with hexachlorobenzene (36) and TCDD (2) represents a valid model for the pathogenesis of organochemical-induced chronic disturbance in human hepatic porphyrin metabolism. As TCDD induces various microsomal enzymes (14, 16,21,23,27), the toxic response to TCDD will depend on its affinity to the receptor protein. Genetically different receptors could also be a reason for the different effects of TCDD on porphyrin and heme synthesis in persons with and without UD deficiency.

The fact that subject 5 of the porphyria family as well as the children of subject 1's brother, all with subnormal UD, did not suffer from PCT, but only developed a subclinical CHP type A (Table 5), could be due to a relationship between the induction of aryl hydrocarbon hydrocarboxylase and the decreased UD activity under the influence of chlorinated aromatic hydrocarbons: Animal experiments showed a dcrease of UD activity only if a genetic responsiveness existed to the induction of aryl hydrocarbon hydroxylase by dioxin (16). A genetically decreased UD will not lead to PCT by itself. Other factors are necessary (Table 1): The enzyme defect is metabolically compensated and clinically unobtrusive. Perhaps genetically fixed differences in the metabolism of the organochemical determine whether or not PCT will evolve in hereditary UD deficiency under dioxine (23). As there exist genetic differences in the metabolism of foreign compounds between strains of laboratory animals (21), we may suppose genetic factors altering the individual response to xenobiotics.

Liver damage and alcohol influence are the most probable reasons for the persistence of porphyrinuria in 3 of the 13 persons with secondary coproporphyrinuria in 1977. To what extent TCDD can be held responsible for the porphyrinuria in 1977 is ultimately not evaluable retrospectively. A frequence of 22% porphyrinopathias in one population group is far above the average (< 5%) and most likely due to a similar organochemical exposure of all people. Our results do not contradict those of the Wageninger group, who also found secondary porphyrinurias with partial transition to CHP type A in TCDD exposed people (3). Yet family investigations show that an incipient CHP type A in subjects 2 and 5 (Table 5) with different erythrocyte UD activity (Table 3) may normalize completely without therapy (Table 4). Thus, a mild form of CHP type A will remain without clinical relevance if alcohol, estrogens and certain foreign chemicals are avoided.

The different sensitivity of persons with subnormal erythrocyte UD activity to TCDD exposure may be due to diverse reasons: Individual differences in exposure amount and duration, in TCDD intake, and in the responsiveness of the toxic agent's hepatic metabolism. A comparison of family members to the sporadic cases shows clearly that a unique TCDD exposure can only lead to a PCT, if a predisposition is given in form of an inherited UD defect. Animal experiments support the thesis that in non-susceptible persons a unique acute TCDD intake will not result in a porphyria (15), whereas an unspecific transitory coproporphyrinuria may be triggered (2). Own studies show that urinary porphyrins are relevant biological indicators of human toxicity of dioxin.

ACKNOWLEDGEMENT

This investigation has been supported by the Deutsche Forschungsgemeinschaft (Grant Do 134), Bonn - Bad Godesberg, Federal Republic of Germany. The authors thank Ms. Angelika Nitz for typing the manuscript.

REFERENCES

1. Bickers DR (1982) Environmental and drug factors in hepatic porphyria. Acta Dermatovener (Stockholm) Suppl 100: 29-41
2. Cantoni L, Salmona M, Rizzardini M (1981) Porphyrogenic effect of chronic treatment with 2,3,7,8-tetrachlorodibenzo--p-dioxin in female rats. Dose - Effect relationship following urinary excretion of porphyrins. Toxicol & Appl Pharmacol 57: 156-163
3. Centen AHJ, Strik JJTWA, Colombi MD (1979) Coproporphyrinuria and chronic hepatic porphyria type A found in people

from Seveso (Italy) exposed to 2,3,7,8-tetrachlorodibenzo--p-dioxin (TCDD). In: Strik JJTWA, KOeman JH (eds) Chemical porphyria in man. Biomedical Press, Elsevier/North-Holland, pp 75-81
4. Colombi AM (1979) Subjective symptomatology prevalence as an additional criterion to define riskgroups exposed to TCDD in the Seveso area, Italy. In: Strik JJTWA, Koeman JH (eds) Chemical porphyria in man. Biomedical Press, Elsevier/North-Holland, pp 83-106
5. Doss M (1980) Pathobiochemical transition of secondary coproporphyrinuria to chronic hepatic porphyria in humans. Klin Wochenschr 58: 141-148
6. Doss M, Sauer H, Sixel-Dietrich F, Tiepermann R v (1984) Different types of porphyria cutanea tarda. Arch Dermatol Res 276: 207-208
7. Doss M, Tiepermann R v, Look D, Henning H, Nikolowski J, Ryckmanns F, Braun-Falco O (1980) Hereditäre und nicht--hereditäre Form der chronischen hepatischen Porphyrie: Unterschiedliches Verhalten der Uroporphyrinogen-Decarboxylase in Leber und Erythrozyten. Klin Wochenschr 58: 1347-1356
8. Elder GH, Evans JO, Matlin SA (1976) The effect of the porphyrinogenic compound, hexachlorobenzene, on the activity of hepatic uroporphyrinogen decarboxylase in the rat. Clin Sci Mol Med 51: 71-80
9. Elder GH, Lee GB, Tovey JA (1978) Decreased activity of hepatic uroporphyrinogen decarboxylase in sporadic porphyria cutanea tarda. N Engl J Med 200: 274-278
10. Elder GH, Sheppard DM, Salamanca RE de, Olmos A (1980) Identification of two types of porphyria cutanea tarda by measurement of erythrocyte uroporphyrinogen decarboxylase. Clin Sci 58: 477-484
11. Elder GH, Sheppard DM, Tovey JA, Urquhart AJ (1983) Immunoreactive uroporphyrinogen decarboxylase in porphyria cutanea tarda. Lancet I: 1304-1304
12. Felsher BF, Carpio NM, Engleking DW, Nunn AT (1982) Decreased hepatic uroporphyrinogen decarboxylase activity in porphyria cutanea tarda. N Engl J Med 306: 766-769
13. Goldstein JA, Hickman P, Bergman H, Vos JG (1973) Hepatic porphyria induced by 2,3,7,8-tetrachlorodibenzo-p-dioxin in the mouse. Res Comm Chem Pathol Pharm 6: 919-928
14. Goldstein JA, Friesen M, Scotti TM, Hickman P, Hass JR, Bergman H (1978) Assessment of the contribution of chlorinated dibenzo-p-dioxins and dibenzofurans to hexachlorobenzene-induced toxicity, porphyria, changes in mixed function oxygenases, and histopathological changes. Toxicol Appl Pharmacol 46: 633-649
15. Goldstein JA, Linko P, Bergman H (1982) Induction of porphyria in the rat by chronic versus acute exposure to 2,3,-7,8-tetrachlorodibenzo-p-dioxin. Biochem Pharmacol 31: 1607-1613
16. Jones G, Sweeney GD (1977) Association between induction of aryl hydrocarbon hydroxylase and depression of uroporphyri-

nogen decarboxylase activity. Res Comm Chem Path Pharm 17: 631-637
17. Kawanishi S, Seki Y, Sano S (1981) Polychlorobiphenyls that induce δ-aminolevulinic acid synthetase inhibit uroporphyrinogen decarboxylase in cultured chick embryo liver cells. FEBS LETTERS 129: 93-96
18. Kushner JP (1982) The enzymatic defect in porphyria cutanea tarda. N Engl J Med 306: 799-800
19. Kushner JP, Barbuto AJ, Lee GR (1976) An inherited enzymatic defect in porphyria cutanea tarda. Decreased uroporphyrinogen decarboxylase activity. J Clin Invest 58: 1089-1097
20. Lehr PA, Doss M (1981) Chronische hepatische Porphyrie mit Uroporphyrinogen-Decarboxylase-Defekt in vier Generationen. Dtsch med Wschr 106: 241-245
21. Nebert DW, Felton JS (1976) Importance of genetic factors influencing the metabolism of foreign compounds. Fed Proc 35: 1133-1141
22. Poland A, Glover E (1973) 2,3,7,8-tetrachlorodibenzo-p--dioxin: A potent inducer of δ-aminolevulinic acid synthetase. Science 179: 476-477
23. Poland AP, Glover E, Robinson JR, Nebert DW (1974) Genetic expression of aryl hydrocarbon hydroxylase activity. J Biol Chem 249: 5599-5606
24. Poland AP, Smith D (1971) A health survey of workers in a 2,4,-D and 2,4,5-T-plant with special attention to chlorance, porphyria cutanea tarda, and psychologic parameters. Arch Environ Health 22: 316-327
25. Sassa S, Anderson KE, Kappas A (1986) Uroporphyrinogen decarboxylase: Genetic defects and inhibition by environmental chemicals. In: Nordmann Y (ed) Porphyrins and porphyrias. Coloque INSERM/J Libbey Eurotext Ltd, Vol 134, pp 45-53
26. Schmid R (1960) Cutaneous porphyria in turkey. New Engl J Med 263: 397-398
27. Sinclair PR, Granick S (1974) Uroporphyrin formation induced by chlorinated hydrocarbons (Lindane, polychlorinated biphenyls, tetrachlorodibenzo-p-dioxin). Requirements for endogenous iron, protein synthesis and drug-metabolizing activity. Biochem Biophys Res Comm 61: 124-133
28. Sixel-Dietrich F, Doss M (1985) Hereditary uroporphyrinogen--decarboxylase deficiency predisposing porphyria cutanea tarda (chronic hepatic porphyria) in females after oral contraceptive medication. Arch Dermatol Res 278: 13-16
29. Smith AG, Francis JE (1981) Increased inhibition of hepatic uroporphyrinogen decarboxylase by hexachlorobenzene in male rats given the oestrogenic drugs diethylstilboestrol and chlorotrianisene. Biochem Pharmacol 30: 1849-1953
30. Smith AG, Francis JE, Kay SJE, Greig JB (1981) Hepatic toxicity and uroporphyrinogen decarboxylase activity following a single dose of 2,3,7,8-tetrachlorodibenzo-p-dioxin to mice. Biochem Pharmacol 30: 2825-2830
31. Swain MG, Follows SB, Marks GS (1983) Inhibition of uroporphyrinogen decarboxylase by 3,3', 4,4'-tetrachlorobiphenyl in chick embryo liver cell culture. Can J Physiol Pharmacol

61: 105-108
32. Stonard MD (1978) Disturbances of porphyrin metabolism. In: Slater TF (ed) Biochemical mechanisms of liver injury. Academic Press, London, pp 443-468
33. Strik JJTWA (1978) Porphyrinogenic action of polyhalogenated aromatic compounds, with special reference to porphyria and environmental impact. In: Doss M (ed) Diagnosis and therapy of porphyrias and lead intoxication. Springer, Berlin Heidelberg New York, pp 151-164
34. Strik JJTWA, Janssen MMT, Colombi AM (1980) The incidence of chronic hepatic porphyria in an italian family. Int J Biochem 12: 879-881
35. Strik JJTWA, Koeman JH (1979) Chemical porphyria in man. The diagnosis and occurrence of chronic hepatic porphyria in man caused by halogenated aromatics (polybrominated biphenyls, polychlorinated biphenyls and 2,3,7,8-tetrachlorodibenzo-p-dioxin). Porphyrinogenic action of halogenated aromatics in experimental animals. Elsevier/North-Holland, Amsterdam
36. Tiepermann R v, Koss G, Doss M (1980) Uroporphyrinogen decarboxylase deficiency in experimental chronic hepatic porphyria. Hoppe-Seyler's Z Physiol Chem 361: 1217-1222
37. Verneuil H de, Nordmann Y, Phung N, Grandchamp B, Aitken G, Grelier M, Noire J (1978) Familial and sporadic porphyria cutanea: Two different diseases. Int J Biochem 9: 927-931
38. Verneuil H de, Sassa S, Kappas A (1983) Effects of polychlorinated biphenyl compounds, 2,3,7,8-tetrachlordibenzo--p-dioxin, phenobarbital and iron on hepatic uroporphyrinogen decarboxylase. Implications for the pathogenesis of porphyria. Biochem J 214: 145-151

27

The use of urinary porphyrins to monitor occupational and environmental exposure to chemicals

A. Ferioli – Institute of Occupational Health Clinica L. Devoto, University of Milan, via S. Barnaba 8, 20122 Milano, Italy

SUMMARY

The occupational and environmental exposure to some polyhalogenated hydrocarbons has been related to the occurrence of chemical porphyria, a disease with clinical and biochemical features resembling Porphyria Cutanea Tarda. The qualitative and quantitative evaluation of urinary porphyrins has been proposed as a valuable indicator of early effects due to exposure to porphyrinogenic chemicals. A review of the literature is made in order to evaluate the validity of this bioindicator for the toxicological assessment of exposure and its usefulness as a screening test in epidemiological investigations. The needs for future research are also discussed.

INTRODUCTION

The occurrence of porphyrin disturbances in subjects exposed to suspected porphyrinogenic agents has been investigated by a number of Authors (Table 1).
As can be seen from this table, the great part of the available studies have dealt with halogenated aryl hydrocarbons. However, it is interesting to note that cases of porphyrin disorders have also been reported following

Table 1: Investigations documenting porphyrin disorders by polyhalogenated arylhydrocarbons in humans.

AGENT	SOURCE OF EXPOSURE	INVOLVED SUBJECTS	REF.
HCB	food contamination	general population	4
	industrial production	manufactoring workers	15
	use as pesticide	agricultural workers	22
TCDD	herbicide production	workers	2
			24
			16
	environmental contam.	general population	5
PBBs	environmental contam.	general population	26
PCBs	industrial use	electrical workers	7
	food contamination	general population	8

occupational exposure to other kinds of chemicals (e.g., vinyl chloride) (12) or after treatment with various drugs (e.g., oestrogens) (27).
Examining each study in detail, it is apparent that most of these investigations have reported sporadic cases or have dealt with rather complex and not well defined exposure conditions.

HEXACHLOROBENZENE (HCB)

The first evidence of chemical porphyria in man was obtained after the mass poisoning by HCB in Turkey in the 1950's, where at least 3000 people developed toxic porphyria after the consumption of wheat treated with HCB (23). It was estimated that the daily intake in these subjects was between 0.05 and 2 grams of HCB for periods ranging from 1 mounth to 3 years (4). Recovery was very slow and some patients showed a marked excretion of porphyrins (mainly uroporphyrin), accompanied by many of the clinical features, even twenty years later (10).
After the Turkish episode, the oldest reported study is that of Gombos et al. (15), who investigated industrial workers manufacturing various cyclic chlorinated hydrocarbons. They found that one worker in contact with HCB developed chloracne and symptomatic porphyria.
Later, Morley et al. (22) in a study on 54 workers who

prepared and distributed HCB-treated grain found, one subject with a latent form of porphyria characterized by a small increase in uroporphyrin and coproporphyrin urinary excretion.

2,3,7,8–TETRACHLORODIBENZO–(p)–DIOXIN (TCDD)

One of the most notable observations on porphyrin changes was reported in workers engaged in the manufacture of chlorinated phenoxy acids. This industrial process has been investigated in different countries and factories, and very similar situations have been described in the subjects involved.
In one American group (2) workers manufacturing 2,4,5-trichlorophenoxy acids had chloracne, liver disorders and a skin disease wholly resembling porphyria cutanea tarda.
The diagnostic investigations performed confirmed the nature of the disease, showing a marked increase in the urinary excretion of porphyrins. Remission of the clinical signs and symptoms was achieved by withdrawal from exposure.
Those workers were checked again six years later after substantial changes had been made in the production process, and consequently of exposure had occurred. Improvement of hygiene conditions in the factory was seen to drammatically reduce the incidence of new cases of the disease, although some workers still showed chloracne and/or mild alterations in porphyrin excretion (24).
Similar findings were obtained in an industrial plant in Czechoslovakia (16). In this case, the investigations originated from an accident that occurred in the main reactor which caused widespread chemical contamination of the factory. The workers affected by the contamination soon afterwards showed clinical signs of chloracne and peripheral neuropathy, associated in some cases with chemical porphyria.
In both these studies, the causative agent implicated in the genesis of porphyria was believed to be TCDD, which occurred as a by-product in the 2,4,5-trichlorophenoxy acid synthesis, expecially in case of uncontrolled thermic conditions (21). Of course, this substance was not measurable in the subjects but only in the working environment at that time and thus the evidence of its responsability for porphyria was only indirect and a calculation of a dose-effect relationship was not possible.
This aspect should be borne in mind when comparing these findings with others obtained in different condition of exposure such as, for example, those from the Seveso area after the 1976 accident, that was similar to the one which

occurred in Czechoslovakia. In the Seveso population, no cases of porphyria nor other chronic diseases were noticed; only mild alterations in urinary porphyrin excretion (mainly coproporphyrinuria) were found in some inhabitants of the polluted areas (5). It is reasonable to assume that in the Seveso area TCDD exposure was probably far less than in the above mentioned cases.
On the other hand cases of porphyria were also reported in the Seveso group when concurring porphyrinogenic factors (i.e. genetic predisposition) were present (13).

HALOGENATED BIPHENYLS

Occupational or environmental exposure to halogenated biphenyls is another important area where porphyrin modifications have been investigated in man. This topic will be dealt with by other speakers in detail, therefore I will only say that the overall evidence of porphyrinogenic effects in humans by these chemicals is rather weak.
The available studies on this subject come from very different exposure situations, different chemical mixtures, various routes of absorption and presumably quite wide dose ranges (7, 8, 18, 26). This make a comparison of the studies very difficult. In any case, however, the fact that mass intoxications such as the Yusho, the Taiwan and the Michigan episodes, have occurred without a single case of chemical porphyria being reported, makes the porphyrinogenic potency of those polyhalogenated biphenyl exposures rather doubtful.

OTHER SUBSTANCES

Besides the major areas so far discussed, sporadic cases of porphyria are present in the literature, which have been connected with various chemical exposures
These cases should be considered with caution because most of them lack documentation on the exact nature and intensity of related exposure, while others were sporadic, so their relevance to a chemical risk cannot be reasonably maintained. Notwithstanding all these limitations, these case reports should merit our attention, especially when experimental data on the porphyrinogenic potential of the implicated chemicals are lacking. In this respect, interesting cases of porphyria have been reported in subjects presumably exposed to methyl chloride, chlorophenol derivates, diazinon (organophosphate pesticide) and caulking fumes (Table 2).

Table 2: Cases of porphvria connected with various chemical exposures.

SUBSTANCE(S)	AUTHORS
Methylchloride	Chalmers et al., 1940 Leurini et al., 1982
Chlorophenol derivatives	Lynen et al., 1975
Diazinon	Bopp and Kosminsky, 1975 Bleakely et al., 1979
Caulking fumes	Conte and Ferioli (not published)

SOME CONSIDERATIONS

The experience so far available of testing porphyrin metabolism in humans leads to some considerations which may be of particular importance for the use of this test in biological monitoring of chemical exposure in man.

The substances of industrial or environmental interest which can be positively defined as porphyrinogenic in man are so far very few. TCDD and HCB are probably the most potent. Other substances such as polychlorinated biphenyls, polybrominated biphenyls and other halogenated hydrocarbons are effective porphyrinogenic agents in animals but do not produce striking effects in man, at least at the dose levels which are more commonly encountered. Some substances show porphyrinogenic potential in experimental models but for many of them there is no human experience. Since not all the experimental or field studies are comparable or concordant, the above conclusions should not be interpreted too rigidly.

Although halogenated aryl hydrocarbons seem to be the most frequent area of application of porphyrins in biological monitoring of exposure, many questions are still unsolved concerning feasibility and interpretation of such monitoring.

First of all, our experience so far is limited to the study of urinary excretion of porphyrins, but recent clinical evidence derived from patients with renal failure demonstrate that the kidney is actively involved in porphyrin metabolism (11) and

therefore, the pattern of urinary excretion of porphyrins may not be closly related to porphyrin liver metabolism.
Another area of uncertainty regards the relative importance of qualitative versus quantitative modifications of the pattern of porphyrin metabolism and excretion.
The urinary excretion of porphyrins can be measured as total concentration of all the homologues by using a simple, fast and accurate spectrophotometric method, based on the second derivative of absorbance (17). The separate determination of all the single homologues, that is, the so called "pattern", requires the use of chromatographic techniques that are in any case time consuming and sophisticated (9, 13).
If it is very clear that the presence of high amounts of porphyrins (above 400-500 mcg/l) with inversion of the copro/uro ratio is a pathological finding, an inversion of the copro/uro ratio in the presence of low or normal total porphyrin excretion is of uncertain meaning.
Similarly, there is uncertainty about the meaning and the prognostic significance of low to moderate increase in total porphyrin excretion (e.g., 200-300 mcg/l) with normal ratio maintained among all the homologues, or with only a slight modification. This uncertainty has so far prevented us from using the assessment of total porphyrin excretion as a screening test, to be followed by the more detailed pattern determination only in selected cases. On the other hand, the laboriousness of pattern measurement has been a serious obstacle to the use of this test in wide-scale human investigations.

CONCLUSION

In order to overcome these limitations, we believe that the main objective for the near future is still the validation of the use of this test in occupational and environmental medicine. This requires that future research activities be oriented toward three main targets: analytical tools, dose-effect studies, mechanism of toxicity.
New analytical procedures for determining porphyrins in urine and other biological media should be explored. The suitable procedures are those that can be automatized and easily used on a large scale. This in turn would result in improved facilities for performing widescale field studies in human populations.
These future studies in humans, should however, be designed so as to allow a very precise qualitative and quantitative

assessment of exposure, whenever possible with the inclusion of internal dose monitoring. These efforts are essential in order to establish dose-effect relationships, which are the only means of providing indisputable evidence of the cause-effect association between a given chemical exposure and porphyria.

The difficulties in interpretation of these results, obtained when measuring porphyrins, will be overcome by a better knowledge of the intrinsic biochemical mechanisms underlying the alterations. This may also lead to the development of other tests related to porphyrin metabolism, such as, the measurement of Uro-decarboxylase in erythrocytes.

Experimental models, and in particular the chick embryo hepatocyte cultures (14, 25), can contribute substantially to progress in understanding, and also represent a valid and cost-effective instrument for identifying the porphyrinogenic properties of new chemicals.

This study was in part supported with the contract no. 85.00581.56 from the National Research Council (C.N.R.) of Italy - Target Project "Preventive Medicine and Rehabilitation"; Subproject "Toxicological Risk".

REFERENCES

1. Bleakely P., Nichol A.W., Collins A.G. (1979): Diazinon and porphyria cutanea tarda. Med. J. Aust. 1: 314.
2. Bleiberg J., Wallen M., Brodkin R., Appelbaum I.L. (1964): Industrially-acquired porphyria. Arch. Dermatol. 89: 793-797.
3. Bopp C., Kosminsky B. (1975): Porfiria hepato-cutanea toxica. Med. Cut. I.L.A., 4:271-280.
4. Cam C., Nigogosyan G. (1963): Acquired toxic porphyria cutanea tarda due to hexachlorobenzene. JAMA, 183: 88-91.
5. Centen A.H., Strik J.J.T.W.A., Colombi A. (1979): Coproporphyrinuria and chronic hepatic porphyria type A found in people from Seveso (Italy) exposed to 2,3,7,8-tetrachlorodibenzo-(p)-dioxin (TCDD). In: Chemical Porphyria in Man, Strik J.J.T.W.A. and Keoman J.H. eds., Elsevier, Amsterdam, 75-81.
6. Chalmers, J.H., Gillian A.E., Kench J.E. (1941): Porphyrinuria in a case of industrial methylchloride poisoning. Lancet ii: 806-807.

7. Chang K.J., Lu F.J., Tung T.C. (1980): Studies on patients with polychlorinated biphenyl poisoning. Res. Commun. Chem. Pathol. Pharmacol. 30: 547-553.
8. Colombi A., Maroni M., Ferioli A., Castoldi M.R., Liu Ke Jun, Valla C., Foà V. (1982): Increase in urinary porphyrin excretion in workers exposed to polychlorinated biphenyls. J. Appl. Toxicol. 2: 117-121.
9. Colombi A., Maroni M., Ferioli A., Valla C., Coletti G., Foà V. (1983): Liquid-chromatography of urinary porphyrins for the biological monitoring of occupational exposure to porphyrinogenic substances. Am. J. Ind. Med. 4: 551-564.
10. Cripps D.J., Goeman A., Peters H.A. (1980): Porphyria turcica. Twenty years after hexachlorobenzene intoxication. Arch. Dermatol. 116: 46.
11. Day R.S., Eales L., Disler P.B. (1981): Porphyrias and the kidney. Nephron. 28: 261-267.
12. Doss M., Lange C.E., Veltman G. (1979): Toxic porphyrinuria and chronic hepatic porphyria after vinyl chloride exposure in humans. In: Chemical Porphyria in Man. Strik J.J.T.W.A and Koeman J.H. eds., Elsevier, Amsterdam, 107-111.
13. Doss M., Sauer H., Von Tiepermann R., Colombi A.M. (1984): Development of chronic hepatic porphyria (Porphyria Cutanea Tarda) with inherited uroporphyrinogen decarboxylase deficiency under exposure to dioxin. Int. J. Biochem. 4: 369-373.
14. Ferioli A., Harvey C., De Matteis F. (1984): Drug-induced accumulation of uroporphyrin in chicken hepatocyte cultures. Biochem. J., 224: 769-777.
15. Gombos B., Pechanova A., Koziak B., Moscovicova E. (1969): Porphyria cutanea tarda and acne chlorica at the production of cyclic chlorinated hydrocarbons. Bratisl. Lek. Listy 51: 640-645.
16. Jirasek L., Kalensky J., Kubec K. (1973): Acne chlorica porphyria cutanea tarda pri vyrobe herbicid. Ceska Dermatol. 48: 306-317.
17. Jones K.G., Sweeney G.D. (1979): Quantitation of urinary porphyrins by use of second-derivative spectroscopy. Clin. Chem. 25/1: 71-74.
18. Kimbrough R.D. (1974): The toxicity of polychlorinated polycyclic compounds and related chemicals. CRC Crit. Rev. Toxicol. 2: 445-498.
19. Leurini D., Giannini M., Riboldi L., Alessio L. (1982): Hepatic cirrohosis and porphyria cutanea tarda in a subject occupationally exposed to methylchloride. Med.

Lav. 73: 571-574.
20. Lynen R.E., Lee R.G., Kushner J.P. (1975): Porphyria Cutanea Tarda associated with disinfectant misuse. Arch. Int. Med. 135: 549-552.
21. May G. (1973): Chloracne from the accidental production of tetrachlorodibenzodioxin. Brit. J. Ind. Med. 30: 276-283.
22. Morley A., Geary D., Harben F. (1973): Hexachlorobenzene pesticides and porphyria. Med. J. Aust. 1:565.
23. Peters H.A., Johnson S.A.M., Cam S., Oral S., Müftü Y., Ergene T. (1966): Hexachlorobenzene-induced porphyria: effect of chelation of the disease, porphyrin and metal metabolism. Amer. J. Med. Sci. 251: 314-322.
24. Poland A.P., Smith D., Metter G., Possick P. (1971): A health survey of workers in a 2,4-D and 2,4,5-T plant. Arch. Environm. Hlth 22: 316-327.
25. Sinclair P.R., Granick S. (1974): Uroporphyrin formation induced by chlorinated hydrocarbons (lindane, polychlorinated biphenyls, tetrachlorodibenzo-p-dioxin). Requirements for endogenous iron, protein synthesis and drug-metabolizing activity. Biochem. Biophys. Res. Commun. 61: 124-133.
26. Strik J.J.T.W.A., Doss M., Schraa G., Robertson L.W., Tiepermann R., Harmsen E.G.M. (1979): Coproporphyrinuria and chronic hepatic porphyria type A found in farm families from Michigan (USA) exposed to polybrominated biphenyls (PBB). In: Chemical Porphyria in Man. Strik J.J.T.W.A. and Koeman J.H. eds., Elsevier, Amsterdam, 29-53.
27. Taylor J.S., Roenigk H.H. (1976): Estrogen-induced porphyria cutanea tarda symptomatica. In: Porphyrins in Human Diseases, Doss M. ed., Basel: Karger, 328-335.

28

Toxicological aspects of exposure to chlorinated hydrocarbons with special reference to chemical porphyria

J.J.T.W.A. Strik – Quality Assurance Unit, National Institute of Public Health and Environmental Hygiene, Bilthoven, The Netherlands

INTRODUCTION

In the past much progress has been made in the ability to detect smaller amounts of chemicals in our environment. In addition, the safety testing of chemicals as a part of toxicology has made enormous progress. Years ago, the acute toxicity was determined. Gradually chronic toxicity testing and reproduction studies were required for many compounds. Mutagenic effects, effects on the immune system and the nervous system, including effects on behaviour became a major concern.

CHLORINATED HYDROCARBONS

Chlorinated hydrocarbons have a wide spread and essential role in the chemical industry and in a variety of manufacturing operations, e.g. plastics, pesticides, solvents. Toxicological effects of exposure consist in general of damage to skin, central nervous system and liver (1,2).

CHEMICAL PORPHYRIA

Chronic hepatic porphyria in man is a disorder of hepatic porphyrin metabolism which can either be inherited as a congenital anomaly or be caused by exposure to certain chemical compounds. Among the so-called porphyrinogenic chemicals which

mostly belong to the group of halogenated hydrocarbons one finds vinylchloride, hexachlorobenzene, certain brominated and chlorinated biphenyls, tetrachlorodibenzodioxin, chlorinated naphthalenes, some organophosphorus - and organochlorine pesticides. The mechanism of chemical hepatic porphyria is based on inhibition of the enzym uroporphyrinogendecarboxylase, which belongs to the heme synthesizing pathway. This results in an increased accumulation and excretion of uro- and heptacarboxylic porphyrins.
Populations and workers exposed to different halogenated hydrocarbons were monitored for urinary porphyrins in order to detect any induction of hepatic porphyria or to find out if this parameter could be relevant in assessing possible exposure. Measured were total porphyrins and the porphyrin composition. This consists of uro-, heptacarboxylic-, hexacarboxylic-, pentacarboxylic- and coproporphyrin. Compounds not known but suspected for their porphyrinogenic potential were screened by use of chick embryo liver cell culture. These studies are presented in detail elsewhere (3,4).

TOTAL URINARY PORPHYRINS

From different studies performed e.g. a PBB exposed population in Michigan, versus a control group from Wisconsin, brominated and chlorinated hydrocarbon exposed workers, versus unexposed workers from the same plant, Yusho patients (PCB exposed; Japan), versus controls from Japan, inhabitants of Seveso and workers of the Icmesa plant (TCDD exposure), versus controls of Seveso region, State Foresty Management workers exposed to TCDD contaminated 2,4-D/2,4,5-T, versus unexposed workers from State Foresty, workers from the Hembrug affair exposed to TCDD, and different control groups, it appeared that a total porphyrin excretion of up to 200 µ g/l is normal. Higher levels indicative for diseased status (5) were not found in the above mentioned studies.

PORPHYRIN PATTERN

In the control groups as mentioned above porphyrin patterns in urine were analysed. In all these different control groups an incidence ranging from a few up to 19% of "increased" uro- and heptacarboxylic porphyrin levels were found. The term "increased" is based on the interpretation by Doss (5) of the urinary porphyrin excretion of patients in different stages of hepatic porphyria. In young people (students) and children the incidence of increased uro- and heptacarboxylic porphyrin was much lower or absent. A higher than normal incidence of the above mentioned porphyrins is considered to be indicative for a porphyric prestage of chemical induced porphyria. This prestage is without any indications for symptoms or illnesses attributable to hepatic porphyria as described in detail by Doss (5). Also other factors and possible age, sex, alcohol, or oestrogens play a role in evoking the increase in uro- and heptacarboxylic porphyrins as found in the control groups studied (5). A higher incidence of these porphyrins was found in the PBB and Seveso study. No differences in incidence of the above mentioned porphyrins were found in the studies with the workers exposed to chlorinated and brominated hydrocarbons, yusho patients, State Foresty Management and Hembrug workers.

CONCLUSION

Qualitative changes in urinary porphyrin patterns are an early indication for a possible chemically induced disturbance in the heme synthesis and may be useful in assessing exposure to porphyrinogenic chemicals as demonstrated in the PBB and Seveso case studies.

REFERENCES

1. HAMILTON, A., and HARDY, H.L. (1974). Industrial Toxicology, third edition. Publishing Sciences Group, Inc.

2. NICHOLSON, W.J., and MOORE, J.A. (1979). Health effects of halogenated aromatic hydrocarbons. The New York Academy of Sciences, New York.

3. STRIK, J.J.T.W.A. (1980). Health status of factory workers with long-term exposure to chlorinated hydrocarbons. Presented before the annual meeting of the Division of Environmental Chemistry of the American Chemical Society, Las Vegas, August 27-29.

4. STRIK, J.J.T.W.A., and KOEMAN, J.H. (1979). Chemical porphyria in man. Elsevier/North-Holland Biomedical Press, Amsterdam.

5. DOSS, M. (1978). Diagnosis and therapy of porphyrins and lead intoxication. Springer, Berlin.

29

Metabolism of steroids in hepatic porphyria

S.W. Golf, V. Graef – Institute of Clinical Chemistry and Pathobiochemistry, University Medical School, D-6300 Giessen, West Germany

SUMMARY

The porphyrias are caused by metabolic events, which are based on acquired or inherited abnormalities of regulation of heme synthesis. The most frequent form is the porphyria cutanea tarda, which may be caused or amplified by environmental chemicals, for example by chlorinated hydrocarbons, or by commercial drugs. The porphyria in the rat, which is induced by hexachlorobenzene (HCB), may serve as a model for this chronic hepatic disease. In these porphyric rats, specific changes in the metabolism of steroids are observed after application of HCB, which lead to an accumulation of porphyrogenic steroid metabolites in rat liver. A porphyria might be the consequence.

The events, which are observed in steroid metabolism, are characterized by a decrease in NAD(P)H-5α-reductase activity and by an increase in NAD(P)H-5ß-reductase activity. In addition, hexachlorobenzene feeding decreases $NAD(P)^{+}$-3-hydroxysteroid-dehydrogenase activities in rat liver . As a result, steroids with 5ß-configuration are accumulating in rat liver cells. Application of the antiandrogen flutamide to porphyric rats causes a decrease in porphyrin excretion in urine, and a corresponding readaptation of activities of 5α(ß)-reductases in rat liver to normal levels.

INTRODUCTION

The heme is metabolized by microsomal heme oxygenase to bilirubin and biliverdin. This reaction is not reversible. Therefore, a continuous heme synthesis in bone marrow and liver is necessary. It is estimated, that 85% of total heme synthesis is located in bone marrow; this process must occur at maximal velocity. 15% of human heme synthesis is located in liver (1). The heme, which is synthesized in liver, is needed as prothetic group for several apoproteins, including cytochrome P-450 (40%), mitochondrial enzymes (30%), cytochrome b_5 (20%), and catalase (10%). Most of these cytochromes are characterized by a very short half-live; in case of cytochrome P-450, the half-live in the cell is two hours. For a continuous presence of these proteins in the cell, a rapid hepatic porphyrin synthesis is necessary.

REGULATION OF HEME SYNTHESIS IN LIVER

While the rate of heme synthesis in bone marrow is rather stable, this process in liver is fluctuating in response to exposition of the organism to environmental factors. It is known that cytochrome P-450 metabolizes specific substrates to reaction products which are more or less biological active compounds. Treatment of experimental animals with these substrates, reagents or drugs causes an induction of hepatic heme synthesis and consequently the biological transformation of these compounds.

This cellular induction of cytochrome P-450 as a reaction of the cell to foreign substrates is one of the reasons for the development of a porphyria and its further amplification.

The biochemical control of hepatic heme synthesis is based on the repression of the first, rate limiting step in heme formation by the end product, the so-called "free heme" (Fig. 1). A complex consisting of heme (corepressor) and an aporepressor (2) inhibits de novo synthesis of δ-aminolevulinic acid synthetase (ALAS) in liver mitochondria and thus decreases the rate of heme formation. The lowered heme concentration, in turn, causes a decrease of active repressor concentration, followed again by an increased heme synthesis.

Fig. 1. – Regulation of hepatic heme synthesis.

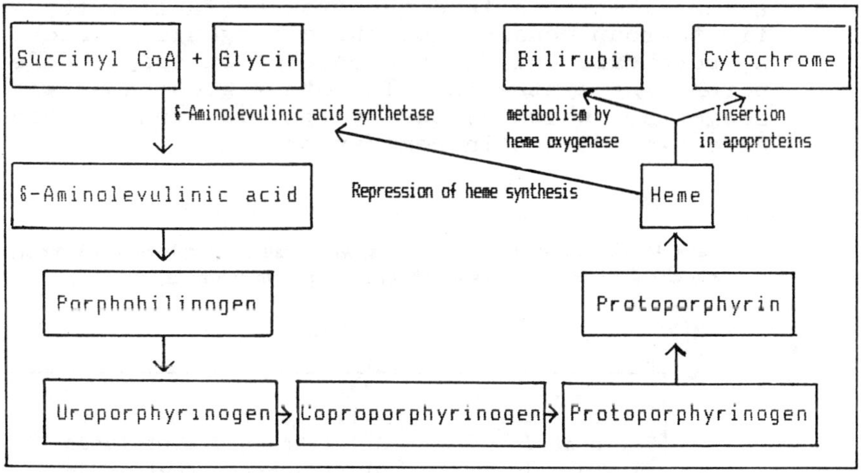

METABOLIC BASIS OF PORPHYRIAS

The maintenance of heme concentration in liver cells is effectively controlled, even under conditions of congenital or acquired defects of intermediary enzymes of heme synthesis. Such specific defects of intermediary enzymes of heme synthesis are observed in the porphyrias (3). Since hereditary transmission of these enzyme defects occurs autosomal dominantly, such persons exhibit half of the enzyme activity, usually observed in healthy persons (3). If the potentially porphyric person has had contact with specific drugs, toxic environmental reagents, or other substances which induce hepatic porphyrin synthesis, the partially labile regulation of heme synthesis in liver is further impaired (4). As a reaction to induction of ALAS, the intermediary steps of heme synthesis are accelerated with the exception of the steps, where the impaired enzymes are located. These steps characterize the type of porphyria. At this location in the pathway from δ-aminolevulinic acid to heme, the porphyrin substrates and products are accumulating, and they are diffusing from tho mitochondria to the cell cytosol and other body fluids and organs. There, the typical symptoms of a porphyria are observed, which range from neurological-visceral signs to the cutaneous photosensitivity in chronic types of porphyria.

It is presumed that most manifestations of porphyrias might be avoided by prevention of contact with

porphyrogenic drugs or environmental reagents.
The porphyrogenic reagents must be subdivided with a few exceptions into two groups (Table 1)(6,7). The first group consists of those reagents, which induce a porphyria only in presence of an enzyme defect of porphyrin synthesis. The other group contains those reagents, which cause a porphyria even in absence of an enzyme defect in heme synthesis.

Table 1: Drugs and chemical reagents, which cause a porphyria in (A) presence, and (B) absence of an enzyme defect in porphyrin synthesis.

A	B
Barbiturates	Allylisopropylacetamide
Sulfonamide	Allylisopropylacetylurea
5ß-Steroids	Allylisopropylacetcarbamide
Ethanol	Diethylcarbonyl-dihydrocollidin
Estrogens	Griseofulvin
Fe-ions	Hexachlorbenzene
Dapson	Tetrachlorbenzo-p-dioxin
Pyrazinamide	chlorinated Phenols
Phenytoin	Pb-ions
Mesantoin	
Ethotoin	
Carbamázepin	
Succinimide	
Gluthethimide	
Meprobamate	
Isopropyl-meprobamate	
Sulfonylurea	
Pentazocin	
Dipyron	
Amidopyrin	
Antipyrin	
Dichloralphenazon	
Phenylbutazon	
Dipyron	
Avapyrazon	
Phenyl-dimethyl-pyrazalon	

Two events must have occured for manifestation of a porphyria:

1. an induction of ALAS,
2. a decrease of one or more enzyme activities, which are observed at specific loca-

tions in heme synthesis.
a. This might be caused by an inhibition of the enzymes by toxic reagents,
b. or it is caused by a congenital enzyme defect.

ROLE OF STEROID HORMONES IN PORPHYRIA

There are several findings which suggest a participation of steroids in development of porphyria. Among the factors which are connected with the onset of a porphyria, are puberty and variations in hormone production during menstruation (8,9). It is also known, that patients suffering from acute intermittend porphyria, are able to metabolize steroids to 5α-reduced metabolites only to a limited extent (10). It was supposed that activity of liver 5α-reductase, which is responsible for synthesis of 5α-reduced metabolites, is decreased in porphyric patients (11). As a cause, 5β-dihydrosteroids are increasingly synthesized by cytosolic 5β-reductase (12). 5β-Dihydrosteroids are potent inductors of ALAS (13). The accumulation of 5β-dihydrosteroids during specific situations, such as puberty, or 5α-reductase deficiency, might therefore lead to an induction of ALAS and consequently to a manifestation of porphyria (12).

In this connection, several questions must be asked:

1. Which factors are responsible for the deficient 5α-reductase activity?
2. Is 5α-reductase the only enzyme of steroid metabolism, which is affected?
3. Are there factors present, which modify both porphyrin synthesis and 5α-reductase activity?

For a clarification of these points, following experiments with rats were carried out.

MODIFICATION OF METABOLISM OF STEROIDS

HCB is a chlorinated hydrocarbon, which causes a porphyria closely related to porphyria cutanea tarda in experimental animals and in man as well. The best known example is the epidemic, which was observed in Turky in 1956, where more than 5000 people suffered from a porphyria after ingestion of seed mixed with HCB (14). When rats are fed for 60 days with a diet containing 0.05% HCB, a porphyria is observed in these animals (15,16). Excretion of porphyrins and porphyrin precursors increased during the treatment of rats with HCB. At the same time, 5α-reductase activity in rat liver decreased by 21%. A second effect, which might be significant for the porphyria, was observed in case of the 3-hydroxysteroid dehydrogenases. This group of enzymes is responsible for the reduction of metabolic products of 5α- and 5ß-reductases (5α- and 5ß-dihydrosteroids) to 3-hydroxysteroids, which are then glucuronidated and excreted. HCB treatment of rats caused a 28% decrease of activity of 3-hydroxysteroid dehydrogenases in rat liver (Table 2).

Table 2: Porphyrin and porphobilinogen excretion in a 24 hour urine and activities of enzymes of steroid metabolism in rat liver after HCB-treatment.

	Control	HCB-treatment 10 days	HCB-treatment 60 days
Porphyrins (μg/24 h)	1.12±1.10	1.72±2.02	163±75
Porphobilinogen (μg/24 h)	5.50±4.20	6.60±2.50	146±63
5α-reductase activity (mU/mg)	24.6±9.50	17.4±4.60	19.5±4.2
5ß-reductase activity (mU/mg)	1.19±0.38		1.76±0.3
Reduction of 5α-dihydrotestosteron in 3-hydroxysteroids (mU/mg)	7.50±1.08	9.28±1.01	5.73±1.0
Reduction of 5ß-dihydrotestosteron in 3-hydroxysteroids (mU/mg)	38.3±7.44	20.6±3.36	27.8±2.9

The changes described for enzyme activities in rat liver after treatment with HCB must lead to an accumulation of steroids in the liver cell, and consequently to an induction of porphyrin synthesis. ROBEL (17) and GHRAF (18) discussed the metabolism of testosterone to products, which are excreted into urine (Fig. 2). Testosterone is first glucuronidated (I) and subsequently reduced to 5ß-dihydrotestosterone (II). After hydrolytic cleavage of the ester bond, 5ß-dihydrotestosterone is oxidized (III). In a healthy rat liver, the 3-oxo group is subsequently reduced by 3-hydroxysteroid dehydrogenases, and again glucuronidated. The water soluble and biologically inactive metabolite (IV) is finally excreted. In porphyric rats, the steps from testosterone to 5α-dihydrotestosterone and from metabolite III to metabolite IV are blocked. As a consequence of these HCB induced changes of steroid metabolism, steroids with a configuration of metabolite III are accumulating in the liver cell. These metabolites are potent inductors of porphyrin synthesis in rat liver (13).

Fig. 2. – Metabolism of testosterone in rat liver after HCB-treatment.

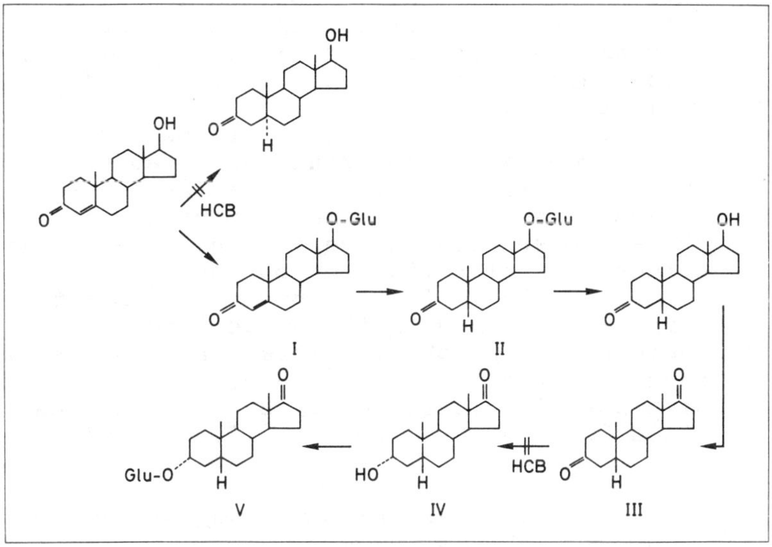

During the 60-day treatment of rats with HCB, these changes in metabolism of steroids are already observed prior to the elevation of porphyrin excretion (19). 10 days after the beginning of treatment, 5α-reductase activity was decreased and 5β-reductase activity was increased. From the 20th day on, the fraction of 5β to 5α dihydrosteroid metabolites was elevated by a factor of 10. This ratio of enzyme activity remained stable until the 40th day of treatment, while the 3-hydroxysteroid-dehydrogenase and UDP-glucuronyl-transferase activities stayed decreased (19).

Our observations suggested, that the period, which is located prior to the onset of increased porphyrin excretion, was most crucial for the development of a porphyria. The metabolic pathways were programmed irreversibly, so that later, after the 60th day of treatment, the cellular mechanisms of regulation of heme synthesis were impaired.

The observed variations in steroid metabolism are

closely connected to the induction of a porphyria.

1. Reagents related to HCB, such as pentachlorphenol, pentachlorbenzene, and 2,4,5-trichlorphenol were not able to induce a porphyria. At the same time, no changes in steroid metabolism were observed (20).

2. Allylisopropylacetamide and allylisopropylacetylurea (Sedormid) are potent inductors of porphyrin synthesis (21). Application of these drugs to rats caused the known changes in steroid metabolism: decrease of activity of 5α-reductase, 3-hydroxysteroid-dehydrogenases, and increase of 5ß-reductase activity (22).

3. The antiandrogen flutamide (4'-nitro-3'-trifluormethyl-isobutyranilide) caused in man a reduced excretion of 5ß-steroids (24). A similar effect on steroid metabolism in porphyric rats should decrease porphyrin excretion in these animals. The effect of flutamide on porphyric rats is seen in Figure 3 (23). It was observed, that treatment of porphyric rats with flutamide lead to a significant decrease of porphyrin excretion in urine within 14 days. At the same time, 3-hydroxysteroid dehydrogenase activity increased by 22%. Flutamide also inhibited 5ß-reductase from rat liver in vitro (23).

Fig. 3. – Porphyrin excretion in urine of porphyric rats without (● – ●) and after (● – – ●) flutamide treatment.

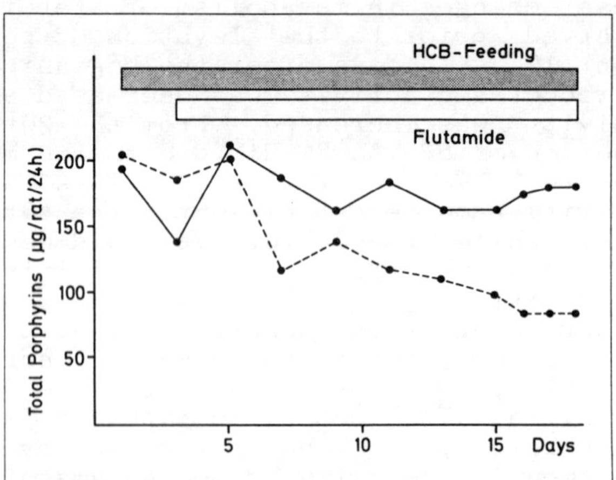

We must presume, that flutamide effectively limits production of 5ß-steroids in rat liver and decelerates porphyrin synthesis.

Our observations have proven, that 5ß-steroids play an important role in the development of a porphyria. The increased porphyrin synthesis is caused by 5ß-steroids via an induction of ALAS. The 5ß-steroids are accumulating in the cell because of specific changes in steroid metabolism. These changes in steroid metabolism are based on the effect of specific drugs on activities of steroid metabolizing enzymes. These modifications in combination lead to the highly significant amplification of porphyrin synthesis and to the porphyria.

REFERENCES

1. Schmid, R., Schwartz, C., Watson, J. (1954). Porphyrin content of bone marrow and liver in the various forms of porphyria. Arch. Int. Med. **93**, 167-167.
2. Granick, S. (1966). The induction in vitro of the synthesis of delta-aminolevulinic acid synthetase in chemical porphyria: a response to certain drugs, sex hormones, and foreign chemicals. J. Biol. Chem. **241**, 1359-1375.
3. Brodie, M.J., Moore, M.R., Goldberg, A. (1977). Enzyme abnormalities in the porphyrias. Lancet **II**, 699-699.
4. Maxwell, J.D., Meyer, U.A. (1976). Effect of lead on hepatic delta-aminolevulinic acid synthetase activity in the rat: a model for drug sensitivity in intermittent acute porphyria. Eur. J. Clin. Invest. **6**, 373-373.
5. Bickers, D.R. (1982). Environmental and drug factors in hepatic porphyria. Arch. Derm. Venerol. Suppl. **100**, 29-41.
6. Disler, P.B., Blekkenhorst, G.H., Eales, L., Moore, M.R., Straughan, J. (1982). Guidelines for drug prescription in patients with acute porphyrias. S. Afr. med. J. **61**, 656-660.
7. Ippen, H., Pierach, C.A. (1983). Verhütung und Behandlung von Attacken induzierbarer Porphyrien. Dtsch. Ärztebl. **80**, 39-47.
8. Meyer, U.A., Schmid, R. (1978). In Metabolic Basis of Inherited Disease (Stanbury, J.B., J.B. Wyngaarden, D.S. Frederickson, Eds.), New York, Mc Graw Hill, p1166.
9. Tschudy, D.P. (1974). In Duncans Diseases of Metabolism (Bondy, P.K., L.E. Rosenberg, Eds.) Philadelphia, W.P. Saunders Co., p775.

10. Anderson, K.E., Bradlow, H.L., Sassa, S., Kappas, A. (1979). Studies in porphyria, VIII. Relationship of the 5alpha-reductive metabolism of steroid hormones to clinical expression of the genetic defect in acute intermittent porphyria. Am. J. Med. **66**, 644-650.
11. Kappas, A., Bradlow, H.L., Gillette, P.N. (1972). Studies in porphyria.I. A defect in the reductive transformation of natural steroid hormones in the hereditary liver disease, acute intermittent porphyria. J. Exptl. Med. **136**, 1043-1043.
12. Bradlow, H.L., Gillette, P.N., Gallagher, T., Kappas, A. (1973). Studies in porphyria. II. Evidence for a deficiency of steroid-delta4-5alpha-reductase activity in acute intermittent porphyria. J. Exptl. Med. **138**, 754-763.
13. Granick, S., Kappas, A. (1967). Steroid induction of porphyrin synthesis in liver cell culture. I. Structural basis and possible physiological role in the control of heme formation. J. Biol. Chem. **242**, 4587-4593.
14. Schmid, R. (1960). Cutaneous porphyria in Turkey New Engl. J. Med. **263**, 397-397.
15. Goerz, G., Krieg, Th., Bolsen, K., Lissner, R. (1977). Langzeit-Exposition von Ratten mit Hexachlorbenzol (HCB): Einfluß auf die Porphyrie-Ausscheidung im Urin und auf das Cytochrom-P450 in der Leber. Arch. Derm. Res. **259**, 199-206.
16. Graef, V., Golf, S.W., Goerz, G. (1979). Effect of hexachlorobenzene on steroid metabolism in rat liver. Archiv für Toxikologie **43**, 115-120.
17. Robel, P., Emiliozzi, R., Baulieu, E.E. (1966). Studies on testosterone metabolism. III. The selective "5ß-metabolism" of testosterone glucuronide. J. Biol. Chem. **241**, 20-29.
18. Ghraf, R., Hoff, H.H., Lax, E.R., Schriefers, H. (1973). The 5ß-metabolites of testosterone: mode and sex-specificity of their formation in the rat liver. Acta Endocrinol. (Kbh.) **73**, 577-584.
19. Lotz, H.E., Singer, I., Golf, S.W. Graef, V. (1984). Steroid-metabolizimg enzyme activities in porphyric rats (Abstract). Hoppe Seyler's Z. Physiol. Chem. **365**, 1028-1028.
20. Graef, V., Golf, S.W., Goerz, G. (1980). The involvement of porphyrogenic steroids in the development of experimental porphyria. Experientia **36**, 1090-1091.
21. Schmid, R. Figen, J.F., Schwartz, S. (1955). Experimental porphyria. IV. Studies on liver catalase and other heme enzymes in sedormid

porphyria. J. Biol. Chem. **217**, 263-274.
22. Graef, V., Golf, S.W., Jung, H. (1984). Effect of allylisopropylacetamide and sedormid on enzymes of steroid metabolism in rat liver. Steroids **44**, 267-273.
23. Graef, V., Golf, S.W., Tyrell, C. (1982). Further evidence for the participation of 5ß-steroids in the development of a porphyria induced by hexachlorobenzene. Archiv für Toxikologie **50**, 233-239.

30

Porphyrinogenesis study in workers exposed to polychlorinated biphenyls (PCBs)

M. Maroni, E.A. Emmett*, A. Fait – Institute of Occupational Health Clinica del Lavoro L. Devoto, University of Milan, via S. Barnaba 8, 20122 Milano. *Center for Occupational and Environmental Health, Division of Occupational Medicine, Department of Environmental Health Sciences, The Johns Hopkins University, Baltimore, U.S.A.

Polychlorinated biphenyls (PCBs) show porphyrinogenic activity in experimental animals as well as in in vitro models (1). This paper reports on investigations on urinary elimination of porphyrins performed in two independent studies involving workers exposed to PCBs in the Washington D.C. area, and in northern Italy. The complete reports of these studies, which comprised many other investigations, have been published in details elsewhere (2, 3) and only data relevant to the porphyrin studies are excerpted and summarized here.

METHODS

American study: The workers exposed to PCBs consisted of 55 transformer repairmen who had been predominantly in contact with Aroclor 1260 and, to a smaller extent, with Aroclor 1254 and 1242. An industrial hygiene survey performed in their working environment had shown very low air concentrations of PCBs but a widespread PCB contamination of the workplace surfaces. 2,3,7,8-tetrachlorodibenzofuran was detected in some transformer oils sampled at the workplace (highest concentration: 31 ppb) (4).

Some of the workers (n = 38) were currently exposed to PCBs at the time of the study, the others had been exposed to PCBs only in the past, the time elapsed from last exposure ranging from 6 months to 5 years. The average lenght of employment with PCB exposure was 7 years.
In order to compare the results obtained in the exposed workers, a reference group (n = 56) was selected among workers employed at the same employer with jobs not involving exposure to PCBs. The demographic characteristics chosen to match the comparison subjects were: sex, age, smoking habits, alcohol intake, place of residence, level of education and marital status. The resulting group did not differ significantly from the exposed group for any of the above parameters.
The internal levels of PCBs in the exposed and comparison subjects were assessed by measuring serum and adipose-tissue-lipid PCB concentrations by packed column GC/ECD (3).
Urinary elimination of porphyrins was determined on 24 hr urine samples by thin layer chromatography (5, 6).

Italian study: The workers exposed to PCBs consisted of capacitor manufacturers, who were exposed to Aroclor 1254 in the past and to Pyralene 3010 (equivalent to Aroclor 1242) at the time of the study. Also in this study, an industrial hygiene survey in the working environment had shown low level of air PCB contamination but very remarkable PCB pollution of the workplace surfaces (7).
The study group was selected so as to include only those subjects who were non-smokers or smoked less than 10 cigarettes per day and who did not consume alcoholic beverages in excess of 30 g alcohol per day. The resulting group consisted of 51 subjects (25 males and 26 females), with a mean age of 39 years and a mean lenght of employment on PCB exposure of 10 years (range: 1 to 30 years).
The comparison group was identified among electronic-component assembly workers and matched with the exposed workers for age, sex, and place of residence, after excluding the subjects who did not meet the smoking and alcohol criteria adopted for the selection of the exposed workers. The resulting comparison group consisted of 67 subjects (30 males and 37 females).
Blood PCB concentration measured by packed column GC/ECD was available for 28 of the exposed workers (14 males and 14 females) and for none of the comparison subjects (3).
Urinary porphyrin concentration was determined in morning spot urine specimens by thin layer chromatography, with

quantification of all the porphyrinic homologues from 8 to 4 carboxylic groups (8).

RESULTS

Table 1 shows the internal levels of PCBs and the urinary excretion of aminolevulinic acid (ALA), coproporphyrin, and uroporphyrin, observed in the American workers. When tested with the analysis of variance, the PCB concentrations were remarkably different between the exposed and comparison workers ($p < 0.01$), although the mean values in the exposed group suggested a low level of exposure.

Table 1: Serum and adipose tissue PCB concentrations and urinary excretion of ALA porphyrins in the American workers.

	CURRENTLY EXPOSED N = 38	PAST EXPOSED N = 17	COMPARISON N = 56
Serum PCBs (ppb)	12.2 (nd–300)	5.9 (nd–30)	4.6 (nd–15)
Adipose tissue lipid PCBs (ppm)	2.1 (nd–33)	0.8 (0.3–5.1)	0.6 (nd–3)
ALA (mg/24 hrs)	2.9 (0.9–4.6)	2.6 (1.1–4.7)	2.5 (0.2–5.0)
Coproporphyrin (µg/24 hrs)	79.5 (26.2–142.5)	70.0 (7.2–139.5)	92.6 (5.5–246.6)
Uroporphyrin (µg/24 hrs)	10.7 (1.7–29.4)	9.9 (3.9–24.7)	10.9 (2.0–28.9)

Values are geometric means for PCB concentrations and arithmetic means for ALA and porphyrins. Maximum and minimum in brackets.

The urinary excretion of ALA and porphyrins did not show significant differences between the groups.

Serum and adipose tissue PCB congeners detected in these subjects were mostly highly chlorinated, reflecting the PCB pattern of the Aroclor 1260 mixture.

In the Italian workers exposed to PCBs, the total blood PCB concentrations ranged from 88 to 1359 ppb with a geometric mean of 308 ppb. When the blood PCB peaks were divided into low and high chlorinated PCBs, according to whether the peaks corresponded to an Aroclor 1242 or an Aroclor 1254 mixture respectively, the geometric means were 87 ppb for the lower chlorinated PCBs and 211 for the higher chlorinated PCBs. Blood PCB concentrations were not available for comparison subjects, but we assume that their concentrations were of the same order of magnitude as those observed in the comparison group in the American study (see Table 1). In fact, the subjects of the two comparison groups live in comparable areas and limited data on human PCB concentrations collected in northern Italy support this assumption.
Urinary porphyrin concentrations are shown in Table 2.

Table 2: Urinary porphyrin excretion (μg/1) in the PCB workers and the comparison group in the Italian study.

	PCB WORKERS (N = 51)		COMPARISON GROUP (N = 67)	
	Range	Geometric Mean	Range	Geometric Mean
Total Porph.	7.1-273.7	74.4	9.1-127.3	36.5
8-COOH Porph.	0.1-45.2	8.2	0.1-16.5	4.3
7-COOH Porph.	0.1-21.7	2.2	0.1-3.4	0.8
6-COOH Porph.	0.1-10.4	0.5	0.1-2.9	0.2
5-COOH Porph.	0.1-5.6	1.2	0.1-6.0	0.4
4-COOH Porph.	6.7-237.0	59.1	7.8-111.0	25.7

All the differences are significant at $p < 0.05$ when tested with the Student t test on log-converted values.

Workers exposed to PCBs showed an excretion of porphyrins about doubled as compared with the non-exposed workers. The enhanced total porphyrin elimination was the result of a generalized increase in every urinary fraction. The comparison

of the porphyrin excretion values corrected for urine creatinine concentration between the groups (not shown in the Table) suggested that the observed difference was not related to an urine dilution effect.
Where blood PCB concentrations were available, correlation coefficients were calculated between respectively total PCB, low and high chlorinated PCB concentrations and the urinary excretion of total porphyrins and single porphyrinic homologues. None of the Pearson's correlation coefficients was higher than 0.2 or statistically significant.

DISCUSSION

Polychlorinated biphenyls are porphyrinogenic agents in animals and in in vitro experimental models (1). Experience in man is limited to the two poisoning episodes occurred in Japan in 1968 and in Taiwan in 1979, where thousands of subjects were intoxicated by cooking oils contaminated with PCBs. In the Yusho epidemics, no reports were made of clinical porphyria or increased urinary excretion of porphyrins in the poisoned patients (9). In the Yu-Cheng accident, porphyrinogenesis alterations were investigated in a limited cohort of patients and only slight increases in the mean excretion of ALA and uroporphyrin were found (10). These findings are markedly different from those observed in patients intoxicated with chlorinated hydrocarbons such as hexachlorobenzene or 2,3,7,8-tetrachlorodibenzodioxin, where marked biochemical modifications of porphyrin metabolism progressing up to clinically overt porphyria were documented in several cases.
In our studies, the porphyrinogenic effects found were very slight in the Italian group and totally absent in the American group.
The absence of correlation observed in the Italian workers between blood total PCB concentration and porphyrin excretion could indicate that a contaminant, rather than PCBs themselves, may have been responsible for the observed effects. An alternative explanation could be that only a few PCB congeners, not proportionally represented in the workers' blood PCBs, were the effective porphyrinogenic chemicals.
So far available evidence indicates that type of PCBs, internal concentration, presence of contaminants and, possibly, route of exposure are all determinant factors in human porphyrinogenesis by PCBs. Human porphyrinogenic effects by PCBs, on the whole, seem to be rather slight. This is not

necessarily in contrast with experimental evidence.
In the American study, serum and adipose tissue samples were also analyzed by capillary column GC/ECD in order to identify and quantitate individual PCB congeners (11). The analysis identified more than 80 different PCB congeners both in serum and adipose tissue. Six major chromatographic peaks accounted for more than 50% of the total PCB concentration. Table 3 illustrates the congeners identified in each peak and the experimental evidence of porphyrinogenic activity available for them. As indicated in the Table, only two congeners have been tested and only one is frankly positive as a porphyrinogenic substance.

Table 3 Major PCB congeners in serum and adipose tissue determined in workers exposed to Aroclor 1260

PEAK RANK (1)	CONGENERS (2)	EXPERIMENTAL PORPHYRINOGENIC ACTION (3)
1°	2,3,5,6,3',4',5' 2,3,4,5,2',4',5' } HEPTA-CB	not tested
2°	2,3,4,2',4',5'-HEXA-CB	+ liver of rodents and birds
3°	2,4,6,3',4',5' 2,4,5,2',4',5' 2,3,4,5,2',5' } HEXA-CB	- rodents/+chick embryo
4°	2,3,4,5,2',3',4'-HEPTA-CB	not tested
5°	2,3,4,5,2',3',5',6' 2,3,4,5,6,2',3',5' } OCTA-CB	not tested
6°	2,4,5,3',4' 3,4,5,2',3' } PENTA-CB	not tested

(1) In order of decreasing concentration
(2) Some chromatographic peaks are not resoluble in a single PCB congener
(3) See Reference 1.

In conclusion, porphyrinogenic activity of industrial PCB mixtures in man appears to be rather weak and not uniform for different PCBs.
Further toxicological studies are needed on the porphyrinogenic power of the single PCB congeners more represented in human tissue.

ACKNOWLEDGEMENTS

The experimental work has been in part supported with the contracts n. 83.02646.56, 84.02332.56 and 85.000581.56 from the National Research Council of Italy (C.N.R.) - Target Project "Preventive Medicine and Rehabilitation"; Subproject "Toxicological Risk", and with a grant of Ministry of Education.

REFERENCES

1. J.J.T.W.A. Strik, F.M.H. Debets (1980): Chemical porphyria. In: Halogenated biphenyls, terphenyls, naphthalenes, dibenzodioxins and related products. Kimbrough R. ed., Amsterdam Elsevier/North Holland 191: 239.
2. M. Maroni, A. Colombi, A. Ferioli, V. Foà (1984): Evaluation of porphyrinogenesis and enzyme induction in workers exposed to PCB. Med. Lav. 75, 3: 188-199.
3. E.A. Emmett, M. Maroni, J.M. Ferrara, B.K. Levin, J. Jeffreys: Studies of transformer repair workers exposed to PCBs. I & II. J. Occup. Med., in press.
4. C.L. Mosely, C.L. Geraci, J. Burg (1982): Polychlorinated biphenyl exposure in transformer maintenance operations. Am. Ind. Hyg. Assoc. J. 43:170-174.
5. C.R. Scott, R.F. Labbe, J. Nutter (1967): A rapid assay for urinary porphyrins by thin layer chromatography. Clin. Chem. 13: 493-500.
6. J. Nutter, R.F. Labbe (1976): Improved screening tests for porphyrins. Letter to the editor. Clin. Chem. 18: 739.
7. M. Maroni, A. Colombi, S. Cantoni, A. Ferioli, V. Foà (1981): Occupational exposure to polychlorinated biphenyls in electrical workers. I. Environmental and blood polychlorinated biphenyl concentrations. Brit. J. Industr. Med. 38: 49-54.
8. A. Colombi, M. Maroni, A. Ferioli, M.R. Castoldi, Liu Ke Yun, C. Valla, V. Foà (1982): Increase in urinary porphyrin excretion in workers exposed to polychlorinated

biphenyls. J. Appl. Toxicol. 2: 117-121.
9. M. Okumura (1984): Past and current medical status of Yusho patients. Amer. J. Ind. Med. 5: 13-18.
10. Yau-Chin Lu, Pin-Nan Wong (1984): Dermatological medical, and laboratory findings of patients in Taiwan and their treatments. Amer. J. Ind. Med. 5: 81-116.
11. A. Fait, E.A. Emmett, E. Grossman, E.D. Pellizzari: PCB congeners in adipose tissue lipid and serum of past and present transformer repair workers and a comparison group. This book.

31

Erythrocyte ZnPP in occupationally unexposed populations and relationship of erythrocyte ZnPP and blood lead in lead battery factory workers

P.K. Suma'mur – National Centre of Industrial Hygiene and Occupational Health, Jakarta, Indonesia

INTRODUCTION

Erythrocyte ZnPP or porphyrin concentrations in occupationally unexposed populations have been reported by Piomelli et al., Hotz and Tomokuni and Ogata (1). Relationships between erythrocyte ZnPP or protoporphyrin and blood lead levels have been reported by Alessio et al. (1, 2) as well as others (1, 2). Despite the facts, more studies are needed to provide better information on the normal levels of erythrocyte ZnPP and the relationship of erythrocyte ZnPP and blood lead.

The objectives of this study were:

1) To find the values of erythrocyte ZnPP in healthy populations occupationally unexposed to lead.
2) To shed light on the relationship between erythrocyte ZnPP and blood lead in occupationally exposed populations.

METHODS AND MATERIALS

1. Methods

Venous blood amounting to 3 ml had been taken from each person in the occupationally unexposed populations or each worker in the exposed ones for the laboratory examinations on erythrocyte zinc-protoporphyrin concentrations, haemoglobine concentrations, haematocrites and also blood lead concentrations. Fluorescence spectrophotometric apparatus Hitachi 650 - 10S had been the chosen instrument to determine the concentrations of zinc-protoporphyrin in erythrocytes. The laboratory examinations of haemoglobine had been done spectrophotometrically with cyan-methaemoglobine method. The micromethod had been adopted to measure haematocrite. For the examination of blood lead, Atomic Absorption Spectrophotometer Polarized Zeeman Effect had been used with flameless spectrophotometric procedures.

2. Materials

The occupationally unexposed groups consisted of 44 female and 97 male employees. These populations had not been exposed to lead, were healthy and aged more than 20 and less than 55 years. As exposed populations, workers in four lead battery factories had been used. These consisted of 153, 112, 41 and 111 or total of 417 males and 43, 20 and 38 or total of 101 females with actual occupational exposure to lead and having worked for 1 year or more.

FINDINGS

Table 1 presents age, length of employment, Hb level, haematocrite, erythrocyte ZnPP and blood lead in occupationally unexposed groups by sex.

Table 1: Age, length of work, HB level, haematocrite, erythrocyte ZnPP and blood lead in control groups by sex.

	Male (n=97)			Female (n=44)		
	Mean	Range	Standard deviation	Mean	Range	Standard deviation
1. Age (yrs)	33.1	21–52	8.3	27.6	21–40	5.4
2. Length of work (yrs)	7.7	1–32	8.2	4.0	1–20	5.1
3. Hb level (g/dl)	14.7	12.1–17.0	1.0	13.4	11.2–15.2	0.9
4. Haematpcrote (%)	44.3	36–55	3.3	38.4	29–48	4.6
5. Erythrocyte ZnPP						
a. µg/l whole blood	534.2	116–1119	174.2	571.5	305–1038	173.6
b. µg/l erythrocytes	1212.8	398–2664	415.6	1504.3	693–3139	539.0
c. µg/g Hb	36.0	12–74	10.8	42.8	21–92	13.6

Table 2 shows correlation coefficients between log ZnPP and age, length of employment, haemoglobine level, haematocrite, and blood lead in male and female groups. The statistically significant correlations have been found for the relationship of log ZnPP and haemoglobine level in men ($p < 0.05$) and women ($p < 0.001$).

Table 2: Correlation coefficients of the relationships between ZnPP in whole blood with age, length of employment, Hb level, haematocrite and blood lead.

	Correlation coefficient (r)	Degree of statist. significance (p)
Male:		
Log ZnPP - age	- 0.03	0.05
Log ZnPP - length of employment	- 0.06	0.05
Log ZnPP - Hb level	0.21	0.05
Log ZnPP - haematocrite	0.13	0.05
Log ZnPP - blood lead	- 0.12	0.05
Female:		
Log ZnPP - age	- 0.08	0.05
Log ZnPP - length of employment	0.10	0.05
Log ZnPP - Hb level	0.42	0.01
Log ZnPP - haematocrite	0.16	0.05
Log ZnPP - blood lead	- 0.07	0.05

Expressed in µg/g of Hb and µg/l whole blood, ZnPP concentrations in men and women were of non-significant differences (each $p > 0.05$). In µg/l erythrocytes, however, they revealed strong statistically significant difference of $p < 0.001$. Table 3 shows age, length of work, Hb level, haematocrite, erythrocyte ZnPP and blood lead in occupationally exposed populations by sex and lead battery factory.

Table 1: Age, length of work, Hb level, haematocrite, blood lead and erythocyte ZnPP in occupationally exposed populations by sex and lead battery factory.*

Lead battery factory	Item	Male			Female		
		Mean	Range	Standard deviation	Mean	Range	Standard deviation
1	a. Age (yrs)	23.7	18–30	3.0	23.7	20–28	2.5
	b. Length of work (yrs)	2.8	1–6	1.7	3.0	1–6	1.6
	c. Hb level (g/dl)	14.7	12.3–16.8	1.0	13.2	10.0–15.2	1.1
	d. Haematocrite (%)	44.5	37–54	2.9	40.0	32–46	3.2
	e. Blood lead (g/l)	359.9	78–932	152.0	249.9	78–778	164.7
	f. Erythrocyte ZnPP (g/l whole blood)	3105.1	524–24448	3326.6	2681.1	267–7561	1450.9
2	a. Age (yrs)	25.7	19–36	4.1	26.0	20–35	3.3
	b. Length of work (yrs)	4.3	1–8	2.4	5.6	3–8	1.5
	c. Hb level (g/dl)	15.2	10.6–17.0	1.4	13.1	11.3–14.6	1.0
	d. Haematocrite (%)	44.7	31–52	4.1	38.6	30–44	3.8
	e. Blood lead (g/l)	488.8	144–1652	200.2	426.6	201–1027	182.0
	f. Erythrocyte ZnPP (g/l whole blood)	4083.8	841–16176	2831.8	3514.1	946–8634	1963.6
3	a. Age (yrs)	26.9	19–42	4.7	-	-	-
	b. Length of work (yrs)	3.4	1–7	1.1	-	-	-
	c. Hb level (g/dl)	15.1	13.3–17.3	1.0	-	-	-
	d. Haematocrite (%)	46.3	41–52	2.8	-	-	-
	e. Blood lead (g/l)	364.1	85–665	143.8	-	-	-
	f. Erythrocyte ZnPP (g/l whole blood)	1925.8	515–7762	1661.4	-	-	-
4	a. Age (yrs)	25.1	17–43	5.3	20.2	16–30	3.8
	b. Length of work (yrs)	4.4	1–12	3.6	2.9	1–10	1.9
	c. Hb level (g/dl)	13.8	9.0–16.0	1.5	11.6	9.0–14.3	1.5
	d. Haematocrite (%)	42.2	30–54	4.7	36.0	28–45	4.6
	e. Blood load (g/l)	690.6	280–2204	348.2	618.0	242–1650	280.0
	f. Erythrocyte ZnPP (g/whole blood)	6524.6	1006–30967	4990.8	10184.6	1766–45296	9104.0

* The sizes of male working populations were 153, 112, 41 and 111 and female were 43, 20, 0 and 38 persons for each lead battery factory 1, 2, 3 and 4.

The values of the correlation coefficients of the relationships between erythrocyte ZnPP expressed in µg per liter of erythrocytes, ug per g of Hb or ug per liter of whole blood were almost similar. As an example, these were found respectively 0.531, 0.527 and 0.493 in lead battery workers of factory 3. To avoid a lot of calculations, ZnPP concentrations in term of µg/l of whole blood have been used for further purposes.

This study has produced 7 regression line equations on the relationship between erythrocyte zinc-protoporphyrin and blood lead concentrations by sex and factory, 31 regression line equations on the same relationship by sex, occupation and factory, and 4 such regression line equations by combinations of male workers of lead battery factories 2, 3 and 4. The exponential correlation coefficients between erythrocyte zinc-protoporphyrin and blood lead concentrations in male working populations were found in varying degrees of statistical significance, whereas in female only correlations without statistical significance were shown. Besides the correlation coefficients were positive or negative reflecting the degrees of exposure. The correlation coefficients ranged from very low to 0.86.

At blood lead concentrations of around 400 µg/l whole blood, the correlations were statistically significant of very significant. At the exposure level above 400 µg/l up to the blood lead level a little above 600 µg/l, the correlation coefficients could still be found of statistical significance. At higher blood lead concentrations the correlation coefficients became statistically unsignificant and finally negative.

By using important regression lines, the hypothetical curve on the overall relationship between erythrocyte ZnPP and blood lead concentrations could be drawn. The curve goes up steeply at the initial stage of exposure until blood lead concentrations between 700-800 µg/l and further levels off (Fig. 1).

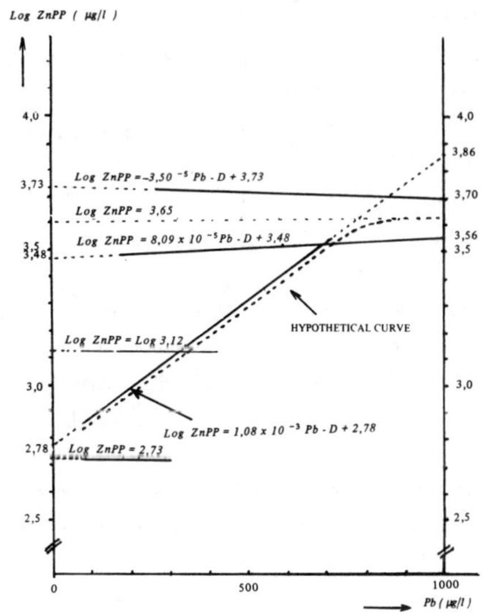

Fig. 1. – Important regression lines of the relationship between erythrocyte zinc-protoporphyrin and blood lead in male working populations.

The results of analysis of variance of erythrocyte ZnPP as well as blood lead in occupationally unexposed populations and workers of lead battery factories 1, 2, 3 and 4 by sex have clear by revealed that either erythrocyte ZnPP or blood lead concentrations in male or female occupationally unexposed and exposed populations were of significantly statistical differences (each $p < 0.01$).

The exposure - effect relationships for blood lead concentrations as indicators of exposure and erythrocyte ZnPP of equal to or more than 3000 µg/l as indicators of effect in male and female workers of each lead battery factories 1, 2, 3 and 4 have been analized. In male workers of lead battery factory 3, reaction was found in 0% of those having blood lead less than 400 µg/l and 14.6% for exposure levels of above 400 µg/l. This difference in degree of reaction for the two groups was statistically significant ($p < 0.01$).

Further analysis indicated that the reactions of $\geq$ 3000 µg/l ZnPP were 14.6% (6 of 41 male workers) in lead battery factory 3, 56.2% (63 of 112 male workers) in lead battery factory 2 and 80.2% (89 of 111 male workers) in lead battery factory 4. By

using Chi-square test, the difference in reaction of lead battery factory 2 with respect to that of lead battery factory 3 and the difference of reaction of lead battery factories 4 and 2 were statistically very significant (respectively $p < 0.001$).
Studying the percentages of male workers of lead battery factories 2, 3 and 4 having erythrocyte ZnPP $\geq$ 3000 µg/l according to blood lead levels, the reactions in form of increased erythrocyte ZnPP of $\geq$ 3000 µg/l were 32.4% (11 of 34 male workers), 41.9% (18 of 43 male workers) and 70.1% (43 of 61 male workers) respectively at blood lead concentrations of $<$ 300, 300-399, 400-499 and 500-599 ug/l. By the use of Chi-square test on the differences of the reaction at blood lead concentrations of 300-399, 400-499 and 500-599 µg/l with respect to that of blood level of $<$ 300 µg/l, these differences gave respectively $p > 0.10$, $p. > 0.05$ and $p < 0.001$.
These findings of dose-effect relationship have emphasized the observation that at blood lead concentrations of 500-599 µg/l the increase in reaction with erythrocyte ZnPP of $\geq$ 3000 µg/l occurs.

CONCLUSIONS

This study has resulted the following conclusions:

1. Erythrocyte ZnPP concentrations in occupationally unexposed men and women were respectively 534.2 (SD = 174.2) µg/l whole blood, 1212.8 (SD = 415.6) µg/l erythrocytes and 360.0 (SD = 10.8) µg/g haemoglobine and 517.5 (SD = 173.6) µg/l whole blood, 1504,3 (SD = 579.0) µg/l erythrocytes and 42.8 (SD = 13.6) µg/g haemoglobine.
2. Normal erythrocyte concentrations did not correlate with age, length of employment, haematocrite or blood lead but weakly correlated with Hb level in men and strongly correlated with Hb level in women.
3. Normal erythrocyte ZnPP levels in men and women were not of significantly statistical differences if such levels were expressed in µg per liter of whole blood or µg per gram of haemoglobine. They were only of statistically significant difference if expressed in µg per liter erythrocytes.
4. Correlation coefficients of the relationship between erythrocyte ZnPP based on µg per liter whole blood, µg per liter erythrocytes or ug per gram of haemoglobine and blood lead were almost similar and have been found for a group of workers of 41 persons respectively 0.493, 0.531 and 0.527.
5. Such correlation coefficients were found in varying degrees

of magnitude and statistical significance and positive or negative. At blood lead concentrations of around 400 µg/l, the correlations were statistically significant or very significant. At blood lead level above 400 µg/l, up to the blood leat level a little above 600 µg/l, they could still be found of statistical significance. At higher level of exposure, they became statistically unsignificant and finally negative.

6. The statistically significant correlation coefficients ranged from very low to 0.86.
7. The hypothetical curve on the overall relationship between erythrocyte ZnPP and blood lead concentrations shows the steep increase at the initial stage of exposure until blood lead concentrations between 700-800 µg/l and this curve then levels off.
8. Analyses of variance of erythrocyte ZnPP or blood lead in occupationally unexposed and exposed groups have been for the statistical significances of the differences in the levels of those parameters.
9. The clear increase in reaction with erythrocyte ZnPP $\geqslant$ 3000 µg/l whole blood occured at blood lead concentrations of 500-599 µg/l.
10. The highest erythrocyte ZnPP concentrations in the exposed workers respectively 30697 and 45296 µg per liter whole blood in men and women.

REFERENCES

1. WHO Technical Report Series, No. 647 (1980) Recommended Health Based Limits in Occupational Exposure to Heavy Metals, Geneva.
2. Yamada, Yuichi (1981), Analysis of interrelationship of some biochemical parameters for lead poisoning in low level longterm lead exposure. The Journal of Science of Labour, Vol. 57, No. 1: 257-277.

32

Effect of mercury exposure on heme biosynthesis and on some enzymatic activities in chloroalkali workers

P. Bonetti, G. Vaudagna – Medical Service – Soc. Solvay & C.ie S.A., Rosignano, Italy

Forty-nine male workers exposed to mercury vapors in a chloroalkali processing plant were studied. They were divided into two groups (A and B) on the basis of their estimated exposure.

Group A was made up of 33 individuals with a mean age of 32±7.8 years and a duration of exposure of 5±5.5 years.

Group B was made up of 16 individuals with a mean age of 47±14.5 years and a duration of exposure of 5±4.6 years. The airbon mercury concentration on each floor of the cell-room was determined through 20 absorbers equipped with a fritted-glass bubbler tube connected in series, and containing a solution of potassium permanganate and sulfuric acid. The mercury content of the solution was measured by flameless atomic absorption. Over the last few years the average mercury vapor level detected has been slightly lower than the maximum permissible concentration of 0.05 mg/cu m (see Fig. 1).

These two groups of exposed workers were compared with a control group (Group C).

Group C was made up of 16 subjects with a mean age of 44±11.4 years and never exposed to mercury or other toxic metals. Each subject provided a 24-hour urine specimen and a single blood sample.

The following parameters were detected: urine and blood mercury (HgU and HgB), delta-aminolevulinic acid (ALA-U) and coproporphyrin (Copro-U) in urine, delta-aminolevulinic acid dehydrase activity in erythrocytes (ALA-D), cholinesterase (ChE) and lactic-dehydrogenase (LDH) activities in serum.

RESULTS AND COMMENT

The means and standard deviation of the different biological parameters in the exposed and control groups, compared by a t-test, are reported in Table 1.

The mean mercury concentrations in blood and urine were significantly higher in the exposed groups than in the control group, and in the mercury-exposed group A were lower than in group B. Neither of the exposed groups exceeded the proposed biological limit values.

The other parameters were within the normal range in all investigated subjects and the t-test showed a level of probability considered to be not significant.

The relationships between some parameters are reported in Table 2.

Only the correlation between levels of mercury in urine and blood was statistically significant. No correlation was observed between blood-mercury, ALA-D, ChE or LDH, and no correlation was found with the same parameters and lenght of exposure (Table 3).

The results obtained in this study show no disturbance in heme byosynthesis and in the other parameters considered in a group of chloroalkali-workers with moderate exposure to inorganic mercury.

Our data suggest that workers exposed to mercury vapor at concentrations lower than the actual TLV-TWA of 0.05 mg/m^3, should not show any detectable biological changes.

In addition, this study seems to confirm the validity of the proposed biological threshold limit values for mercury concentration in blood (3 µg/100 ml) and in urine (50 µg/g creatinine).

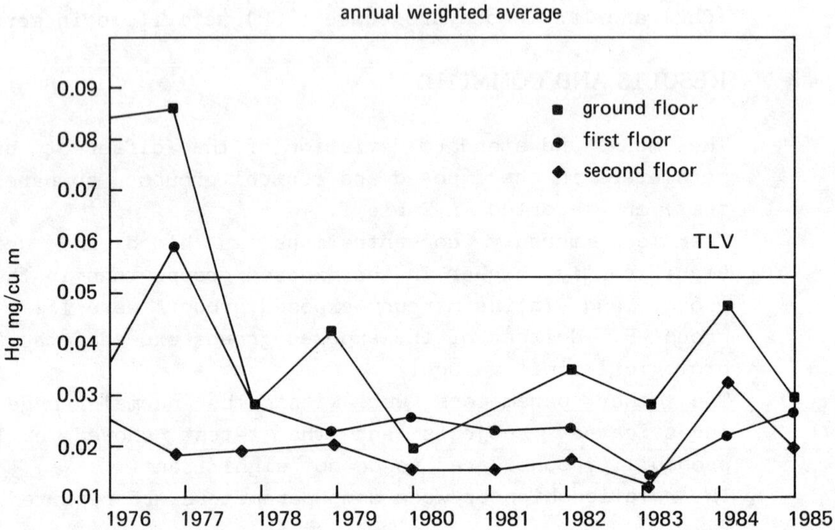

Table 1: Biological determinations in exposed and control group.

PARAMETERS	Group A mean (sd)	Group B mean (sd)	Controls mean (sd)
HgU mcg/g.creat.	13 (7) +	23 (14) +	3 (1) +
ALA-U mcg/g. creat.	1.6 (0.6)*	1.3 (0.2)*	1.5 (0.5)*
Copro-U mcg/g. creat.	37 (22)*	38 (10)*	34 (12)*
HgB mcg/100 ml	0.8 (0.3)+	1.2 (0.4)+	0.6 (0.1)+
ALA-D U.I./ml.	40 (13)*	45 (11)*	42 (7)*
ChE U.I./ml.	2945 (448)*	2879 (470)*	2800 (610)*
LDH U.I./ml.	175 (27)*	185 (26)*	152 (21)*

+ $p < 0.01$

* n.s.

Table 2: Correlation between HgB and the other blood parameters.

PARAMETERS	Group A	Group B
HgB - HgU	r = 0.476 $p < 0.01$	r = 0.204 n.s.
HgB - ALA-D	r = 0.039 n.s.	r = 0.379 n.s.
HgB - ChE	r = 0.194 n.s.	r = 0.04 n.s.
HgB - LDH	r = 0.06 n.s.	r = 0.190 n.s.

Table 3: Correlation between length of exposure and blood parameters.

PARAMETERS	Group A	Group B
ALA-D	r = 0.155 n.s.	r = 0.411 n.s.
ChE	r - 0.091 n.s.	r - 0.021 n.s.
LDH	r = 0.198 n.s.	r = 0.332 n.s.
HgB	r = 0.123 n.s.	r = 0.150 n.s.

33

A high performance liquid chromatography technique for separation of porphyrins from porphyric hepatocytes

F. De Matteis, M. Manno, C. Harvey & A. Ferioli* – MRC Toxicology Unit, Carshalton, Surrey, U.K. *Institute of Occupational Health, Clinica L. Devoto, University of Milan, Italy

INTRODUCTION

Chicken hepatocytes cultured *in vitro* are very sensitive to the porphyrogenic effect of drugs and can be used as a convenient test system to screen drugs for porphyrogenic potential. Different types of porphyria can be produced by different chemicals and the type of porphyrins accumulating (whether of I or III isomeric type, and varying in the number of carboxylic acid functions) depends on the enzymatic block in haem biosynthesis which is a characteristic of each variety of porphyria. Previous work (Sinclair *et al.*, 1983; Swain *et al.*, 1983; De Verneuil *et al.*, 1983; Ferioli *et al.*, 1984) has shown that in this cell culture system polyhalogenated chemicals induce accumulation of uroporphyrin (URO) and heptacarboxylate porphyrin (HEPTA), giving rise to a type of porphyria closely resembling human *porphyria cutanea tarda* (PCT). In both PCT and the uroporphyria induced in culture a block of haem biosynthesis at the level of uroporphyrinogen decarboxylase has been documented and this readily accounts for the accumulation of URO and HEPTA. However, patients with PCT also excrete increased amounts of the unphysiological URO I isomer and an additional defect at the level of the cosynthetase has been postulated (see Doss *et al.*, 1976; Meyer & Schmid, 1977). We have now investigated the isomeric type of URO and HEPTA accumulating in cultures of hepatocytes made porphyric by drugs.

THE HPLC TECHNIQUE

We have developed a reversed-phase gradient elution h.p.l.c. technique for simultaneous separation of type I and type III isomers of 8, 7, 6, 5 and 4 - carboxylated porphyrins as well as of protoporphyrin. The porphyrins, extracted from cellular monolayers (through contact with Methanol/$HClO_4$ for 5 min) or excreted in the medium, are first concentrated on Sep Pak C 18 cartridges, eluted with methanol containing 2.5 mM tetrabutyl ammonium hydroxide buffered with phosphate (Fisons) and after further concentration separated by h.p.l.c. The h.p.l.c. technique makes use of a high concentration of ammonium acetate to separate isomeric porphyrins (Lim et al., 1983) but the mobile phases are now modified so as to extend the range of porphyrin separation to dicarboxylate porphyrins. The solvents for the gradient elution were a) 13% acetonitrile in 1 M ammonium acetate buffer pH 5.2 containing EDTA 10 mg/100 ml; b) 95% methanol containing 2.5 mM tetrabutyl ammonium hydroxide.

Separation was carried out on columns of 25 cm x 5 mm internal diameter packed with μ Bondapak C 18 (Waters). A 22.5 min linear gradient from 10 to 72% B (in the A + B mixture) was followed by a second gradient reaching 100% B over the next 2 minutes; this concentration was then kept constant for a further 10 minutes.

EXPERIMENTS WITH HEPATOCYTES

Hepatocytes prepared from 16 days-old chicken embryos were cultured and incubated with drugs as described (Ferioli et al., 1984). After incubation with excess 5-aminolaevulinate (ALA, 25 μg/ml) for 19 h, the hepatocytes accumulated mostly protoporphyrin (PROTO), but URO and HEPTA of both isomeric types were also found in the cellular extracts (Table 1). When 3,3',4,4'-tetrachlorobiphenyl (TCBP, 0.25 μg/ml) was added together with ALA, PROTO markedly decreased, while URO and HEPTA of both isomeric types accumulated in excess. In contrast, an iron chelator, desferrioxamine (DES, 1.5 mM), which limits the supply of iron for incorporation into PROTO, caused marked accumulation of PROTO from exogenous ALA as expected, but also markedly decreased both isomers of URO and HEPTA (Table 1).

Table 1: Porphyrins extracted from cell monolayer (Units of absorbance at 400 nm/10 cm plate).

Treatment	URO I	URO III	HEPTA I	HEPTA III	PROTO
ALA	753	619	137	604	1430
ALA + TCBP	1856	3053	699	1643	147
ALA + DES	95	73	62	79	4495

The porphyrins excreted in the culture medium were also resolved by h.p.l.c.: in contrast to URO and HEPTA, which were mostly demonstrable in the cells, 5 and 4 carboxylated porphyrins of both isomeric types were only present in the medium, while PROTO was present in both cells and medium (results not shown).

COMMENT

A h.p.l.c. technique capable of separating isomeric free-carboxylated porphyrins, ranging from URO to PROTO, has been described. The technique has been applied to cultures of hepatocytes made porphyric by drugs, but could easily be used to monitor urinary and faecal excretion of porphyrins in people exposed to environmental chemicals.

Chicken hepatocytes cultured in presence of ALA accumulated some URO and HEPTA of both isomeric types, even in the absence of a porphyria-inducing drug. TCBP, a chemical which induces uroporphyria, markedly increased the accumulation of both isomers of URO and HEPTA (as compared with values seen with ALA alone), suggesting that in this experimental uroporphyria - like in human PCT and in murine hexachlorobenzene porphyria (Smith & Francis, 1986) there may be a defect of both uroporphyrinogen decarboxylase and uroporphyrinogen III cosynthetase. DES, an iron chelator, given together with ALA, decreased the accumulation of both isomer I and III of URO and HEPTA: this finding, which requires further study, may indicate that an iron-dependent mechanism is involved in decreasing the activity of both enzymes in the intact cells.

Support of the Italian Min. of Public Educ. is acknowledged.

by A. Ferioli

REFERENCES

De Verneuil, H., Sassa, S. & Kappas, A. (1983) Biochem. J. 214: 145-151.

Doss, M., Schermuly, E., Look, D. & Henning, H. (1976) In: Porphyrins in Human Disease, Doss, M. (ed.) p.286, Basel, Karger.

Ferioli, A., Harvey, C. & De Matteis, F. (1984) Biochem.J. 224: 769-777.

Lim, C.K., Rideout, J.M. & Wright, D.J. (1983) Biochem. J. 211: 435-438.

Meyer, U.A. & Schmid, R. (1977) In: The Metabolic Basis of Inherited Diseases, Stanbury, J.B. et al. (eds.) p.1203, 4th Edit. New York, McGraw-Hill.

Sinclair, P.R., Elder, G.H., Bement, W.J., Smith, S.G., Bonkowsky, H.L. & Sinclair, J.F. (1983) FEBS Lett 152: 217-221.

Smith, A.G. & Francis, J.E. (1986) In: Porphyrins and Porphyrias, Nordmann, Y. (ed.) p.127 INSERM/John Wiley Eurotext Ltd.

Swain, M.G., Follows, S.B. & Marks, G.S. (1983) Can. J. Physiol. Pharmacol. 61: 105-108.

Porphyrins as biological indicators of exposure to environmental chemicals. Concluding remarks for session III

F. De Matteis* & L. Alessio† – *MRC Toxicology Unit, Carshalton, Surrey, U.K. and
†Istituto di Medicina del Lavoro, Università di Brescia, Italy.

INTRODUCTION

The main purpose of this paper is to summarise, in the light of evidence discussed in the Porphyrin Session of the Symposium or reported elsewhere in the literature (see Marks, 1985 for a recent review), the mechanisms by which environmental pollutants stimulate porphyrin biosynthesis in the liver and to discuss the toxicological significance of these effects in man. The use of porphyrins as biological indicators of exposure will therefore be discussed briefly and also the main research priorities in this particular area of Environmental Medicine.

MECHANISMS AND BIOLOGICAL SIGNIFICANCE OF INCREASED LIVER PORPHYRIN FORMATION

Two main mechanisms can be distinguished for the stimulation of hepatic porphyrin biosynthesis by foreign chemicals (Smith & De Matteis, 1980).

A) The first involves increased activity of liver 5-amino-laevulinate synthase without loss of activity of an intermediary enzyme of the pathway. Drugs which operate by this mechanism are very many and include: 1) inducers of cytochrome P-450; b) stimulators of haem oxygenase and 3) "suicide" substrates of cytochrome P-450 (alkenes and alkynes). The increased liver haem biosynthesis which they produce – intended to provide for increased cytochrome P-450 biosynthesis or to replenish the loss of haem due to accelerated haem turnover - is accompanied by accumulation of the intermediary porphyrins, which are excreted in excess. The increase in porphyrin excretion (involving mostly Coproporphyrin and Protoporphyrin) is usually modest and reversible and is not accompanied by symptoms which are attributable to porphyrin accumulation. The biological significance of such increased excretion is probably that of an adaptive response with limited toxicological importance. When properly validated, such increase in porphyrin excretion may be useful as an indicator of exposure to foreign chemicals.

B) The second mechanism also involves increased activity of the first enzyme, 5-aminolaevulinate synthase, but in this case a partial block in haem biosynthesis is also present, due to inhibition of one of the intermediary enzymes of the pathway. Two major groups of drugs can be distinguished, depending on the particular enzyme which is inhibited:
1) 4-Alkyldihydropyridines, griseofulvin (De Matteis et al., 1982) and sydnones (Ortiz de Montellano et al., 1986), where the last enzyme of the pathway, ferrochelatase, is inhibited and Protoporphyrin accumulates in vast excess. So far, this type of disorder (drug-induced PROTOPORPHYRIA) has been clearly documented only in experimental animals.
2) Polyhalogenated aromatic hydrocarbons, which inhibit the enzyme Uroporphyrinogen decarboxylase and lead to accumulation of Uroporphyrin, with the development of a chronic toxic porphyria in the most severe cases. This type of metabolic alteration (drug-induced UROPORPHYRIA) has been described not only in the laboratory animal but also in man, where it represents by far the most important drug-induced porphyrin disorder of the liver. The syndrome gradually evolves from an initial coproporphyrinuria into a massive uroporphyrinuria accompanied by other symptoms of hepatic-cutaneous porphyria (PCT), such as photosensitive skin lesions, liver damage and, possibly, increased incidence of hepatoma (Doss & Martini, 1978). As these toxic symptoms are closely associated with (or directly attributable to) the accumulation of Uroporphyrin, the increased urinary excretion of porphyrins, when the porphyrin profile is also altered in favour of Uroporphyrin and other highly carboxylated porphyrins, should be regarded as an indicator of a toxic response capable of evolving into a chronic porphyria. The significance of the initial coproporphyrinuria is more difficult to assess. It is not yet known whether it represents the first stage of the same process ultimately leading to Uroporphyria (for example by a modest increase in hepatic synthesis of uroporphyrinogen, which is then decarboxylated by the kidney; or whether an unrelated drug action is involved similar to that described in mechanism A, above, for inducers of cytochrome P-450.

ROLE OF PORPHYRINS IN BIOLOGICAL MONITORING OF EXPOSURE

Porphyrins can be assayed in urine, faeces and liver tissue.

a) Assay of urinary porphyrins has the advantage of being a non-invasive test, simple and relatively rapid to perform (it can therefore be easily carried out on a routine basis). It is also sensitive and reproducible, as the urinary porphyrins can be first concentrated, if necessary, on Sep Pak C18 cartridges before being resolved, for example, as the free carboxylic acid by reversed-phase h.p.l.c. (see De Matteis et al. 1986). It should be noted, however, that the urinary levels of porphyrins can be influenced by alterations in kidney and liver function and that any increased Protoporphyrin formation would be missed in a urine test, as this porphyrin is excreted in the faeces.

b) The assay of faecal porphyrins is more difficult to organize on a routine basis, but is essential to detect an increase in protoporphyrin excretion which is a characteristic of protoporphyria (see above). A qualitative screening test (Elder, 1980) for porphyrins in faeces should first be performed, and if this is positive, a suitable faecal extract can be prepared and the porphyrins can then be resolved and identified by h.p.l.c., as for urines.

c) Studies on liver samples can obviously be performed only in exceptional cases. The main advantage of such studies lies in confirming the hepatic origin of the increased prophyrin formation and in defining, by enzymatic studies, the precise type of the metabolic disorder. Some information on these two points can also be obtained - less directly - by measuring the concentration of porphyrins in the excreta, so in most cases studies on bioptic material in patients are not necessary.

RESEARCH PRIORITIES

Further work is necessary, in our view, on the following points.

A) In the short term, the methods for porphyrin estimation in urine and faeces should be further improved and validated. It would be highly desirable if comparable techniques were adopted in different laboratories, so that inter-laboratories variations could be more easily assessed and results obtained in different human populations compared more easily. A reversed-phase h.p.l.c. technique would seem to be the method of choice as the porphyrins could then be separated as the free carboxylic acids and no intermediary step of derivatization would therefore be necessary.

B) In the medium term, work is required to investigate further the significance of increased porphyrin excretion as an indicator of exposure of man to environmental chemicals. How rapidly are changes in porphyrin excretion reversed when exposure is discontinued and are various environmental chemicals different in this respect? Is it possible to predict, on the basis of an abnormal porphyrin test, that a chronic hepatic porphyria is gradually becoming established? Can the results of porphyrin tests help us to distinguish different stages or phases (as outlined by Alessio et al., 1984) in the establishment and evolution of a drug-induced liver disorder? The answer to these questions will only be provided by future work.

C) Finally, in the long term, more efforts are needed to elucidate the mechanisms by which foreign chemicals stimulate liver porphyrin biosynthesis and - in particular - the mechanism of induction of Uroporphyria by polyhalogenated compounds. These mechanistic studies may well lead to the discovery of new biological indicators, closer to the "primary" molecular events and therefore more specific in their diagnostic value.

REFERENCES

Alessio, L., Buratti, M., Bertelli, G. & Dell'Orto, A. (1984) In: Monitoraggio biologico negli ambienti di lavoro, Bertazzi, P.A. et al. (eds.) pp. 23-56, Milano, Angeli.

De Matteis, F., Gibbs, A.H., Farmer, P.B., Lamb, J.H. & Hollands, C.(1982) Adv. Pharmacol. Ther. Proc. Int. Congr. 8th 5, 131-138.

De Matteis, F., Manno, M., Harvey, C. & Ferioli, A. (1986) In: Proceedings of This Symposium.

Doss, M. & Martini, G.A. (1978) In: Falk Symposium No. 25, H. Remmer et al. (eds.) pp. 409-420, Lancaster, MTP Press Ltd.

Elder, G.H. (1980) Clin. Haemat. 9, 371-398.

Marks, G.S. (1985) Critical Rev. Tox. 15, 151-179.

Ortiz de Montellano, P.R., Costa, A.K., Grab, A., Sutherland, E.P. & Marks, G.S. (1986) Collque. INSERM, 134, 109-117.

Smith, A.G. & De Matteis, F. (1980) Clin. Haemat. 9, 399-425.

Part IV

Biochemical and cellular indices of renal changes induced by exogenous chemicals

Coordination:
R. Lauwerys (Bruxelles, Belgium)
I. Franchini (Parma, Italy)

34

Biological indices of renal disturbances induced by exogenous chemicals

R. Lauwerys, A. Bernard – Industrial Toxicology Unit, Catholic University of Louvain

This paper reviews the tests currently available for detecting the early renal changes induced by long term exposure of humans to nephrotoxic industrial chemicals. Some tests are already used routinely. Others are still at a stage of development or validation. We are not considering in this paper the biological signs (e.g. increased serum creatinine) which usually appear at an advanced stage of renal insufficiency.

1. DETECTION OF EARLY EFFECTS AT THE GLOMERULAR LEVEL

1.1. Measurement of high molecular weight proteins in urine

An increased urinary excretion of specific proteins may precede the detection of increased total proteinuria by commercial dipsticks (Barrat 1983). An increased urinary excretion of high molecular weight (HMW) proteins may result from different types of renal changes :

(a) loss of proteins from the cells of the urinary pathway. This phenomenon does not apply to albumin, the main component of HMW proteinuria.

(b) defective tubular reabsorption of filtered plasma proteins. Competition experiments with HMW and low molecular weight (LMW) proteins have demonstrated that both types of proteins are reabsorbed by the same process but albumin has much less affinity for the reabsorption sites on the tubular membrane than β_2-microglobulin (Bernard et al. 1986a). It can, therefore, be concluded that an isolated proteinuria without concomitant increased urinary excretion of LMW proteins is the consequence of an increased glomerular permeability.

(c) loss of nephrons. This mechanism only plays a role in the advanced stage of renal insufficiency.

(d) increased glomerular permeability. Reduction of the anionic charge of the glomerulus by certain xenobiotics may cause an increased filtration of plasma proteins.

(e) modification of the physico-chemical characteristics of the filtered proteins. Modifications of the charge and/or the structure of plasma proteins (e.g. through binding of endogenous or exogenous substances) may favorize their glomerular filtration.

Evaluation of the reserve glomerular filtration capacity.

The increase of the glomerular filtration rate after an acute oral load of proteins might be a reflection of the reserve glomerular filtration capacity (Rodriguez-Iturbe et al. 1985). It has not yet been assessed whether this reserve capacity can be reduced at an early stage of intoxication by nephrotoxic chemicals.

Circulating antiglomerular basement membrane antibodies.

An immune type glomerulonephritis associated with the occurrence of antiglomerular basement membrane antibodies may be induced by some inorganic or organic xenobiotics (Druet et al. 1982). Anti-laminin antibodies were found in higher titer in a group of workers exposed to cadmium (unpublished results). The significance of this test requires further assessment.

Glomerular basement antigens in serum and in urine.

Glomerular basement membrane (GBM) antigens can be detected in human serum and urine of normal subjects and in patients with various types of glomerular diseases (Huttinen et al. 1979). Is has not yet been investigated whether nephrotoxic chemicals may exacerbate the release of some GBM components in blood or in urine.

Determination of red blood cells and/or platelets membrane negative charges.

The interest of this test (Levin et al., 1985) to detect chemicals which may reduce the density of the glomerular anionic sites needs further assessment.

DETECTION OF EARLY EFFECTS AT THE TUBULAR LEVEL

Measurement of low molecular weight proteins in urine.

The increased urinary excretion of low molecular weight (LMW) proteins such as β_2-microglobulin, retinol binding protein, is a sensitive index of impaired proximal tubular function at least as long as the glomerular filtration rate has not dropped below 30 ml/min. An increased urinary excretion of LMW proteins may not always be associated with the existence of tubular cell lesions since competition between certain xenobiotics and LMW proteins for the same reabsorption process may also lead to the same effect (Bernard et al. 1986b).

Enzymuria.

The determination of various enzymes (e.g. β-N-acetylglucosaminidase, adenosine deaminase, alkaline phosphatase, β-D-galactosidase, -glutamyltransferase, lactate dehydrogenase, leucine aminopeptidase) has been proposed for assessing the integrity of different sites along the nephron. The place of enzymuria as an indicator of early effects of nephrotoxic chemicals is discussed in another chapter of this volume.

Renal tubular antigens.

Tubular proteins can be shed in greater quantity in urine following absorption of nephrotoxic chemicals (Mutti et al. 1985) but further work is needed to determine their interest as a marker of early toxic effects.

Kallikrein and prostaglandin excretion.

Kallikrein and prostaglandin are synthesized in the kidney (Nasjletti and Melik 1979) and seem to play a role in the regulation of the renal flood flow. Reduction of their urinary excretion could represent an early manifestation of exposure to some chemicals.

Other parameters.

Disturbances of renal tubular function may lead to other biological changes such as glucosuria, aminoaciduria, hyperphosphaturia, hypercalcemia, hypouricemia, hypophosphatemia, hypokalemia. With the possible exception of increased urinary excretion of glucose without concomitant hyperglycemia, the other changes do not necessarily represent an early sign of chemically induced tubular impairment.

REFERENCES

Barrat M., Proteinuria
Brit. Med. J. 287, 1489, 1983.

Bernard A., Viau C., Ouled A., Lauwerys R., Competition between low and high molecular weight proteins for renal tubular uptake.
Nephron 1986a (under press).

Bernard A., Viau C., Ouled A., Tulkens P., Lauwerys R.,Effects of gentamicin on the renal uptake of endogenous and exogenous proteins in conscious rats.
Tox. Appl. Pharm. 1986b (under press).

Druet P., Bernard A., Hirsch F., Weening J.J., Gengoux. P., Mahieu P., Birkeland S. Immunologically mediated glomerulonephritis induced by heavy metals.
Arch. Toxicol. 50, 187, 1982.

Huttunen N.P., Turner M.W., Barratt T.M., Physico-chemical characteristics of glomerular basement membrane antigens in urine.
Kidney Int. 16, 322, 1979.

Levin M., Smith C., Walton M.D., Gascoine P., Barratt T.M. Steroid-responsive nephrotic syndrome : a generalised disorder of membrane negative charge.
Lancet ii, 239, 1985.

Mutti A., Valcari P., Fornari M., Lucertini S., Neri T.M., Alinovi R., Franchini I., Urinary excretion of brush-border antigen revealed by monoclonal antibody : early indicator of toxic nephropathy.
Lancet ii, 914, 1985.

Nasjletti A., Malik K.V., Relationships between the kallikrein-renin and prostaglandin system.
Life Sciences 25, 99, 1979.

Rodriguez-Iturbe B., Herrera J., Garcia R., Response to acute protein load in kidney donors and in apparently normal postacute glomerulonephritis patients : evidence for glomerular hyperfiltration.
Lancet ii, 461, 1985.

35

Urinary enzymes as early indicators of renal changes

R.G. Price, J. Halman, C.T. Yuen – Biochemistry Department, King's College (Kensington Campus), Campden Hill, London W8 7AH, U.K.

SUMMARY

The potential of urinary enzymes as indicators of renal damage and disease is discussed. A series of non-invasive tests for nephrotoxicity are evaluated using cis-dichlorodiammine platinum (II) cis-DPP and propyleneimine as nephrotoxins. The potential of N acetyl-β-D-glucosaminidase (NAG) as a screening procedure and prognostic indicator of renal disease in man is also considered. Interest has recently developed into the possibility that the determination of NAG isoenzyme will provide additional diagnostic information and some preliminary results are discussed. The recent improvements in the methodology used in the assay of urinary enzymes, particularly the availability of new chromogenic substrates suggests that these methods will be more widely used in the future.

INTRODUCTION

Although the presence of enzymes in urine has been known for a considerable time (Raab, 1972) only in the last decade has their potential as diagnostic indicators been realised (Price, 1979; 1982). Before urinary enzymes could be used in the assessment of toxicity and renal disease a number of inherent problems had to be overcome. A major difficulty is the effect of variation in urine flow on enzyme activity. In man factoring enzyme activity by creatinine concentration is convenient when random urine samples are assayed. Experience has shown that the second sample of the day is preferred unless carefully timed samples can be collected. The accuracy of twenty-four hour samples are notoriously unreliable. These problems are not encountered in animal toxicity studies but care must be taken to ensure that animal urine is not contaminated by food or faeces. Steps should be taken in any assay procedure to ensure that any endogenous inhibitors including urea (Bondiou et al., 1985) or drug induced inhibition is eliminated by filtration, dialysis or dilution. Dilution of the urine in the assay is the most convenient procedure but the effectiveness of this depends on the sensitivity of the assay used. Initially the sensitivity of fluorimetric assays allowed the urine to be diluted sufficiently to eliminate the effects of inhibitors (Price et al., 1970). More recently two colorimetric substrates based on ω-nitrosyrol (Yuen et al., 1982) or m-cresolsulphonphthalein (Noto et al., 1983) have become available, both chromophores have high extinction coefficients which give sufficient sensitivity to avoid gel filtration or dialysis.

The nephron is divided into discrete morphological and functional regions each with its characteristic complement of enzymes (Guder and Ross, 1984) and by careful selection of urinary enzymes it may be possible to identify the primary site of damage induced by a nephrotoxin or the region of the nephron initially affected in renal disease. However, urine is a hostile environment for many enzymes and the number of enzymes which are stable in urine is limited (Price, 1982). Urinary enzymes provide an early indication of renal damage and a rise in enzyme excretion can be used to determine the most relevant time to undertake pathology.

The methodology currently available allows urinary enzyme assays to be used as a general screen for renal damage or disease. They are particularly valuable if carried out in conjunction with other simple non-invasive tests of renal function. Kit procedures are now available which can be used in conjunction with tests for specific proteins for screening industrial workers (Yuen et al., 1984). Finally the availability of colorimetric substrates will enable the routine assay of urinary enzymes in the clinical chemistry laboratory where they can provide valuable prognostic indications (Wellwood et al., 1975). The possibility that the separation of isoenzymes may provide additional diagnostic information is currently the subject of intensive research in a number of laboratories (Yuen et al., 1986; Vigano et al., 1983; Kind, 1982; Bourbouze et al., 1984).

In the present report the relative value of a number of non-invasive tests for the assessment of nephrotoxicity and NAG isoenzymes are compared in the rat using a proximal tubular toxin and a papillotoxin propyleneimine. The value of the determination of the isoenzymes of NAG in man is also considered.

MATERIALS AND METHODS

Experimental animals

Male Sprague-Dawley rats (220-225 g) bred in-house were maintained in metabolism cages for several days prior to administration of the nephrotoxins, cis-DDP (5 mg/kg body weight i.p.) or propyleneimine (20 μl/kg body weight i.p.). Rats were fed for three hours a day outside the metabolism cages and 21 h urine volumes collected. NAG, β-D-galactosidase and β-D-glucosidase were assayed using the appropriate ω-nitrostyryl substrate+, the conditions used and those for leucine aminopeptidase and alkaline phosphatase were as described previously (Halman et al., 1984). NAG isoenzymes profiles were determined by semi-automated ion-exchange chromatography (Ellis et al., 1975; Yuen et al., 1986). Urinary protein was determined by the Coomassie Blue G-250 dye binding procedure (Bradford, 1976). Urine specific gravity was determined with a hand refractometer (Urian-N, Atago, Japan) and osmolality using a vapour pressure osmometer (5100B, Wescor Inc., USA). Urinary electrolyte excretion (sodium and potassium) was measured using a 543 flame photometer (Instrumentation Laboratory, Milan) while urinary pH and glucose were determined with a Hema-combistix (Ames Ltd., Slough) and creatinine by the method of Bonsnes and Taussky (1945).

Normal rat or human kidney was homogenised in a Potter-Elvehjem homogeniser (0.5 mm diameter clearance) and a 10% homogenate in 10 mM disodium phosphate buffer pH 6.8 prior to the separation of NAG isoenzymes. All samples were dialysed, using Visking tubing (4-22/32", Medicell Int. Ltd., London) and concentrated to an activity of approximately 8 mmol MNP released/h/l using a Minicon B-15 concentrator (Amicon Corp., Lexington, USA).

Normal urine and serum samples were obtained from students and staff at our Institution. Pathological urine samples were freshly voided mid-stream samples collected following permission of both the patients and consultants in charge.

RESULTS

The administration of the tubular toxin cis-DDP to rats resulted in an immediate small rise in urine volume which was accompanied by a 4-fold rise in lysosomal (NAG) and brush border marker (leucine aminopeptidase, alanine aminopeptidase and alkaline phosphatase) enzymes but no change in the cytosolic marker of β-glucosidase was found. In contrast only NAG was elevated in urine immediately after the administration of propyleneimine and no significant increase in the activity of the brush

---

+Further information on these substrates can be obtained from Dr. R.G. Price.

border membrane marker enzymes was observed. However an increase in β-D-galactosidase and β-D-glucosidase did occur ten days after the injection. The administration of a higher dose of propyleneimine (30 μl/kg body weight) resulted in a marked elevation of all the enzymes, indicating enzyme excretion is dose-related.

An increase in urine volume (3-fold) lasted for three days following the administration of cis-DDP. In contrast a sustained increase in urine volume (4-fold) was maintained for 16 days following the administration of the papillotoxin. Urinary S.G., osmolality and creatinine concentration altered in parallel with changes in urine volume. Electrolyte depletion resulted from papillary damage and the rate of excretion of sodium and potassium increased gradually over 16 days by which time the urinary concentration was four times the normal rate. The changes in electrolyte excretion following cis-DDP injection were transient and less marked. Glucose excretion was high in cis-DDP toxicity but not in rats given the papillotoxin, whereas urinary protein increased in both models but the rise was sustained over a longer period in papillary damage.

Urinary NAG Isoenzymes

Marked differences were found in the isoenzyme profiles of rat urine following the injection of propyleneimine. The major change was an increase in the intermediate form. This finding may be explained by the greater amount of this form present in rat medulla and papilla, Table 1. In contrast little change was observed in the NAG isoenzyme profile in rats given cis-DDP.

Urinary NAG and renal damage in man

The assay of NAG activity in urine provides a sensitive indicator of drug nephrotoxicity (Wellwood et al., 1976). The assay of this enzyme also provides a good prognostic indicator and this is illustrated in Table 2. Both patients a. and b. had acute renal failure but patient b recovered more quickly with urinary levels of NAG falling relatively rapidly into the normal range. Patient a. gradually recovered renal function and this was paralleled by a gradual fall in NAG activity. If NAG activities remain elevated as occurred in the patient with diabetic nephropathy then the prognosis is poor; this particular patient eventually required a transplant. Transplant rejection is often preceded by an increase in the intermediate form of NAG as well as increased NAG activity. Whether this finding has diagnostic potential is currently under study.

DISCUSSION

The most useful non-invasive procedures used to assess renal damage are listed in Table 3. When renal damage is localised

Table 1: NAG isoenzymes in normal and pathological tissue.
Number of samples are indicated in parenthesis.

		Isoenzymes (% of total recovered activity)	
	B	I	A
Normal Rat			
Kidney (20)	0.8 ± 0.12	18.4 ± 1.7	80.8 ± 2.4
cortex (5)	0.7 ± 0.18	8.5 ± 1.3	90.8 ± 1.5
medulla (5)	0.8 ± 0.15	13.2 ± 1.4	86.0 ± 2.1
papilla (5)	0.7 ± 0.14	17.8 ± 1.8	81.5 ± 2.1
Urine (20)	8.1 ± 1.6	14.2 ± 2.1	77.7 ± 2.6
Propyleneimine Treated Rat			
Urine day +1	2.8	26.8	70.4
day + 5	2.4	50.8	46.8
day + 9	3.2	69.6	27.2
Cis-DDP Treated Rats			
day + 1	3.8	16.6	79.6
day + 5	2.8	17.7	79.4
day + 9	2.7	23.8	73.5

Table 2: Variation of urinary NAG in renal disease.

	NAG Activity (μmol MNP/h/mmol creatinine) normal - 13.3 ± 6.3			
Patient	Months after 1st Visit			
	0	5	10	15
Acute renal failure a	600	360	200	110
b	110	25	-	-
Diabetic nephropathy	160	270	250	160
NAG Isoenzymes				
	B	I_1	I_2	A
normal	8.9	3.2	0.1	87.8
rejection	14.9	9.1	21.1	55.0

to specific regions of the nephron then marked differences are found with some but not necessarily all of the tests. This is illustrated by the comparison of a tubular and papillotoxin discussed in the results section. In our laboratory urinary NAG activity and volume are measured as initially screening procedures. If changes occur then the agent may be nephrotoxic while an elevation of glucose and brush border enzymes would suggest proximal tubular damage which can be confirmed by demonstrating the presence of low molecular weight proteinuria. If there is a sustained increase in protein, sustained polyuria and increase in NAG intermediate form papillary damage should

Table 3: Non-invasive tests.

1. Urine volume (S.G., osmolality), creatinine.
2. Na^+, K^+, pH.
3. Glucose protein.
4. Enzymuria - NAG, alanine aminopeptidase, alkaline phosphatase, LDH, β-galactosidase.
   Isoenzymes - NAG, LDH.
5. Proteinuria - lysozyme, β_2-microglobulin, retinol binding protein, Tam-Horsfall glycoprotein, albumin, α_1-microglobulin, SDS-PAGE.

be suspected. Distal damage is more difficult to establish non-invasively and the assay of LDH and alkaline phosphatase together with Tam-Horsfall glycoprotein (Sikri et al., 1985) may be indicators of damage to this region. The site of damage can be confirmed using histopathological procedures. The assay of urinary enzymes provides an excellent screening procedure for renal damage and their value is enhanced if the assays are carried out with a number of other tests.

Many of the tests listed in Table 3 can be carried out with dipsticks, minimal equipment or test kits. They therefore form the basis of procedures for screening for renal disease or damage in man. The methodological problems which previously limited the use of urinary enzymes in screening procedures have been overcome and dipstick and kit procedures will increase their use. There are a number of areas where the assay of urinary enzymes particularly NAG promises to be valuable, these include detection of renal involvement in hypertension (Johnson et al., 1983), transplant rejection (Wellwood et al., 1978), diabetic nephropathy (Whiting et al., 1979), urinary tract infections (Kiker et al., 1982). Urinary enzymes have potential in monitoring the nephrotoxicity of drugs including the aminoglycosides (Luft et al., 1978), cyclosporin (Finn and Gitelman, 1985), cis-DDP (McAllister et al., 1985) since the enzyme excretion related to the damage induced in the kidney drug therapy can be modified. Now that assay procedures have been simplified urinary enzyme assays can be used to screen 'at risk' industrial workers and populations exposed to potential nephrotoxins. It now seems probable that urinary enzymes will find a role in the assessment of nephrotoxicity, renal disease and in environmental medicine.

ACKNOWLEDGEMENTS

This study was partly supported by the SERC and MRC. We are grateful to Dr. J.S.L. Fowler for helpful discussions.

REFERENCES

Bondiou, M-T., Bourbouze, R., Bernard, M., Percheron, F., Perez-Gonzalez, N. and Cabezas, J.A. (1985) Clin. Chim. Acta 149, 67.

Bonsnes, R.W. and Taussky, H.H. (1975) J. Biol. Chem. 158, 581.

Bourbouze, R., Baumann, F-C., Bonvalet, J.P. and Farman, H. (1984) Kid. Int. 25, 636.

Bradford, M.M. (1976) Analyt. Biochem. 72, 248.

Ellis, R.B., Ikonne, J.U. and Mason, P.K. (1975) Analyt. Biochem. 63, 5.

Guder, W.G. and Ross, B.D. (1984) Kid. Int. 26, 101.

Halman, J., Price, R.G. and Fowler, J.S.L. (1984) in Selected Topics in Clinical Enzymology (Ed. D.M. Goldberg and M. Werner) Walter de Gruyter & Co., Berlin, 2, pp 435.

Johnson, I.D.A., Jones, N.E., Scoble, J.E., Yuen, C-T. and Price, R.G. (1983) Clin. Chim. Acta 133, 317.

Kind, P.R.N. (1982) Clin. Chim. Acta 119, 89.

Noto, A., Ogawa, Y., Mori, S., Yoshioka, M., Kitakaza, T., Hori, T., Nakamura, M. and Miyake, T. (1983) Clin. Chem. 29, 1713.

Price, R.G., Dance, N., Richards, B. and Cattell, W.R. (1970) Clin. Chim. Acta 27, 65.

Price, R.G. (1979) in Diagnostic Significance of Enzymes and Proteins in Urine (Ed. Dubach, U.C. and Schmidt, U.) Hans Huber, Berne, pp 150.

Price, R.G. (1982) Toxicology 23, 99.

Raab, W.P. (1972) Clin. Chem. 18, 5.

Sikri, K.L., Foster, C.L. and Marshall, R.D. (1985) Biochem. J. 225, 481.

Vigano, A., Assael, B.M., Villa, A.D., Gagliardi, L., Principi, N., Ghezzi, P. and Salmona, M. (1983) Clin. Chim. Acta 130, 297.

Wellwood, J.M., Ellis, B.G., Price, R.G., Hammond, K., Thomson, A.E. and Jones, N.E. (1975) Brit. Med. J. 3, 408.

Wellwood, J.M., Lovell, D., Thompson, A.E. and Tighe, J.R. (1976) J. Pathol. 118, 171.

Yuen, C-T., Price, R.G., Chattagoon, L., Richardson, A.C. and Praill, P.F.G. (1982) Clin. Chim. Acta 124, 195.

Yuen, C-T., Kind, P.R.N., Price, R.G., Praill, P.F.G. and Richardson, A.C. (1984) Ann. Clin. Biochem. 21, 295.

Yuen, C-T., Corbett, C.R.R., Kind, P.R.N., Thompson, A.E. and Price, R.G. (1986) Clin. Chim. Acta in press.

36

Separation and quantification of proteins to detect effects of chemical exposure

O. Vesterberg – Chemistry Division, National Board of Occupational Safety and Health, S-171 84 Solna, Sweden

BACKGROUND

Synthesis and concentration regulations of many proteins are influenced by environmental factors. The proteins of the cells account for their life manifestations as well as their individuality. Through studies of proteins, information can be obtained on the functions of active genes, regulation of protein concentrations and also on postsynthetic changes of proteins. We are now entering an era when this can be studied efficiently by using high resolution separation methods such as isoelectric focusing in gels, especially using this method as the first separation dimension in two-dimensional electrophoresis. This means that proteins are first separated according to charge and in the second dimension in respect to molecular size differences. Using < 1 µl of a serum sample several hundred proteins can be separated and displayed in a gel. The method can be said to give in one shot an overview of most of the proteins present in a sample. It therefore has many advantages to currently used procedures depending on, e.g. enzyme activity determination or use of antibodies, which are limited firstly by requiring knowledge about many detailed procedures, secondly for practial reasons allowing studies of only a few proteins at a time and thirdly demanding knowledge about what to assay before you know what has happened i.e. the effects of exposure to the chemicals. Two-dimensional electrophoresis can thus provide a detailed picture allowing simultaneous examination of a large number of proteins in the sample of biological interest.

The staining or visualization of the separated proteins by sensitive procedures is very important. Detection after separation of proteins occuring at low concentrations require sensitive staining procedures. It is possible to visualize in a gel using the dye Coomassie brilliant blue about 100 ng of protein in one spot [33]. For a few years it has been possible, with improved silver staining or silvering, to detect as little as 0.1 ng [15]. Silvering, thus being more one thousand times more sensitive than the usual Coomassie staining, allows visualization of several proteins which are only minor components of biological fluids, and permits direct analysis of biological fluids with low protein concentration, such as cerebrospinal fluid and urine without preceeding concentration [18]. This not only facilitates work but also obviates problems with the concentration procedures such as uncontrolled losses. A low detection limit is also of importance in analysis of blood serum and cellular proteins because it allows a limited total protein load in isoelectric focusing and/or electrophoresis thus avoiding overloading effects. This may be understood if it is recognized that e.g. in a serum sample of less than 1 µl about 1200 polypeptide spots can be separated [16]. If there were equal amounts of all these a dection limit of 1 ng would require a total load of 1200 ng. However, in reality the concentration ratios of minor to major components may well be one to thousand. Studies of such components would then require a substantial increase of the total load. With Coomassie staining which has a detection limit about 1000 times higher than silvering a total load of some hundred µl of serum would be required. But this would give overloading problems because usually more than one µl of serum per gel rod used in the isoelectric focusing step of two-dimensional electrophoresis causes trailing and other adverse effects. Sensitive silvering abolishes such problems. It must be recognized that there are many examples where minor components i.e. proteins occuring in a concentration of only one thousandth of the major ones may have significant biological functions, e.g. if they have regulatory or catalytic functions as enzymes. In this context it is not only of interest to detect concentration changes of a certain protein, e.g. an enzyme, but changes of an isoform of a protein e.g. an isoenzyme may also have biological significance [30]. Isoelectric focusing, which separates molecules by charge differences, is outstanding for this purpose. Such fundamental facts and some understanding of cell and body functions have stimulated us to devote much time and effort to lower the detection limits and to increase the reproducibility of the procedures for protein visualization. This has resulted in methods which are well suited to agarose gels [33] or polyacrylamide gels [15] after isoelectric focusing or electrophoresis, respectively. Very recent improvements are included in a manuscript in preparation. The reproducibility is also a very important parameter especially when examining many samples over a time period and almost mandatory if using computerized scanning evaluation of the stained gels.

The rationale behind the use of the mentioned methods to detect effects of chemical esposure may be described as follows. Proteins make up and are essential for the function of most elements of cells such as structure proteins, membranes and enzymes. Normally the amount of each protein is kept within quite narrow limits, by controlled rates of synthesis and catabolism. These two phenomena and especially the synthesis has been found to be composed of many steps in sequence, which may be regarded as chainlinks.

Proteins in blood plasma can be subdivided into two categories. Firstly, we think of the proteins secreted into the blood plasma to make up its original composition characteristic of the normal, "healthy" situation, the "true plasma proteins". The majority of them are produced in the liver. Secondly, there are a number of conditions, e.g. diseases, cancer and other malfunctions, where proteins, normally present in cells or on cellsurfaces, can be desquamated, secreted or come into the blood plasma by partial or complete failure of cell membranes anywhere in the body. The concentrations in blood plasma are regulated by the rates of supply balanced with the rates of removal. The reticuloendothelial system is dedicated to the later function. Proteins in blood plasma are also affected by passage through the kidneys. They have a cathabolic function and proteins are also lost at filtration without reabsorption, into the urine. The normal dayly output in urine amounts to about 150 mg of total protein. In a number of conditions this loss may increase. In e.g. glomerulonephitis and nephrosis the loss may increase by more than tenfold, but it is usually more moderate, e.g. in tubular kidney malfunction. To get a better understanding of changed concentrations of specific protein in blood plasma or urine it is advisable to make such quantification in plasma and urine.

The concentration of proteins in blood is usually in the normal healthy state kept within certain limits. It is quite astonishing that the relative variation may be as low as 0.3 even in a large multiracial group of individuals [26]. It is remarcable that the cells and the human body can often maintain a sufficient constancy not only over a certain day but for almost its whole life opon. This is achieved for each protein by several regulatory finely tuned systems. The halflife times for proteins in blood plasma span over a wide range, e.g. from hours for certain protein hormones as insulin to many days for several proteins e.g. ten days for transferrin in humans [26]. Toxic chemicals may excert their effects through interferences with such systems or influence on cell membranes e.g. causing decreased rates of removal, or the opposite, protein leakage. Furthermore, toxic effect may be caused by e.g inhibition of certain enzymes or interference with ion flux through membranes. Such disturbances may directly or indirectly, by compensatory activities of cells or the body, cause changes of protein concentrations. Because so many regulatory functions are involved for each protein it is most probable that the concentrations of specific proteins are sensitive to disturbances caused by adverse exposure to agents such as chemicals and radiation. From a theoretical point of view protein studies are therefore highly justified.

In addition protein molecules may be modified during their life span in the body, e.g. by loss of carbohydrate or amide groups [30]. This often results in isoproteins, which may be separated and studied by electrophoretic methods. Such modifications may be normal molecular ageing phenomena or effects of malfunctions [30]. Changed concentrations of such isoforms may be due to disturbances of various kinds e.g. of membrane function or regulatory mechanisms. From the above text it can be understood why studies of proteins can give information not only on toxic effects of chemicals but also on other forms of malfunction such as diseases and malignancy [30]. Nowadays several such examples are known and many more will certainly be discovered within the next few years. To a large extent this is thanks to recently developed electrophoretic procedures with high resolving power.

STUDIES OF URINARY PROTEINS

It is well known that the kidneys have the important functions of filtration of blood plasma and partial reabsorption of most molecules. Impaired kidney function leads to detoriating life quality and may end life. Already many years ago we could make high resolution separation of proteins in polyacrylamide gels by using isoelectric focusing. We have published results on studies of kidney malfunction after occupational exposure to cadmium causing urinary protein concentrations changes [28]. We observed similar protein patterns when studying samples from Japanese people having the diagnosis Itai-Itai disease as well as several cases suspected for having this condition, which was later officially recognized as being a characteristic kidney tubular malfunction partially due to excessive intake of cadmium via rice grown in a heavily puluted district. Our findings supported the clearing up of this problem. After staining the separated proteins were quantified by scanning densitometry using one of the best densitometers available at that time. We made accurate determinations of β_2-microglobulin [28], which was found to be increased in relation to estimates of preceeding occupational exposure and uptake of cadmium [9, 29]. Increase of over one hundred times the normal were found in some cases. We also found that after cadmium exposure not only the concentration of β_2-microglobulin but also that of many others were increased, notably albumin. This is in accord with later publications by others [2]. From such studies it can be concluded that there is most probably no agent or condition which causes an isolated increase in only β_2-microglobulin [2, 3, 29]. This makes interpretation of results at estimation of different urinary protein concentrations somewhat more complex especially for cases having a moderate increase. It must be recognized that there are a number of factors such as infection, inflamation, fever and heavy work, which may be accompained by slight to moderate elevations of concentrations of many urinary proteins including β_2-microglobulin [10,14,21,22, 24,25]. We showed that isoelectric focusing could be used to detect many changes of the normal protein excretion pattern, including glomerular malfunctions of the kidney [29]. Other methods such as electrophoresis with sodium dodecyl sulphate (SDS) with or without the presence of a gradient of polyacrylamide gel may give similar information [8]. Many assays have been made for β_2-microglobulin, but only a few years ago it became more generally accepted that such assays have some severe limitations and shortcommings. This protein is rapidly destroyed at pH below 5.6 [3,29]. As this can occure already in the urinary bladder it does not help to adjust pH in a sample. Already in 1976 we found, by using isoelectric focusing and immunological methods, urinary retionol binding globulin (RBP) to be a more optimal candidate for estimation of kidney malfunction of mainly tubular type. However, the time was not ripe for that yet. Groups in Belgium and Italy have confirmed the suitability of RBP [3,23].

In later years it was also found that exposure to metals such as chromium could cause slight elevations of concentrations of small size proteins in urine [13]. This was, however, found to be reversible [13] in contrast to the situation for cadmium [4,5]. In addition we showed that the concentration of orosomucoid in urine [5], determined with a simple but very efficient immunoelectrophoretic procedure called zone immunoelectrophoresis assay (ZIA) (for references see 31 and 32), is nearly as predictive as β_2-microglobulin. Cadmium is outstanding in having a biological halflife in humans of decades [4]. This circumstance makes highly relevant monitoring, by determination of suitable protein markers, to avoid adverse effects of excessive cadmium accumulation. Such monitoring may be very rational. Analysis of only a few urine samples a year is much safer than determination of cadmium in air samples even if made e.g. once a week. This is because measurement of substances in air does not tell the adsorbed dose and about possible biological effects for a number of reasons. A comprehensive review of biological monitoring concepts with central refecences have been published recently by Elinder and Vesterberg [6].

One difficulty in using urine stems from the fact the urine flow rate may vary from hour to hour. Thus an assayed substance may either be concentrated or diluted giving higher and lower concentration values (mg/l), respectively. In many places this has been corrected for by using the density or creatinine concentration in the urine sample. However, it has been recognized that this may cause missleading results [34], therefore it is recommended not use adjusted values of creatinine < 0.3 g/l or density < 1.010 [12].

One difficulty for the determination of urinary proteins, especially in the normal or close to normal range, is due to the low concentrations present. This has often necessiated a preconcentration, which means extra work and risks for variable losses of protein. We have devoted much time to study this and how to avoid it by developing sensitive protein staining in gels [18,33] and immunological procedures [31,32].

In order to study more completely the spectrum of proteins excreted in urine, in healthy subjects as well as at minor kidney malfunction, we have used two-dimensional electrophoresis. Numerous proteins can be visualized in unconcentrated samples also from healthy subjects by using an utmost sensitive silverstaining procedure [18].

OTHER APPLICATIONS

The use of two-dimensional electrophoresis is very challenging when striving to detect differences accompained by various diseases, malignancy and toxic effects of chemicals including drugs. For some years we have devoted a lot of effort to this aim, mainly in studies of proteins in blood plasma or serum from humans as well as from experimental animals. Numerous publications have resulted. For space reasons the reader is refered to a few recent publications [16-20] where references to earlier publications also can be found. Here may be mentioned protein studies on rats exposed to various chemicals such as dimethylformamide [19] and on humans after alcohol abuse [16, 17]. Toxic effects in the time course of the exposure were detected as well as in the recovery phase after exposure [17, 19]. Such protein changes are very interesting because they may offer better insight into the specific changes and disturbances on the molecular level. We are involved in numerous different toxicological studies, and reference is given to a recent publication [7].

Although we believe that use of two-dimensional electrophoresis is a very promissing approach few others have published using this methodology. The explaination is probably that, although it may appear simple, it takes a lot of effort and experience to have all the steps under control. Very interesting results on toxic effects by PCB (polychlorinated biphenyls) have been published recently [1]. This is of high importance especially because PCB is widely distributed in large quantities e.g. in transformers for electricity. At repair and replacement significant quantities may be released. PCB is chemicaly very stable and not simply biodegradable. Exposure to low quantities may give serious toxic effects. The effects on animals and microorganisms are alarming [27].

Some leading pharmaceutical industries have in recent years showed increasing interest in using two-dimensional electrophoresis. One rational behind this could be the awareness of it being more logical to detect adverse effects of potential drugs early than after years of use by large population groups. Another explanation could be the outstanding possibilities with two-dimensional electrophoresis to detect new biological phenomena and the study of such in great detail in for example pathology of cancer and diabetes. Similar causes are probably also behind the fact that an increasing number of people (at present estimated to be some hundred) are now applying two-dimensional electrophoresis to the study of proteins in numerous projects ranging from plant genetics, to advanced molecular biology in human cells and diseases such as hepatitis and AIDS (Aquired immune deficiency syndrome).

Many applications using high resolution protein separation methods are so recently published that they have not been dealt with in any overview. The best way at present to find such publications is by computer aided litterature search. Arthrosclerosis and diabetes constitute huge problems for many. These conditions are of growing interest in e.g. health care and preventive occupational medicine. It is known that lipoproteins are important in pathogenesis. We have devoted some research efforts to this field [11, 20]. The mentioned diseases are also of relevance in occupational medicine. Furthermore, influences of various treatments and regimens can be studied. - Simultaneous exposure to two or more chemicals may give additative and sometimes potentiated effects, i.e. significantly more than the sum of each. Detection and study of such effects is of growing importance, partly because of the increasing use of chemicals. Humans are for every year exposed to an increasing number of chemicals. Studies of effects of combined exposure is therefore urgent and of rapidly growing importance. Two-dimensional electrophoresis offers a comprehensive and realistic approach to protein mapping of effects of chemicals exposure. The use of this method in studies of cancerogenesis will be dealt with in fourthcomming publications.

REFERENCES

1. Anderson, N.L. et al. Effectc of Aroclor 1254 on proteins of mouse liver: Application of two-dimensional electrophoretic protcin mapping. Electrophoresis 7(1986)44-48.

2. Bernard, A., Buchet, J.P., Roels, H., Masson, P. and Lauwerys, R. Renal excretion of proteins and enzymes in workers exposed to codmium. European J. Clin. Invest. 9(1979)11-22.

3. Bernard, A.M., Moreau, D. and Lauwerys, R. Comparison of retinol-binding protein and β_2-microglobulin determination in urine for the early detection of tubular proteinuria. Clin. Chim. Acta 126(1982)1-7.

4. Elinder, C.G., Edling, C., Lindberg, E., Kågedal, B. and Vesterberg, O. Assessment of renal function in workers

previously exposed to cadmium. Brit. J. Industr. Med. 42(1985)754-760.

5. Elinder, C.G., Edling, C., Lindberg, E., Kågedal, B. and Vesterberg, O. β_2-Mocroglobulinuria among workers previously exposed to cadmium: Follow-up and dose-response analyses. Am. J. Ind. Med. 8(1985)553-564.

6. Elinder, C.G. and Vesterberg, O. Environmental and biological monitoring. Scand. J. Work. Environ. Health 11(1985): Suppl 1, 91-103.

7. Ghantous, H., Danielsson, B.R.G., Dencker, L., Gorczak, J. and Vesterberg, O. Trichloroacetic acid accumulates in murine amniotic fluid after tri- and tetrachloroethylene inhalation. Acta pharmacol. et toxicol. 58(1986)105-114.

8. Görg, A., Postel, W., Weser, J., Schiwara, H.W. and Boesken, W.H. Horizontal SDS electrophoresis in ultrathin pore-gradient gels for the analysis of urinary proteins. Science Tools, Vol. 32(1985)5-10.

9. Hansén, L., Kjellström, T. and Vesterberg, O. Evaluation of different urinary proteins after occupational Cd exposure. Int. Arch. Occup. Health 40(1977)273-282.

10. Hemmingsen, L. and Skaarup, P. Urinary excretion of ten plasma proteins in patients with febrile diseases. Acta Med Scand 201(1977)359-364.

11. Holmquist, L. and Vesterberg, O. Quantification of human serum apolipoprotein A-I by zone immunoelectrophoresis assay. Clin. Chim. Acta. Accepted for publication 1986.

12. Lauwerys, R.R. Industrial chemical exposure: Guidelines for biological monitoring. Biomedical Publications. 1983 Davis, California.

13. Lindberg, E. and Vesterberg, O. Urinary excretion of proteins in chromeplaters, exchromeplaters and referents. Scand J Work Environ Health 9(1983)505-510.

14. Lizana, J., Brito, M. and Davis, M.R. Assessment of five quantitative methods for determination of total proteins in urine. Clin. Biochem. 10, 2(1977)89-93.

15. Marshall, T. Detection of protein in polyacrylamide gels using an improved silver stain. Anal. Biochem. 136(1984) 340-346.

16. Marshall, T., Williams, K. and Vesterberg, O. Two-dimensional electrophoresis of proteins in human serum: Improved resolution by use of narrow pH gradients and prolonged electrophoresis. Clin. Chem. 30(1984)2008-2013.

17. Marshall, T., Vesterberg, O. and Williams, K.M. Effects of alcohol abuse on human serum proteins revealed by two-dimensional electrophoresis. Electrophoresis 5(1984)122--128.

18. Marshall, T., Williams, K.M. and Vesterberg, O. Unconcentrated human urinary proteins analysed by high resolution two-dimensional electrophoresis with narrow pH gradients: Preliminary findings after occupational exposure to cadmium. Electrophoresis 6(1985)47-52.

19. Marshall, T., Williams, K.M. and Vesterberg, O. A comparison of two-dimensional gel electrophoresis methods for analysis of rat serum proteins following dimethylformamide exposure. Electrophoresis 6(1985)392-398.

20. Marshall, T., Williams, K.M., Holmquist, L., Carlson, L.A. and Vesterberg, O. Plasma apoliproprotein pattern in fisheye disease by highresolution two-dimensional electrophoresis. Clin. Chem. 31(1985)2032.

21. McElderry, L.A., Tarbit, I.F. and Cassells-Smith, A.J. Six methods for urinary protein compared. Clin. Chem. 28/2(1982)356-360.

22. Mogensen, C.E., Gjöde, P. and Christensen, C.K. Albumin excretion in operatin surgeons and in hypertension. The Lancet, April 7(1979)774-775.

23. Mutti, A. et al. Urinary excretion of brush-border antigen revealed by monoclonal antibody: Early indicator of toxic nephropathy. The Lancet, October 26(1985)914-917.

24. Poortmans, J. and Van Kerchove, E. Dosage de la proteiurie: Comparaison de deux méthodes. Clin. Chim. Acta 8(1963)485-488.

25. Poortmans, J.R. and Vancalck, B. Renal glomerular and tubular impairment during strenuous exercise in young women. Europ. J. Clin. Investigation 8(1978)175-178.

26. Putnam, F.W. The plasma proteins. Structure, function, and genetic control. Second ed. 1984, Acad. Press, New York.

27. Safe, S. Polychlorinated biphenyls (PCBs) and polybrominated biphenyls (PBBs): biochemistry, toxicology, and mechanism of action. CRC Crit Rev Toxicol 13/4(1984)319-395.

28. Vesterberg, O. and Nise, G. Urinory proteins studied by use of isoelectric focusing. I. Tubular malfunction in association with exposure to cadmium. Clin. Chem. 19(1973) 1179-1183.

29. Vesterberg, O., Nise, G. and Hansén, L. Urinary proteins in occupotionol exposure to chemicals and in diseases. J. Occupational Medicine 18(1976)473-476.

30. Vesterberg, O. Isoelectric focusing a review of analytical techniques and applications. International Laboratory May/June 1978, pp. 61-68 and American Laboratory June 1978, pp. 13-24 and Japan Journal of Electrophoresis, 1979.

31. Vesterberg, O. Quantification of albumin in urine by a new method: zone immuno-electrophoresis assay (ZIA). Clin.

Chim. Acta 113(1981)305-310.

32. Vesterberg, O. Protein quantification with zone immunoelectrophoresis assay (ZIA). Modern Methods in Protein Chemistry - Review Articles. Ed. H. Tschesche 1983, Walter de Gruyter Berlin, New York pp. 187-206.

33. Vesterberg, O. and Gramstrup-Christensen, B. Sensitive silver staining of proteins after isoelectric focusing in agarose gels. Electrophoresis 5(1984)282-285.

34. Vesterberg, O., Sollenberg, J. and Wrangskog, K. Evaluation of determinations made in urine samples. Adjustments of mandelic acid concentration using creatinine and density. Proc. Int. Symp. on Occup. Exposure Limits, ACGIH, Copenhagen 1985. Ann. Am. Conf. Ind. Hyg. 12(1985)301-304.

37

The study of urinary excretion of kidney antigens to reveal early effects of exposure to exogenous chemicals

A Mutti – Institute of Internal Medicine and Nephrology, Laboratory of Industrial Toxicology, University of Parma, Italy

INTRODUCTION

Urine might also be considered a kidney biopsy, in the form of cells and other tissue constituents, which are released in a freely available suspension. Moreover, very sensitive immunochemical methods, such as latex-, enzyme- and sol particles-immuno assays reveal as few as pico or even femtomoles of a given antigen.

Because of their monospecificity, monoclonal antibodies fired the imagination of investigators in many fields of biology and medicine. Since the publication of Kolher's and Milstein's report in 1975, monoclonal antibodies have been considered as a powerful answer looking for questions and applications rather than being the right solution for a particular problem. This is also why, when planning our study, we considered different goals.

The first one was the demonstration of the presence of common antigenic determinants in different structures: such a result would have supported the view that a primary toxic damage may relase hidden antigens from kidneys or from other organs, and cause an immune reaction resulting in glomerulo- or tubulo-interstitial nephritis.

The second goal was the measurement of the urinary excretion of tissue proteins as an early indicator of toxic nephropathy.
Another objective was the study of the association between microtissue damage and dysfunction.
This paper will discuss some promising findings we obtained along with some problems we encountered when using monoclonal antibodies to reveal kidney antigens in urine from subjects exposed to nephrotoxic chemicals.

PRODUCTION OF MONOCLONAL ANTIBODIES

A detailed description of methods has been reported elsewhere (Mutti et al., 1985). Only a summary of hybridoma technology will be described below (fig. 1). Briefly, splenocytes from mice immunized with a renal cortex (Fx1A) preparation were fused in a ratio 4:1 with HPRGT(Hypoxanthine-Guanine-PhosphoRibosil-Transferase)-deficient non-secreting mouse myeloma cells. Hybridomas were selected out by incubation in a medium with hypoxanthine-aminopterin-thymidine (HAT) additives. Repeated cloning of mutant cells was performed by limiting dilution. If positive for relevant antibody activity, which was checked both by enzyme-linked immunosorbent assay (ELISA) and by indirect immunofluorescence on renal slices, hybridomas were then expanded into flasks, frozen and stored in liquid nitrogen. Cells were also maintained in long-term culture or injected into pristane-pretreated mice for production of ascites antibodies, which were then precipitated and purified.

HISTOCHEMICAL LOCALISATION OF ANTIGENS

From two fusions, 365 growing primary cultures were obtained; 41 secreted antibodies reacting with human renal cortex. Most antibodies reacted with the brush border (fig. 2). A cross-reaction with other structures throughout the nephron, e.g. glomeruli, and the endothelial wall of peritubular capillaries, was however found by using monoclonal antibodies HF5 and CB7, respectively.
This suggest that antigenic determinants may be shared by different kidney structures. It is conceivable that some of such antigens are normally hidden and may be released from the kidney tubule or from other damaged tissues. For instance, HF5 and CB7 monoclonal antibodies show a widespread reactivity, including the nervous system for the latter whereas CG9 is specific for the brush-border of proximal tubules (N. Kirkham, personal communi-

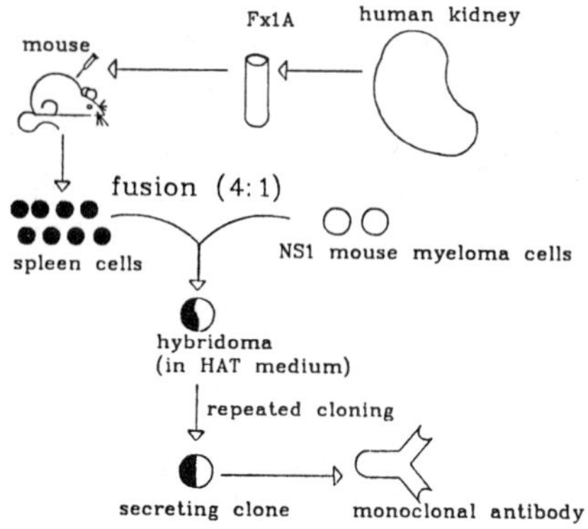

Fig. 1. – Production of monoclonal antibodies to human renal antigens. See text for abbreviations and for a detailed description.

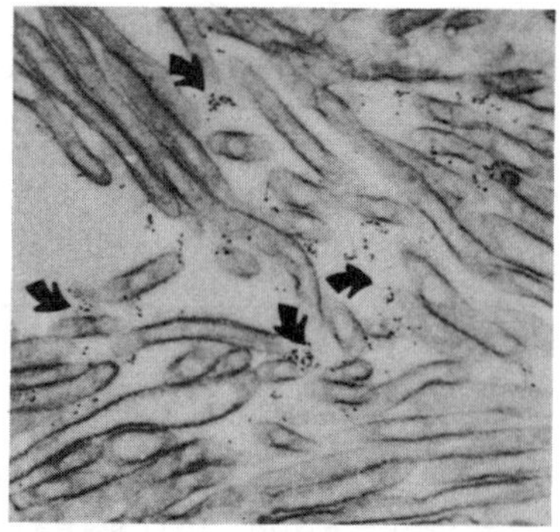

Fig. 2. – Binding of CB7 to the brush-border of renal tubules revealed by immunogold electron microscopy.

cation). Toxic lesions involving kidney tubules or even other tissues might release hidden antigens and lead to immune reactions. Immune-mediated glomerulopathies would in turn require a sort of individual predisposition, e.g. certain HLA antigens (Wooley et al., 1980).

ELISA

ELISA is now used routinely to measure a variety of soluble antigens present at low concentrations in body fluids. Its sensitivity is of the order of picomoles per liter the other analytical features relying entirely on the properties of antibodies. Different ELISAs may be designed, one of these being particularly advantageous. In the indirect "sandwich" ELISA, the solid phase-bound antibody reacts with one of the antigenic sites on the molecule to be analyzed, which can react with a second antibody or with a monospecific antiserum (e.g. from rabbits). An enzyme-conjugated antibody from a different species (e.g. goat) directed towards the second antibody is then allowed to react. Washing out all the unreacted material after each step, it is possible to reveal the antigen which has been bound to the solid phase through the sandwich, the antigen concentration being proportional to enzymic activity. One particular advantage of this technique is its versatility: different antigens may be analysed in parallel within the same run, simply by changing the 1st antibody coating the solid phase.

Similar techniques have been proposed to reveal trace amounts of plasma proteins (Lucertini et al., 1984; Mutti et al., 1986; Valcavi et al., 1986) and have been used to measure urinary albumin and RBP (retinol-binding protein) during cross-sectional surveys of subjects exposed to known nephrotoxins.

CHARACTERIZATION OF ANTIGENS

The antigen recognized by CB7 (the 1st monoclonal antibody we used) was called BB50, since it was identified as a brush-border protein migrating with an apparent m.w. of 50,000 in SDS-PAGE (Mutti et al., 1985). Unfortunately, both the artifactual loss of membrane constituents and the production of new bands resulting from aggregation or fragmentation of nascent proteins can occur with this method. As a result, though it is believed to be a "mild" extraction procedure, SDS-PAGE does not provide information about the true features of tubular antigens in both tissue preparations and urine. An alternative approach not involving denaturing conditions is gel-permeation of concentrated urines. Table 1 summarizes the behaviour of antigens recognized by three different monoclonal antibodies. Owing to their very high apparent m.w., only a range may be specified: the so-called BB-50 antigen recognized by CB7 migrates with an apparent m.w. of between 1.5 and 5 millions, whereas the antigens recognized by HF5 and CG9 a m.w. higher and lower respectively than such an interval.
The antigens also show different behaviours when treated with detergents reacting with hydrophobic segments of membrane proteins. Table 2 shows that the reactivity of tissue constituents recognized by different monoclonal antibodies may be decreased, unchanged, and increased by pre-incubation with Triton X. Such effects of Triton X are due to its interaction with the antigens rather than with antibodies, since no differences were seen when incubation was performed before or after gel permeation.

Table 1: Apparent m.w. of native brush-border antigens isolated from urine and revealed by ELISA.

Gel-permeation with	Apparent m.w. of antigens reacting HF5	CB7	CG9
Sephadex G-100	>100,000	>100,000	>100,000
SephacrylS-300	>1,500,000	>1,500,000	<1,500,000
Superose	>5,000,000	<5,000,000	
Biogel 15 M	<15,000,000		

Table 2: Behaviour of brush-border antigens when incubated with Triton X 100.

Triton X 0.5%	*Relative amount of Ag recognized by HF5	CB7	CG9
Before gel-permeation	0.20	1.00	1.00
After gel-permeation	0.25	0.90	1.00
None	1.00	0.80	0.10

*1.00=highest concentration measured by ELISA

During gel-permeation, all unreacted detergent is removed from samples. Changes in reactivity are thus due to stable modifications of antigens' hydrophobic components interfering with their immunereactivity rather than with their solubilization.
All such antigens also seem to have a very high m.w., when they are extracted from urine under non-denaturing conditions. Such an apparent m.w. might reflect that of microvesicles where antigens are incorporated, or might be artifactually high because of non-proteic components or of aggretions which may occur during isolation procedures. Finally, it might represent the true mass of structural components of the brush-border or of cytoskeleton. Each one of these possibilities has differential likelihood for the different antigens reacting with our monoclonal antibodies.
The above experiments show that the monoclonal antibodies we are using react with proteins which are probably different in nature and have different physicochemical properties. Since they also have different histochemical localization, they seem to constitute a pattern which deserves further investigation in subjects exposed to nephrotoxic agents.

RESULTS FROM FIELD SURVEYS

A summary of cross-sectional investigations dealing with the urinary excretion of the BB-50 antigen recognized by CB7 monoclonal antibody is based on published material (Mutti et al., 1985 a and b; Viau et al., 1986) and on preliminary results from on-going studies.
BB-50 was found to be increased in different groups of subjects occupationally exposed to or treated with potentially nephrotoxic agents (fig. 3), i.e. cadmium, cisplatin, chromium(VI), mercury, and hydrocarbons. A group of patients with histologi-

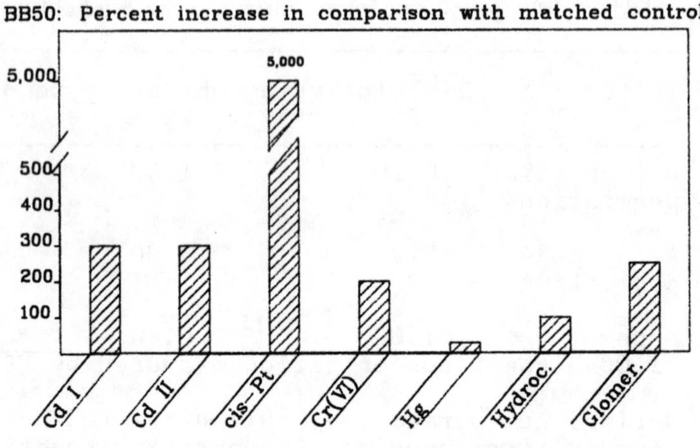

Fig. 3. – Increase in urinary BB-50 in various groups of subjects exposed to nephrotoxins and in patients with chronic glomerulonephritis. All groups but Hg-exposed workers were significantly different from matched controls (not shown here).

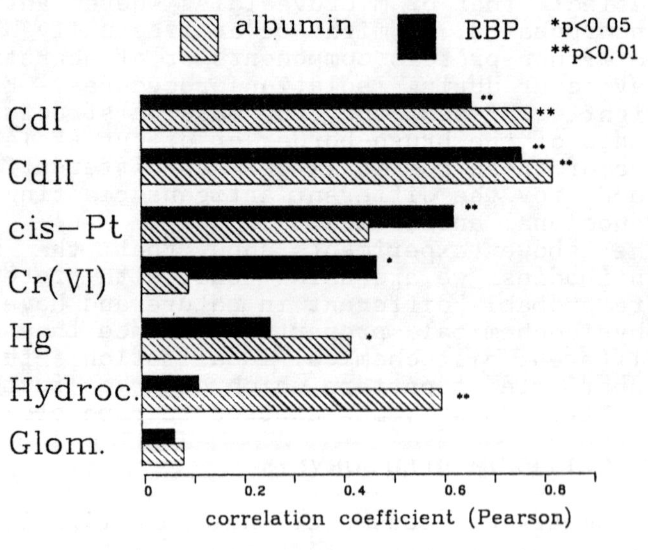

Fig. 4. – Correlations between BB-50 and specific urinary proteins in the same groups as in fig. 3.

cally verified glomerulopathy is also included. The level of BB50 is correlated with those of low and high m.w. proteins, when these are increased, with the exception of patients with glomerulopathy (fig. 4). The urinary excretion of renal antigens is however easier to interpret than that of plasma proteins, since it is an indicator of epithelial turnover throughout the proximal tubule. Increased excretion can only be interpreted as resulting from necrosis, cell breakdown or loss of brush-border vesicles. It represents the direct measurement of microtissue damage, which is revealed only indirectly by enzymuria.

The correlations between BB-50, i.e. a specific indicator of tubular damage, and both low and high m.w. plasma proteins in the urine suggest that tubular impairment might be responsible not only for low m.w. proteinuria, but also for slight increases in albuminuria, even when the latter occur isolated. The urinary excretion of a brush-border antigen and alterations in renal handling of plasma proteins give different pictures in patients with primary glomerulopathies and in subjects exposed to nephrotoxins. The former group shows important changes in albuminuria and slight increases in BB-50, without any relationships between these markers. On the contrary, slight increases in albuminuria among subjects exposed to nephrotoxic agents are always correlated with increased levels of brush-border antigens in the urine. These relationships might be due to independent effects of the same agent. In this case, any pathophysiological interpretation would be misleading. However, until the general meaning of albuminuria as a marker of increased glomerular permeability is not disproved, we should conclude that tubular lesions and glomerular damage may be associated in subjects exposed to a variety of nephrotoxic agents.

As a whole, our findings suggest that tubular impairments leading to increased excretion of plasma proteins present in filtered load are associated with microtissue damage, which may be revealed by the urinary excretion of brush-border antigens. Furthermore, our results are not inconsistent with the hypothesis that a primary tubular lesion may lead to immune-mediated glomerular damage.

LIMITATIONS OF THE TEST

In addition to problems of stability in urine samples, which should be buffered and frozen as soon as possible, renal antigens share with other proteins the main limitation of immunochemical reactions, for which a correspondance is assumed between the amount of a given antigen and its immunological reactivity. This assumption may be misleading in various situations. If serum RBP is used to build-up a calibration curve for urinary

RBP, falsely high values will be found, depending on the fact that serum RBP is complexed with prealbumin and has lower immune-reactivity than urinary RBP, which is mostly free. Albumin may be fragmented or polymerized especially when frozen, with unpredictable consequences on its immunological reactivity (Wiggins et al., 1985).

These possible biases may well occur also for tissue antigens, only limited experiences being available so far in this new field of investigation. The test also requires further clinical validation before it can be routinely used. Meanwhile, any evaluation should be made only at a group level. Even in this case, a rigourous scientific approach, e.g. blind examinations of alternate samples whitin the same series, will help us to avoid the possible transformation of femtomoles we are measuring into phantom-moles, which are certainly not S.I. units.

REFERENCES

Kohler G., Milstein C. Continuous culture of fused cells secreting antibodies of predefined specificity. Nature 1975; 256: 497-97.

Lucertini S., Mutti A., Valcavi PP., Franchini I. Enzyme-linked immunosorbent assay of retinol-binding protein in serum and urine. Clin Chem 1984; 30: 149-51.

Mutti A., Lucertini S., Valcavi PP., Neri TM., Fornari M., Alinovi R., Franchini I. Urinary excretion of brush-border antigen: Early indicator of toxic nephropathy. Lancet 1985; 8461: 914-917.

Mutti A., Lucertini S., Fornari M., Franchini I., Bernard A., Roels H., Lauwerys R. Urinary excretion of brush-border antigen revealed by monoclonal antibodies in subjects occupationally exposed to heavy metals. Proc. Intern. Conf. on Heavy Metals in the Environment. Vol I, 565-7, Edinburgh: CEP, 1985.

Valcavi PP., Lucertini S., Mutti A., Chezzi C. Urinary albumin. In: Bergmeyer HU (ed): Methods of Enzymatic Analysis. Vol. IX, 56-64. Weinheim: VCH, 1986.

Wiggins RC., Kshrisagar B., Kelsch RC., Wilson BS. Fragmentation and polymeric complexes of albumin in human urine. Clin Chim Acta 1985; 149: 155-63.

Supported in part by the Commission of the European Communities (contr. ENV-749-I)

38

Cadmium in blood as a cumulative dose estimate

L. Järup, C.G. Elinder, G. Spang* – Department of Occupational Medicine, National Board of Safety and Health, S-171 84 Solna, Sweden. *Industrial Health Center, S-572 00 Oskarshamn, Sweden

Long term exposure to cadmium may cause kidney damage. The first sign of cadmium-induced renal dysfunction is an increased secretion of low molecular urinary proteins such as ß-2-microglobulin. The biological half-time of cadmium is very long (10–20 years) and thus the metal is accumulated in the body in proportion to cumulative exposure (Friberg et al, 1985).

Air cadmium data are usually used to estimate the exposure of individual workers. Such data are, however, inadequate since the individual internal dose is influenced also by several other factors such as personal hygiene and cigarette smoking.

A recent study has suggested that cadmium in blood may be used as a cumulative dose estimate (Rogenfelt et al, 1984).

This has been evaluated further in the present report, which includes 435 workers in a Swedish battery factory.

The individual exposure varies from little or no exposure through medium to high exposure. Cadmium in blood has been analyzed since 1967 and urinary ß-2-microglobulin since 1972.

Cumulative blood cadmium (expressed in nmol x months/l) and urinary ß-2-microglobulin (µg/mmol creatinine) were calculated for each study member. Cadmium in blood for each individual was cumulated during the exposure period. Tubular proteinuria

was defined as urinary ß-2-microglobulin-levels above 35 µg/mmol creatinine. We determined the response of tubular proteinuria for six dose classes. The resulting dose-response relationship is show in Figure 1.
A dose-response relationship was also computed using cumulative air cadmium as the measure of exposure. The cumulative air cadmium dose was calculated as the product of the number of years of employment and the mean concentration of cadmium dust in the workroom air during different time periods.
Regression analysis showed a good correlation (r = 0.75) between the two different approaches to estimate the dose: cumulative air-cadmium and cumulative cadmium in blood. The corresponding dose-response relationships compare quite well as can be seen in Table 1.

Table 1

DOSE CLASS	CUMULATIVE B-Cd nmol x months/l	RESPONSE %	CUMULATIVE AIR-Cd µg x years/m3	RESPONSE %
1	< 5000	1.3	<359	1.1
2	5000–15000	7.8	359–1710	9.2
3	15000–30000	19.4	1710–4578	23.3
4	30000–50000	24.1	4578–9458	32.3
5	50000–69178	36.4	9458–15000	31.2
6	69178+	40.7	15000+	50.0

A consequence of the dose-response relationship with cumulative air cadmium as the dose estimate is that continuous exposure to the current Swedish TLV of 20 µg/m3 for 20 years would lead to a response of 3%, i.e. very close to the background response rate. Similar dose-response relationships have been found by other (Kjellström et al., 1977; Elinder et al, 1985; Ellis et al, 1985). The lowest air cadmium level corresponds to a mean blood cadmium level of 21 nmol/l for the same exposure period, which also yields a 3% response rate.
Similarly, the current no adverse level for cadmium in blood proposed by Bernard et al (1979), 10 µg/l (89 nmol/l) and

later recommended by the WHO (1981), would correspond to a response of 14% for a 20 year work period. This recommendation is, however, based on cross-sectional data whereas the present study uses cumulative blood cadmium as the dose estimate.
Our data suggest that cumulative blood cadmium may be used as an estimate of the individual cadmium exposure. A cumulative blood cadmium of less than 5000 nmol x months/l corresponds to a mean blood cadmium level of 89 nmol/l for five years or an average blood cadmium of 21 nmol/l for 20 years.
We thus believe that our results indicate that cumulative blood cadmium may be used to identify persons at risk for developing cadmium induced renal damage.

REFERENCES

BERNARD A, BUCHET JP, ROELS H, MASSON P, LAUWERYS R (1979) Renal excretion of proteins and enzymes in workers exposed to cadmium. Eur J Clin Invest 9: 11-22.
ELINDER CG, EDLING C, LINDBERG E, KAGEDAL B, VESTERBERG O (1985) ß2-microglobulinuria among workers previously exposed to cadmium: follow-up and dose-response analyses. Am J Ind Med 8: 553-564.
ELLIS KJ, COHN SH (1985) Cadmium inhalation exposure estimates: their significance with respect to kidney and liver cadmium burden. J Toxicol Environ Health 15: 173-187.
FRIBERG L, ELINDER CG, KJELLSTROM T, NORDBERG GF (1985) Cadmium and Health: a toxicological and epidemiological appraisal, vol 1: 159-160.
KJELLSTROM T, EVRIN PE, RAHNSTER B (1977) Dose-response analysis of cadmium-induced tubular proteinuria. A study of urinary ß2-microglobulin excretion among workers in a battery factory. Environ Res 13: 303-317.
ROGENFELT A, ELINDER CG, JARUP L (1984) A suggestion on how to use measurements of cadmium in blood as a cumulative dose estimate. Int Arch Occup Environ Health 55: 43-48.
WHO (1981) Recommended health-based limits in occupational exposure to heavy metals. Report of a study group. Tech Rep Ser 647. World Health Organization, Geneva.

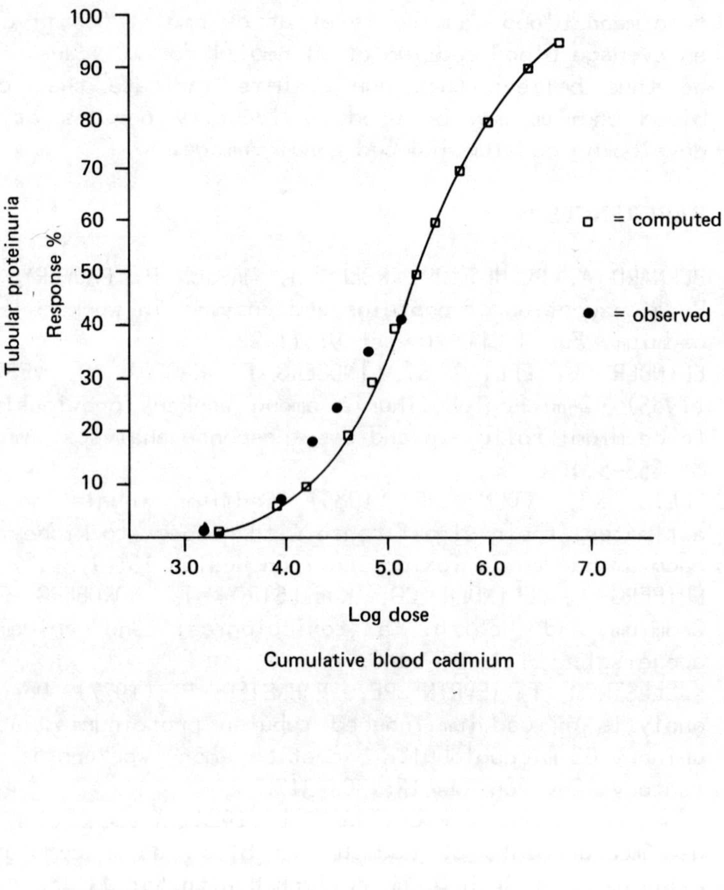

Fig. 1. Response of tubular proteinuria plotted against the logarithms of the cumulative blood cadmium dose. The dose-response relationship was computed using the probit analysis method.

Evaluation of renal function in workers with low blood lead levels

C.N. Ong, G. Endo*, K.S. Chia, W.O. Phoon, H.Y. Ong – Department of Social Medicine and Public Health, National University of Singapore, National University Hospital, Lower Kent Ridge Road, Singapore 0511, Republic of Singapore. *Department of Preventive Medicine, Osaka City University, Japan

INTRODUCTION

In the 60's, there was little concern on lead exposure until the worker's blood lead level (PbB) exceeded 80 or even 100 ug/100 ml. However, during the last 15 years there has been a dramatic reappraisal of the level of lead absorption considered to be acceptable. Today, blood lead levels are often considered elevated if they exceed 40 ug/100 ml (WHO, 1980). Usually, workers exposed to low levels of lead seldom develop clinical manifestations and therefore are difficult to recognise because of the absence of pathognomonic signs or symptoms. However, recent studies have shown evidence of subclinical effects of blood lead concentrations below those at which clinical symptoms are usually manifest. The usual tests showed these subclinical effects are related to neurophysiological or haematological studies. There were relatively few studies on the laboratory parameters that have been shown to be useful and reliable for the monitoring of kidney function on workers exposed to low levels of lead.

The pathological features of lead nephropathy have been controversial. It was first thought that the most common feature of lead intoxication was interstitial nephritis leading to a contracted kidney (Morgan et al, 1966). This has not always been substantiated by later observations. Recent studies have suggested that lead causes damage to proximal tubular cells and affected tubular reabsorption (Weeden et al, 1975; Meyer et al, 1984).

Urinary enzyme analysis has been proven to be an extremely sensitive indicator of renal injury in recent years. Such enzymes had been shown to be elevated in glomerular, tubular and interstitial renal diseases even before the onset of renal failure or any abnormality in other excretory function is detectable (Piperno, 1981; Price, 1982). One of the urinary enzyme is a lysosomal enzyme, N-acetyl-β-D-glucosaminidase. It is present in renal tubular cells. NAG can be assayed accurately and reproducibly and had been shown to be an extremely sensitive indicator of early renal disease and injury (Mansell et al; 1978; Lockwood and Bosmann 1979).

In the present study, the excretion of NAG was measured in over 200 low to moderate lead exposed workers with blood lead (PbB) of 10-80 ug/dl. In addition, three conventional methods for evaluating renal function; serum creatinine (SC), blood urea nitrogen (BUN) and creatinine clearance tests (CCT) were also conducted.

METHODS

Subjects

169 workers from two factories manufacturing lead storage batteries in Singapore and 40 workers from a secondary lead smelting plant in Osaka, Japan participated in this study. Thirty professional and laboratory staff with no history of renal disease or lead exposure and had normal urinary analysis using clinistix were used as controls.

Laboratory measurement

Lead in blood was determined by a Pye-Unicam SP-9 Spectrophotometer with a graphite furnance. The detailed analytical procedures for PbB were as reported earlier (Ong et al, 1985). External quality control was carried out under the National External Quality Assessment Scheme (NEQAS) in U.K. The degree of precision over the concentration range of 1 ug/dl to 90 ug/dl was ± 0.5 ug/100 ml.

To evaluate renal function, serum samples of 0.1 ml were obtained for determination of blood urea nitrogen (BUN).

Serum concentrations of creatinine and BUN were measured by Jaffe alkaline Dicrate and Crocker's Diocetyl Monoxime Methods(Wootton & Freeman, 1982), respectively. A single voided urine sample was collected from each subject for NAG determination. To obtain a more standard sample of urine, they were usually collected in the morning around 10.00 am. Samples were analysed on the same day. N-acetyl-β-D-glucosaminidase (NAG) measurement was performed according to Noto's MCP-NAG method(Noto et al, 1983).

RESULTS

The distribution pattern, standard deviations and renal function are presented in Table 1. Of the 209 workers who took part in this study, 51 were females and 150 were males. The mean age was 34 years old. Less than 25% of them were over 40 years old. Fewer than 25% had been employed for 10 years or longer. The highest PbB detected was 80 ug/100 ml, and only five workers (3%) had PbB over 60 ug/100 ml. The females were noted to be slightly younger and had a shorter history of lead exposure than their male counter parts. For the various parameters tested males had higher values than females except for urinary NAG.

Table 1: Age, years of lead work, concentrations of lead in blood and renal function.

			Male		Female	
	Total n	range	n	mean ± SD	n	mean ± SD
Age	209	17.0-68.0	158	34.9±11.5	51	31.53±10.3
Lead Work (Yrs)	209	1.0-36.0	158	6.77±7.9	51	3.94±6.2
PbB	209	3.0-80.0	158	42.1±16.6	51	31.9±14.3
PbU	202	0.5-49.7	154	6.86±8.2	48	5.98±4.3
SC	206	0.5-1.96	157	1.10±0.2	49	0.93±0.2
BUN	206	4.0-28.0	157	12.73±3.7	49	9.67±3.2
CCT	165	10.0-753	121	104.6±40.9	44	118.2±130.8
NAG	203	0.2-10	154	2.38±1.17	49	2.71±1.53

Table 2 summarises the relationship of PbB and PbU together with the four indicators for renal function. PbB had the highest correlation coefficient with serum creatinine (SC), r=0.26 ($p<0.001$). Lower correlations were found for BUN and creatinine clearance test. No significant association was found between PbB and NAG. However, a significant association was observed for PbU and BUN (r=0.34), whereas, a lesser correlation was noted for PbU and NAG (r=0.23). Close correlation was also observed for SC and CCT (r=0.28).

Table 2: Correlations of various parameters tested.

	PbU	BUN	SC	CCT	NAG
PbB	0.41***	0.16**	0.26***	-0.16**	0.04^{NS}
PbU		0.34***	0.20**	-0.11*	0.23***
BUN			-0.04^{NS}	-0.01^{NS}	0.15**
SC				-0.28***	0.16**
CCT					-0.17**

***p<0.001 **p<0.01 *p<0.05 NS : Not significant

The mean NAG levels for unexposed controls and exposed subjects are shown in Table 3. The results indicate significantly higher urinary NAG amongst the exposed subjects.

Table 3: Mean NAG levels for controls and lead exposed workers.

Age	Non-exposed		Exposed		t-test		
Year	n	Mean ± SD	n	Mean ± SD	T	df	P
< 20	--	--	7	119.7±22.1	--	--	--
20-29	12	84.7±28.9	81	119.4±58.7	2.01	91	0.05
30-39	9	98.4±51.4	60	144.3±69.5	2.69	67	0.01
40-49	9	113.8±50.9	27	184.5±87.8	2.76	33	0.01
> 50	6	162.2±73.6	21	197.2±54.2	2.68	22	0.02
Total	30	86.7±29.2	196	144.4±70.9	4.40	224	0.001

p<0.0001 by analysis of variance

Fig. 1 illustrates the relationship between PbB and NAG. There is an increase in the release of NAG with the rise in the level of PbB. When adjusted for age, this increase of NAG excretion with increasing blood level was statistically significant (F=6.27, p<0.001).

The correlation between PbU and NAG is as shown in Fig. 2. The scattergram suggests an association between the two variables, giving a weak but significant linear correlation (r=0.23, p<0.001). The age-adjusted mean corrected NAG levels indicates a significant increasing trend of NAG excretion with increasing urine lead levels (F=8.1, p<0.0001).

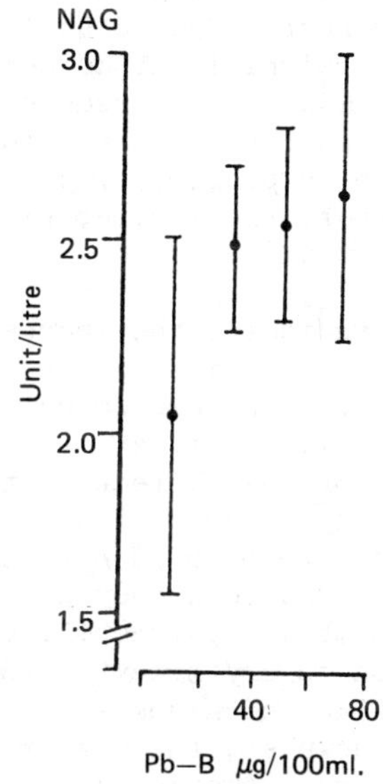

Fig. 1. – Relationship between PbB and NAG.

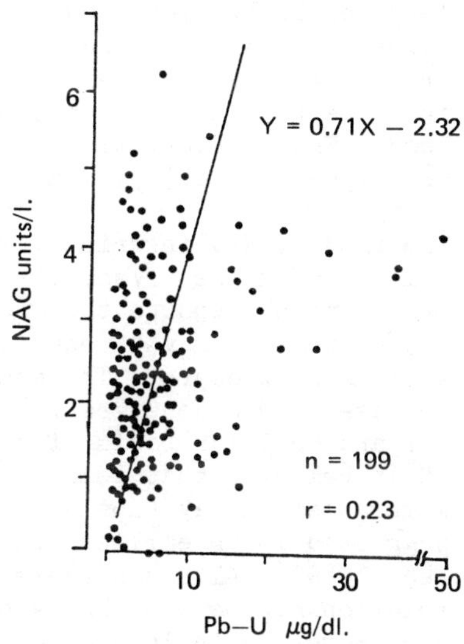

Fig. 2. – Correlation between PbU and NAG.

DISCUSSION

Lead and some other heavy metals can produce acute renal disease(Choie & Richter, 1974). However, the level of occupational and environmental exposure to this toxic substance has been so markedly reduced that exposure to this agent is now a rare cause of renal insufficiency(Weden, 1982). Cramer et al(1975) and Weden(1982) have demonstrated that chronic exposure to lead caused structural changes in renal tissues in workers exposed to lead. Recently Landrigan and his co-workers observed renal impairment in lead workers of PbB between 40 and 80 ug/dl (1983). In a more recent study, an association was seen between urinary enzymes and lead exposure (Meyer et al, 1984).

The assessments of renal function based on creatinine clearance, urinary concentration ability, serum urea and proteinuria have been widely used, but they are technically tedious for screening workers exposed to lead. In addition, several recent studies have shown that changes in serum urea and proteinuria were detected only after higher dose of nephrotoxic substance was given (Price, 1982; Hemino et al, 1985). On the other hand, urinary NAG has been shown in recent years to be a more reliable and useful indicator for renal function(Wellwood et al, 1975; Mansell et al 1978; Hultberg & Ravaskov, 1982).

This study was conducted primarily to evaluate the subclinical changes of renal function in workers exposed to low level of lead. The results suggest significant associations between PbB and SC, PbB and BUN in addition to reduce creatinine clearance with increase in PbB. These findings indicate that kidney function had been possibly impaired by lead. These observations also support the earlier findings of Campbell et al(1977) and Lilis et al(1980).

The rise of NAG in urine (Fig. 2) indicates the increase in release of kidney lysomal enzyme due to lead exposure. This observation supports the view of Meyer and his co-workers (1984) that it may be one of the earliest effect of long-term lead exposure. The mechanism of lead affecting the release of NAG deserves further study. The precise molecular and cellular basis of the enzyme release is not clear. It is believed that lead could be bound to kidney cell membranes and may thus alter the membrane permeability as suggested in an earlier observation on erythrocytes (Ong & Lee, 1980). Although there was a lack of a dose-response relationship between PbB and NAG, however, the higher mean NAG excretion in lead exposed workers (Table 3) suggested the presence of renal impairment in exposed persons. The poor association between PbB and NAG may reflect the fact that PbB is a poor indicator of the renal load of lead.

Further studies are in progress in which workers exposed to different concentrations of lead are being observed. It is hoped that a long-term follow-up of these cohorts will provide more precise information on earlier renal effects of lead exposure.

ACKNOWLEDGEMENTS

We are grateful to B L Lee and L H Chua for their technical help, P C Lim for typing the manuscript. The research project was partly sponsored by International Centre for Medical Research (ICMR) Japan.

REFERENCES

Campbell BC, Beattie AD, Moore MR and Goldberg A (1977). Br. Med. J. 1:482.

Choie & Richter (1974). Lab. Invest. 30, 652.

Cramer K, Goyer RA, Jagenberg R, and Wilson WH (1974). Br. J. Ind. Med. 31, 113.

Himeno S, Watanabe C & Suzuki T (1985). Jap. J. Ind. Hlth. (in press).

Hultberg B, Ravnskov U (1981). Clin. Nephrol. 15, 33.

Landrigan PJ, Baker EL (1983). J. UOEH, 5, 145.

Lilis R, Fischbein A, Valciukas J, Blumberg WE and Selikoff IJ (1980) in Mechanisms of Toxicity and Hazard Evaluation, ed. Holmstedt B, Elseiver.

Lockwood TD, Bosmann HB (1979). Toxicol. Appl. Pharmacol. 49, 323.

Mansell MA, Jones NF, Ziroyanis RN, Marsen WS (1978). Lancet, 2, 803.

Meyer R, Fischbein A, Rosenman K, Lerman E, Drayer E, Reidenberg MM (1984). Am. J. Med. 76, 989.

Morgan JM, Hartley MW and Miller RE (1966). Arch. Intern. Med. 118, 17.

Noto A, Ogawa Y, Mori S, et al. Clin. Chem. 29, 1713.

Ong CN and Lee WR (1980). Br. J. Ind. Med. 37, 78.

Ong CN, Phoon WO, Law HY, Tye CY and Lim HH (1985). Arch. Dis. Child. 60, 756.

Piperno E (1981) in Toxicology of the Kidney, Raven Press, N.Y. pp. 31-35.

Price RG (1982). Toxicology, 23, 99.

Vander AJ, Taylor DL, Kalitis K, Mouw DR and Victery W (1977). Am. J. Physiol. 233, 532.

Weeden RC, Maesaka JK, Weiner B, Lipat GC, Lyons MM, Vitale LF and Joselow, MM (1975). Am. J. Med. 59, 630.

Weden RP (1982). Clin. Exp. Dialysis Apheresis, 6, 113.

Vigano A. Cavanna G, Capodaglio P, Assael BM, Salmona M (1981). Biochem. Med. 25, 26.

Wellwood JM, Eliis BG, Price RG, Hammond K, Thompson AE and Jones NF (1975). Br. Med. J. 3, 408.

World Health Organization (1980). Technical Report Series 647.

Wootton IDP and Freeman H (1982). Microanalysis in Medical Biochemistry, 6th Ed. Churchill Livingston.

40

Studies on the nephrotoxicity of heavy metals in iron and steel industries

G. Triebig, W. Zschiesche, K.H. Schaller, D. Weltle, H. Valentin – Institute of Occupational and Social Medicine, University of Erlangen-Nuremberg (Fed. Rep. Germany)

ABSTRACT

In a cross-sectional field study we examined 230 workers of high alloy steel industries to elucidate possible adverse effects on the kidneys due to long term exposure to various metals. Measurements of external exposure after personal air sampling and of internal exposure by biological monitoring revealed relatively high levels for chromium (air: up to 0.9 mg/m³; urine: up to 79 µg/g creatinine) and nickel (air: up to 6.3 mg/m³; urine up to 65 µg/g creatinine). The assessment of renal effects was based on urinary excretion of total protein (TP), β_2-microglobulin (β_2MG) and the activities of alanine-aminopeptidase (AAP), N-acetyl-β- D-glucosaminidase (NAG), y-glutamyltranspeptidase (y-GT) as well as β-galactosidase (GAL). In comparison to the reference group of 50 healthy men without known exposure to nephrotoxic chemicals, there were no consistent findings indicating a higher prevalence of adverse effects in the occupationally exposed persons. No significant correlations concerning dose-effect/response-relationship were calculated.

INTRODUCTION

It is well known that, depending on kind, concentration and duration of exposure, a renal disorder can occur after acute or chronic intoxication with various metals [1,2]. But there is some lack of information about the nephrotoxicity of heavy metals like chromium and nickel in relatively low concentrations which can be found at workplaces in iron and steel industries. Therefore we studied four groups of workers with different exposure to metals with respect to possible adverse effects of the kidney functions.

METHOD

We investigated 230 men (age 18 to 60 years, median 41 years) of high alloy steel plants and of one plant producing metal-magnets. The workers were employed 7 years in average (range 1 to 32 years). There was an exposure to chromium, nickel and inorganic compounds resp.; in some cases also to other metals like cobalt, copper, vanadium, aluminium, molybdenum and lead.

Regarding the aim of our study and to avoid bias, persons with non-occupational risks of nephropathy were excluded. The criteria were as follows: former renal and/or urological diseases, diabetes mellitus, gout and the intake of analgetic drugs over a longer period. Altogether 74 persons were dropped.
For comparison a reference group of 50 healthy men without renal diseases in the past and without known exposure to nephrotoxic chemicals was investigated.
Regarding the various working conditions and exposure levels the workers were subdivided in the following 4 groups.

Group	All workers	workers without non-occupational risks	Description of of workplace
A	150	107	arc furnace, converters, transport, diecasting and supervision
B	20	13	thermal cutting, hard facing, plasma cutting, arc air gouging
C	56	34	grinding, polishing, fettling
D	4	2	production of metal magnets before sintering

To evaluate the exposure to metals ambient air monitoring (AM) by personal air sampling and biological monitoring (BM) in blood and urine were performed.

The content of chromium and nickel of the filters was measured by electrothermal atomic absorption spectrometry (ET-AAS). For BM the chromium- and nickel-concentrations in post-shift urine specimens were also determined by ET-AAS.

In order to check up kidney functions several biological indicators have been determined (table 1). Furthermore their upper normal limit values and the corresponding morphological-biological basis or the underlying effect are given resp. [3] .

The upper normal limit value corresponds to the 95 percent con-

confidence limits in the reference group.

Table 1: Review of kidney parameters examined.

Paramaeter	Upper normal limit releated to liter	Upper normal limit releated to gram creatinine	effect measured concerns
TP-U	400 mg	250 mg	glomerular membrane
β_2MG-U	400 µg	250 µg	tubular reabsorption
y-GT-U	35 U	25 U	brush border enzyme
APP-U	24 U	12 U	brush border enzyme
β-Gal-U	8 U	5 U	lysosomal enzyme
NAG-U	13 U	10 U	lysosomal enzyme

Total protein was measured according to the Coommassie-method and β_2-MG with an enzyme-immuno-assay. The urinary enzyme activities were determined photometrically according to assays published by MARUHN and JUNG et al. [4, 5].
All concentrations and activities resp. of the kidney parameters in urine were related to liter as well as to gram creatinine in order to compensate actual diuresis variabilities.

STATISTICAL ANALYSES

The results are given as means with standard deviations or ranges. Differences were analysed using the U-Test (Mann-Whitney) or the H-Test (Kruskal-Wallis). Linear regression analyses were performed to examine possible dose-effect/response-relationship.

RESULTS

The results of ambient air monitoring (AM) are shown in table 2, those of biological monitoring (BM) in table 3.

Table 2: External exposure to chromium (calculated as CrO_3) and nickel. Given are mean values and ranges.

Subgroup	Concentrations in air (mg/m³) chromium	nickel
A	0.13 (0.01-0.89)	0.04 (0.01-0.07)
B	0.61 (0.04-2.28)	0.48 (0.04-1.42)
C	0.26 (0.01-1.28)	0.12 (0.01-0.46)
D	-	2.19 (0.13-6.33)

Table 3: Internal exposure to chromium and nickel in urine. Given are mean values and ranges.

Subgroup	Concentrations in urine (µg/g creatinine) chromium	nickel
A	2.1 (0.1-17.7)	9.2 (0.9-89.3)
B	11.7 (0.1-78.9)	32.2 (1.5-97.4)
C	3.2 (0.1-27.2)	6.1 (0.5-28.2)
D	no exposure	21.8 (2.1-64.8)

Regression analyses considering age and duration of exposure as well as urinary proteins and enzyme activities revealed no significant correlations indicating age/exposure-effect/response-relationships. The comparison of enzyme activities in urine between smokers and non-smokers as well as between persons with an admitted alcohol-consumption of less or more than 60 g per day showed no consistent significant differences. There is just a higher excretion of AAP in urine of smokers than of non-smokers, only by relation to liter, but not to creatinine.

In table 4 the prevalence excess rates of the upper normal limits of the kidney parameters are given for all persons examined as well as for men without non-occupational risks of nephropathy. The eximation is based on creatinine related values. The results related to liter urine are alike.

Table 4: Percental prevalence rates of abnormal findings of kidney parameters.

Parameter	prevalence rates of abnormal findings in % All men (N = 230)	Men without non-occupational risks (N = 156)
TP-U	0.3 (N = 229)	0.7 (N = 143)
β2-MG-U	7.3 (N = 201)	6.7 (N = 134)
GT-U	8.0 (N = 213)	3.5 (N = 142)
AAP-U	19.2 (N = 123)	30.9 (N = 94)
GAL-U	9.7 (N = 196)	0.8 (N = 133)
NAG-U	4.9 (N = 223)	1.4 (N = 139)

DISCUSSION

As the results of AM and BM demonstrate, there is a considerable variation of exposure to chromium and/or nickel at the workplaces examined. In the Federal Republic of Germany there exist a threshold limit value or a recommmendation for these metals:
Nickel in air: 0.5 mg/m^3, and in urine 25 µg/l; chromium (VI) in air: 0.1 mg/m^3, and in urine 30 µg/l. The upper normal limits in urine are somewhat lower: 3 µg Cr/l and 4.0 µg Ni/l. With respect to these values in urine, usually there is a tolerable exposure in the average, except for nickel in group B (thermal cutters, welders, etc.).
The prevalence rates of abnormal enzyme activities for all workers ranged between 0.7 % and 6.7 %, only for AAP there is a considerable higher rate of 30.9 %. Because of possible confounding factors (bacterial contamination of urine samples) so far the question cannot be answered at present, if this result actually indicates an adverse effect due to metal exposures. But in summary there are no significant and/or consistent findings in urinary enzyme excretion of workers that could indicate adverse effects on the kidney functions caused by metal exposure found in high alloy steel industry.

Further studies are recommended to substantiate our findings. For this reason, a longitudinal study is in progress in our working group.

ACKNOWLEDGEMENT

This work was supported by grants of the Commission of the European Communities (CEC), Contract No. 7 248-23-006.

REFERENCES

1 Meyer, B.R., A. Fischbein, K. Rosenman, Y. Lerman, D.E. Drayer, M.M. Reidenberg: Increased Urinary Enzyme Excretion in Workers Exposed to Nephrotoxic Chemicals, Am.J. Med. 76, 989-998 (1984)

2 Landrigan, P.J., R.A. Goyer, T.W. Clarkson, D.P. Sandler, H.J. Smith, M.J. Thun, R.P. Wedeen: The Work-Relatedness of Renal Disease, Arch. Environ. Health 11, 83-90 (1985)

3 Price, R.G.: Urinary enzymes, nephrotoxicity and renal disease, Toxicology 23, 99-134 (1982).

4 Maruhn, D.: Rapid colorimetric assy of β-galactosidase and N-acetyl-β-D-glucosaminidase in Human Urine, Clin.Chim. Acta 73, 453-461 (1976).

5 Jung, K. and D. Scholz: An Optimized Assay of Alanine Aminopeptidase Acitivity in Urine, Clin. Chem. 26. 1251 (1980).

Investigations on renal impairments of high alloy steel welders

W. Zschiesche, G. Emmerling, K.H. Schaller, D. Weltle – Institute of Occupational and Social Medicine, University of Erlangen-Nuremberg, Federal Republic of Germany

ABSTRACT

In a cross-sectional field study we examined 131 high alloy steel welders from 19 welding shops. Measurements of external exposure after personal air sampling and of internal exposure by biological monitoring show levels for chromium in air up to 3.3 mg/m^3 and in urine up to 84.0 µg/g creatinine, for nickel in air up to 0.5 mg/m^3 and in urine up to 47.3 µg/g creatinine. The assessment of renal effects was based on urinary excretion of total protein (TP) and the activities of N-acetyl-β-D-glucosaminidase (NAG) and alanine-aminopeptidase (AAP). In comparison to an age-matched reference group there were no consistent findings indicating a higher prevalence of adverse effects on the kidneys in the occupationally exposed persons. There might be a slight positive correlation of kidney impairments with increasing age and time of exposure.

INTRODUCTION

In the last few years some publications showed an increased risk of renal impairments due to heavy metals like cadmium, mercury, etc. But there is still a controversial discussion, whether also chromium (Cr) and nickel (Ni) can affect the kidneys. One of the groups at workplaces with the haviest industrial exposure to these metals are high alloy steel welders. Nevertheless only little data have been published so far on the question, if this group is subject to a major risk of renal dysfunctions.

These data do not show consistent findings of adverse effects of high alloy steel welding fumes on the kidneys [1, 2, 3]. On the other hand these studies partly investigated only small numbers of welders or did not consider the different kinds of welding processes.

METHODS

Therefore we investigated 131 high alloy steel welders and the same number of age-matched (+ 5 years) controls, who were not occupationally exposed to Cr and/or Ni or other substances with a possible kidney affection (e.g. lockmakers, electricans etc.) in 19 factories.
For the present study all persons were excluded who had been suffering from nephrological, urological or internal diseases in the past, which could have affected the kidneys as well as those who declared any consumption of analgetic drugs. Therefore we dropped 13 welders. The number as well as distribution of age and duration of exposure of this collective divided into several subgroups according to the welding processes mainly applied are shown in table 1.

Table 1: Number, age and time of exposure of the welders according to the various welding processes. Median values and ranges.

and ranges.

Group	Main welding process	N	Age	Duration of exposure (years)
I	Manual metal arc welding with stick electrodes (MMA)	51	38 (21-58)	10 (1-33)
II	Shielded gas welding with wires (MAG, MIG)	45	37 (22-55)	9 (3-25)
III	Thermal cutters and sprayers	6	33,5 (20-58)	8 (1-25)
IV	Processes with low exposure (plasma cutting under water, tungsten inert gas welding, submerged arc welding)	16	35 (24-56)	9,5 (2-30)
V	Reference Group	107	37 (21-61)	16 (1-40)

The external exposure of these welders to total fume, Cr and Ni was estimated by personal air sampling behind the welding helmets and masks respectively.
The internal exposure of the welders and controls was assessed by biological monitoring of Cr and Ni in spot urine and blood specimens taken before and after shift of a working day at the end of a working week.
We investigated the kidney function of the welders and the reference group by measuring the excretion of total protein (TP), N-acetyl-β-D-glucosaminidase (NAG) and alanineaminopeptidase (AAP) in the morning urine specimens. All analytical methods are given in the last chapter.

STATISTICAL ANALYSES

The results are given as median values with the central 68 % ranges. Differences have been analysed with the U-test by MANN and WHITNEY and the H-test by KRUSKAL and WALLIS respectively. Linear regression analysis was performed to investigate possible dose-effect/response-relationships of the urinary parameters with age and time of exposure to high alloy steel welding fumes.

RESULTS

So far 47 filters of dust samples have been analysed. The Cr and Ni concentrations in the air are shown in table 2. In general the external exposure to Cr and Ni is considerable. The percental content of Cr VI in welding fumes varies between the different welding processes and consumables applied. Our results are in good agreement with the data given in the international literature. These show Cr VI concentrations of 30 - 90 % of total Cr in fumes of coated stick electrodes (group I) and of 0 - 4 % in fumes of blank wires for metal gas shielded arc welding (group II). In thermal cutting and spraying no Cr VI was detected.

Table 2: External exposure of welders to total fume, Cr and Ni. Median values and central 68% ranges.

Group	Total Fume (mg/m^3)	Total Chromium* ($\mu g/m^3$)	Nickel ($\mu g/m^3$)
I	2.3	79.5	18.5
(N=21)	(1.2 - 6.2)	(11.3 - 258.5)	(2.3 - 70.6)
II	3.4	350.5	66.3
(N=23)	(1.1 - 7.2)	(106.5 - 491.3)	(22.5 -135.2)
III	1.7	23.4	29.4
(N=3)	(1.3 -11.5)	(9.7 - 683.3)	(15.5-223.6)

*calculated as CrO_3

So far the urine specimens of 49 welders and the same number of controls have been analysed. The post-shift urine samples show significantly higher Cr and Ni concentrations in the welders' group than in the reference group. The results of the controls mainly are within the normal range. These data are given in table 3.

Table 3: Concentrations of Cr and Ni in post-shift urines of high alloy steel welders. Median values and central 68% ranges.

Group	Cr-U		Ni-U	
	µg/l	µg/g Creat	µg/l	µg/g Creat
I	11.6	6.7	7.5	4.5
(N=21)	(2.4-42.0)	(2.2-34.8)	(2.5-15.0)	(2.5-12.0)
II	6.3	4.9	11.2	8.1
(N=25)	(3.4-21.9)	(3.2-10.9)	(4.1-28.0)	(3.5-21.4)
III	4.5	2.7	24.0	13.4
(N=3)	(4.1- 8.6)	(2.3- 4.5)	(14.6-24.2)	(9.5-14.6)
V	1.6	1.4	2.3	1.8
(N=45)	(0.8- 3.4)	(0.8- 2.4)	(1.2- 5.1)	(1.1- 5.0)

RENAL PARAMETERS

The percent excess rates of the upper normal limits are shown in table 4.

Table 4: Percental prevalence rates of abnormal findings of renal parameters in welders. UNL = upper normal limit: upper 95% value of the reference group.

Group	NAG-U		AAP-U		TP-U	
	U/l	U/g Creat	U/l	U/g Creat	mg/l	mg/g Creat
I (N=51)	7.8	11.8	5.9	5.9	9.8	7.8
II (N=45)	6.7	6.7	9.9	9.9	6.7	6.7
III (N=6)	16.7	0	0	0	0	16.7
IV (N=16)	12.5	6.2	6.2	6.2	0	6.2
Σ N=118	9.3	8.5	6.8	6.8	6.8	7.6
UNL	11.6	8.4	14.1	10.2	120	79

There are no consistent findings of all parameters neither when related to liter urine nor to gram creatinine nor to a specific gravitiy of 1,018 that would indicate a major renal risk of the welders compared to the reference group. Also the comparison of the distribution of the values revealed no significant differences ($p < 0.05$) for all parameters between all welders and controls as well as between the various subgroups of welders. Also the smoking habits did not influence the results significantly in most of the parameters.
The regression analyses show an increasing excretion of NAG, AAP and total protein with rising age and time of exposure to high alloy steel welding fumes. But these results only stand for NAG - and total protein-excretion related to creatinine and for AAP related to the specific gravity. Furthermore, the coefficients of correlations are small within a range of 0.19 - 0.25.

DISCUSSION

The results of the ambient and the biological monitoring which are available so far show that the exposure to Cr and Ni in general is relatively low compared to the findings in other studies [4]. The German TLV´s for Cr VI and Ni in the air are mainly not exceeded. In biological monitoring usually the upper normal limits of Cr (3 µg/l) and Ni (4 µg/l) in urine are exceeded, but not the values of 30 and 25 µg/l respectively that are discussed as industrially tolerable in the FRG. Under these circumstances the renal parameters show no significant influence of welding fumes on the kidney functions. The various welding processes do not cause different effects on the kidney functions. Especially Cr VI does not seem to influence the tubular or the glomerular cells in another way than the other oxidation stages of Cr.
On the other hand the results show a slight positive correlation of the enzyme and protein excretion with age and duration

of exposure to high alloy steel welding fumes.
So there might be a possible cumulative effect of Cr and/or Ni or an adverse effect due to the age of the welders for long term exposure. Also synergistic effects could be discussed.

But we have to realize that all these results can only be shown when the protein or enzyme excretion is related to particular reference parameters.

In summary there are no consistent findings that would prove an adverse effect of high alloy steel welding fumes on the kidney functions.
Further studies on welders with higher exposure to Cr and Ni should be performed.

ACKNOWLEDGEMENT

This study was supported by grants of the Commission of the European Communities (CEC), Luxembourg (Contract No. 7 248-11-023).

The study was performed in cooperation with the German Welding Society (Deutscher Verband für Schweißtechnik e.V.).

REFERENCES

1 Littorin, M., M. Welinder, B. Hultberg: Kidney Function in Stainless Steel Welders, Int. Arch. Occup. Health 53, 279-282 (1984).

2 Lindquist, B.: Kidney Diseases in Welders, IRCS Med. Sci. 11, 99 (1983).

3 Mutti, A., A. Cavatorta, C. Pedroni, A. Borghi, C. Giaroli, I. Franchini: The Role of Chromium Accumulation in the Relationship between Airborne and Urinary Chromium in Welders, Int. Arch. Occup. Health 43, 123-133 (1979)

4 Sjögren, B:, L. Hedström. U. Ulfvarson: Urine Chromium as an Estimator of Air Exposure to Stainless Steel Welding Fumes, Int. Arch. Occup. Health 51, 347-354 (1983).

42

Assessment of renal function in lead poisoned workers

G. Maranelli, P. Apostoli – Istituto di Medicina del Lavoro, Università Degli Studi di Verona

INTRODUCTION

The association between lead and kidney disease has been known since the end of the past century: clinical and epidemiological reports are at present available to confirm the nephrotoxicity of this metal (1). Renal function disturbances, characterized by aminoaciduria, glucosuria and hyperphosphaturia were evidenced in lead-poisoned children (2). Ultrastructural changes in the proximal tubular cells and tubular dysfunction have been reported in particular in acute or subacute lead poisonings (3). Epidemiologic surveys in lead workers, on the contrary, have provided equivocal evidence of occupational lead nephropathy (4). Whether long-term exposures, similar to those found in occupational exposure, are associated with occurence of chronic irreversible nephropathy is a more controversial question. Furthermore, little is known about dose/effect relationship between lead absorption-effect indicators and kidney damage indicators and whether the impairment of the renal function is reversible.

At present the diagnosis of lead nephropathy may be made by evaluation of indicators of tubular cells damage (urinary enzymes, aminoaciduria and proteinuria) or evaluating the traditional routine renal function tests such as blood urea nitrogen , serum creatinine, serum uric acid and the related clearances.

This research has been carried out in 60 lead poisoned workers to put in evidence some aspects of renal function by verifying the renal tests generally used in clinical practice.

SUBJECTS AND METHODS

60 workers, admitted into our Institute between 1979 and 1985 with diagnosis of "lead poisoning", were studied. They were been employed in various manufactures (pottery, pewter, battery, lead smeltery, plastics). The time after cessation of exposure, the number of acute-subacute intoxications and the chelation therapy periods were extremely variable and, in some cases , difficult to evaluate.

A control group was selected among the patients hospitalized in the same period because of respiratory diseases and certainly not exposed to lead.

Lead in blood (PbB) and in urine after chelation (PbU EDTA) was determined by atomic absorption spectrophotometry; chelatable lead was measured in urine samples collected in the 24 hours after I.V. administration of 1 g of calcium disodium edetate. Erythrocyte zinc protoporphyrin (ZPP) was measured using an haematofluorimeter and urinary δ-aminolaevulinic acid (ALA-U) with the method of Grisler Griffini.

Blood urea nitrogen (BUN), serum-creatinine and serum-uric acid were determined by enzymatic method using a Technicon Auto-analyzer ; renal clearances were evaluated in 24 hours urine samples.The laboratory "normal values" were the following: BUN 8-22 mg/100 mL; S-creatinine: 0.4-1.2 mg/100mL; S-uric acid: 2.5-7 mg/100 mL. For the clearances the lower normal limit were: BUN 40 ml/min; Creatinine 80 ml/min; Uric Acid 7 ml/min.

The statistical analysis on results was carried out by the "t" test and the linear regression test.

RESULTS AND DISCUSSION

Table 1 shows age, exposure length and values of lead absorption and effect indicators in 60 lead patients. The comparison of the renal function tests of the control group and the three clearances values, with "normal values", are reported in tables 2-3. The mean values of BUN and S-uric acid and the percentage of values above the upper normal limit of the three considered tests are higher in lead poisoned individuals ($p<0.01$ at "t" test for mean values). BUN, serum uric acid, and related clearances appear to be the more frequently affected parameters. No significant differences in renal function tests were found in two subgroups of patients with different lead body-burden: in the group with edetate-mobilization test above 3500 µg/24 hours, BUN and uric acid are slightly higher (table 4). A comparison of lead absorption-effect indicators in the workers subgroups, having respectively BUN and s-uric acid above or below the upper normal limit is reported in tab.5 and in tab.6. Patients with outlying values of BUN have significantly higher values of an indicator of effect (ZPP); hyperuricemic patients have higher both an effect indicator (ALA-U), and body burden indicator ($p<0.05$). The linear regression analysis between renal function tests and lead absorption or effect indicators shows correlation only between BUN and PbU-EDTA ($r=0.27;p<0.01$) and between BUN and ZPP ($r=0.38;p<0.005$).

Table 1: Age, exposure length, and biological indicators in 60 lead poisoned workers.

	mean±sd	range
AGE (years)	39.5±10.0	18-57
EXPOSURE LENGTH (years)	10.8± 8.0	1-34
PbB (μg/100mL)	71.9±16.5	50-128
PbU-EDTA (μg/24 hours)	3375±2737	1050-15000
ZPP (μg/g Hb)	12.8± 6.9	2.5-29
ALA-U (mg/24 hours)	1.4±1.2	0.4-6.16

Table 2: Results of renal function tests (bun, s-creatinine and s-uric acid in m mg/100 ml) in 60 lead poisoned workers and in 76 not exposed individuals (controls).

	LEAD POISONED (n=60)		CONTROLS (n=76)	
	mean ± s.d.	outlying values	mean ± s.d.	outlying values
BUN	18.9 ± 5.12*	20%	16.7 ± 3.2	3.9%
S-CREATININE	1.00 ± 0.18	11.5%	0.96 ± 0.15	7.9%
S-URIC ACID	6.00 ± 1.45*	20%	5.5 ± 1.1	10.5%

(*)= $p<0.01$

Table 3: Results of renal clearance of bun, creatinine, and uric acid in 60 lead poisoned workers.

TEST	mean ± s.d.	normal values	outlying values
BUN- CLEARANCE	42.9 ± 20.1	> 40	45%
CREATININE-CLEARANCE	94.9 ± 28.9	> 80	35%
URIC ACID-CLEARANCE	5.8 ± 3.2	> 7	68%

Table 4: Renal function tests in two groups of patients with PbU-EDTA above or below 3500 mg/24 h.

	PbU-EDTA (mg/24 h)	
	3500	3500
	n=41	n=19
BUN mg/dl	18.3±5.2	20.3±4.6
(values $\geq$ 23)	19.5%	21%
S-CREATININE mg/dl	0.99±0.16	0.99±0.21
(values $\geq$ 1.3)	9.7%	15.8%
S-URIC ACID mg/dl	5.9±1.4	6.2±1.6
(values $>$ 7)	14.6%	31.6%

Table 5: Biological indicators of lead poisoning in patients with BUN values above or below the upper normal limit (23 mg/100mL).

	BUN	
	$<$ 23	$\geq$ 23
PbB µg/100mL	72.4±16.8	70.8±16.1
PbU-EDTA µg/24h	3273±2311	3975±4010
ZPP µg/g Hb	11.7±6.8	17.1±6.3*
ALA-U mg/24h	1.4±1.1	1.8±1.8

(*)= $p<0.01$

Table 6: Biological indicators of lead poisoning in two groups of patients with normal or abnormal values of serum uric acid (upper limit: 7 mg/100mL).

	S-URIC ACID (mg/100mL)	
	$\leq$ 7	$>$ 7
PbB µg/100mL	70.4±13.2	77.5±26
PbU-EDTA µg/24h	3109±2469	4541±3340*
ZPP µg/g Hb	12.8±7.0	12.9±6.9
ALA-U mg/24h	1.3±1.1	2.1±1.7*

(*)= $p<0.05$

CONCLUSION

In recent years mortality studies of lead workers have consistently reported excess deaths from "chronic or unspecified nephritis", particularly in heavily or long term exposed workers (3). However in the literature the renal function tests, and the filtration tests in particular, are not reported to be strictly correlated with lead exposure or effect indicators. In fact, some authors, have reported a decrease in glomerular filtration rate (4), others an increase of BUN and serum uric acid (5), others, at least, normal or not lead-correlated findings (6). The degree and statistical significance of functional disturbances are variable and difficult to evaluate, due to lead exposure extent in the examined group and to the analytical and statistical methods employed. In lead poisoned workers examined in the present research a moderate renal function impairment is confirmed by comparison of BUN and serum uric acid mean values with those of the control group. A filtration decrease, may be seen by analyzing creatinine and BUN clearances data, even if for these functional parameters, no control data except the "normal laboratory values" exist. The uric acid clearance appears more altered: this seems to be due to both disturbances of the glomerular filtration and altered tubular transport of uric acid. Our data confirm the lack of a strict correlation between renal damage and values of lead absorption or effect indicators. In clinical casistics, like that presented in this study, important factors which should be probably taken into consideration, are previous episodes of acute or subacute poisoning and previous chelation therapies.

REFERENCES

1. BENNETT W.M : Lead nephropathy. Kidney Int. 1985,28:212-220
2. WEDEEN R.P.-MAESAKA J.K.-WEINER B. et al.: Occupational lead nephropathy. Am.J.Med. 1975,59:630-641.
3 COOPER W.C.-WONG O.-KHEIFETS L.: Mortality among employees of battery plants and lead producing plants,1947-1980. Scand.J.Work Environ.Health 1985,11:331-345.
4. WEDEEN R.P.-MALLIK D.k.-BATUMAN V.: Detection and treatment of occupational lead nephropathy. Arch.Intern.Med. 1979,139:53-57.
5 CAMPBELL B.C.-BEATTIE A.D.-MOORE M.R. et al.:Renal insufficiency associated with excessive lead exposure. Br.Med.J. 1977,I:482-485.
6. POCOCK S.J.-SHAPER A. et al.: Blood lead concentration, blood pressure, and renal function.Br.Med.J.1984, 289:872-4

43

Biological monitoring of silver exposure in gold casting process

C. Minoia, G. Catenacci – Fond. Clinica del Lavoro, Università di Pavia
M.C. Oppezzo – USSL 71 Valenza (AL)

ABSTRACT

The environmental levels of silver and other metals in were determined the particulate fumes produced during gold casting processes. Ag levels in urine of the exposed subjects in jewelry handicraft are also reported. The significance of AgU as an exposure indicator and the relationship between environmental and urinary concentrations of silver are discussed.

INTRODUCTION

In jewelry handicraft the preparation of coloured golds (gold binary or ternary alloys with silver, copper, nickel, zinc and cadmium) is performed by using investment casting. During these processes the crucible is directly heated with an oxy-acetylene flame or alternatively by using an electromagnetic induction oven (1). The high temperature (about 1000° C) may generate particulate fumes that contain appreciable quantities of metals, determining an inhalation risk for the exposed subjects (the size of the inhaled particles is very low). During casting of "red" or "yellow" gold, silver, gold and copper are the prevailing contaminants present in the working area, whereas during the preparation of "white" gold zinc and nickel are also present in the fumes. Occasionally a binary alloy (gold and cadmium) was used. In this study the environmental levels of metal present in the particulate fumes during the investment casting process were determined.

In order to clarify the utility of biological monitoring, Ag levels in urine of the exposed workers were also determined. The significance of AgU as an indicator of exposure and the relationship between exposure level and metal concentration in urine are here discussed.

MATERIALS AND METHODS

During investment casting about 0.5 kg of pure gold was transferred to the crucible and various metal pieces were added in order to obtain the required alloy (e.g., for preparation of "red gold" the alloy was composed of Au 75%, Ag and Cu 25%).

By using an oxy-acetylene flame (or alternatively an electromagnetic induction oven) the metal fragments were directly heated with a melting temperature of around 950°C, varying from 900°C to 1000°C.

When the casting was completed, the crucible was centrifuged and the melted alloy transferred into a metallic cylinder where jewelry articles are obtained by means of wax models.

The average time required for a single casting process is about 15 min., varying from 5 to 15 min.

During this operation the operator works at a distance of 1 to 2 meters from the melting crucible.

Subjects. Six subjects (all males, age ranging from 28 to 51 years) employed in jewelry handicraft were selected for this study; they had been employed in 3 different factories for 7 to 32 years. No albuminuria was found in any of the subjects and no subject had ever suffered from renal disease.

Environmental sampling. The air in the breathing zone of the exposed workers was sampled during five 8-hour shifts. Four samples per shift were collected, each over a period of about 2 hours. The content of the various metals in the particulate fumes was determined by ETA-AAS or ICP-AES (2).

Biological monitoring. Urine samples were taken during a working week. All urinary samples collected for every monitored subject were analyzed for specific gravity and the volume excreted at the time of collection was measured. The determination of Ag in urine was performed by using ETA-AAS after separating the element by liquid anion exchanger (Amberlite LA-2, Bdh) solubilized in MIBK (3-4-5-6-7-8).

RESULTS

Environmental monitoring. The comparison of the data obtained in different environmental conditions is shown in Table 1, where for each investment process the average concentration of metals, before and during casting, is listed.

Biological Monitoring. The urinary Ag concentrations (μg) in before/after shift samples of three subjects working on gold investment casting performed by oxy-acetylene flame, over five consecutive days of sampling are reported in Table 2.

The daily urinary Ag excretion (μg/24 h) of the same workers is listed in Table 3.

The average data of the Ag urinary excretion (before/after shift samples) of all the subjects under study are reported in Table 4.

Table 1: Average environmental concentrations ($\mu g/m^3$) of metals before and during gold investment casting.

Casters		Before casting ($\mu g/m^3$)				During casting ($\mu g/m^3$)				
	Ag	Au	Cu	Ni	Zn	Ag	Au	Cu	Ni	Zn
Oxy-acet.5 flame	1.8 (n88)	0.85 (n88)	0.3 (n88)	0.09 (n28)	0.6 (n56)	318 (n10)	35 (n10)	7.2 (n10)	0.9 (n10)	9.5 (n10)
Electric 1 oven	0.45 (n20)	0.00 (n20)	0.20 (n20)	0.04 (n20)	0.50 (n20)	5.1 (n12)	3.8 (n12)	1.0 (n12)	0.08 (n12)	0.93 (n12)

n = number of samples.

Table 2: Ag(μg) in before/after shift samples of 3 subjects working on gold casting with oxy-acetylene flame on 5 consecutive days.

	AgU μg									
Subjects	1st day		2nd day(c)		3rd day		4th day(c)		5th day	
	*	**	*	**	*	**	*	**	*	**
N.D.	21.4	16	33.7	37.2	38.4	3.7	0.8	0.9	7.3	237.3
N.U.	20.3	10.5	2.4	58.4	47.4	0.9	1.0	18.2	3.2	2.7
G.G.	27.2	5.9	2.0	36.5	43.2	0.9	1.7	2.0	0.8	0.5

* = before shift;
** = after shift;
(c) = casting 2 hours.

Table 3: AgU μg/24 h, on 5 consecutive days, of 3 subjects working on gold casting with oxy-acetylene flame.

	AgU μg/24 h				
Subjects	1st day	2nd day(c)	3rd day	4th day(c)	5th day
N.D.	61.4	102.6	104.6	14.9	261.8
N.U.	57.4	58.2	93.8	20.8	7.7
G.G.	38.6	148.9	47.9	5.2	95.9

(c) = casting 2 hours.

Table 4: Average data of AgU (before/after shift) in 6 subjects working on gold casting processes, on 5 consecutive days.

Casting	Subjects	AgU µg($\bar{x}\pm SD$) Before shift	After Shift
oxy-acet.flame	5	16.8±6.3(n25)	26.7±19.0(n25)
el. induct.oven	1	4.5±1.8(n 5)	5.1± 2.3(n 5)

n = number of urinary samples.

DISCUSSION

The results clearly confirm that Ag is the prevalent environmental contaminant during the investment casting process when gold melting is performed using oxy-acetylene flame (the environmental Ag concentration ranged from 0.27 to 0.60 mg/m^3). Even when Ag is the prevalent metal in the environmental particulate, the melting fumes contain appreciable amounts of Au and small amounts of Cu and Zn. Only the Ni concentration was negligible and always below than 1 µg/m^3.
However, when the electro-magnetic induction heating process is used, the environmental pollution by Ag and other metals is considerably lower.
This is confirmed by the variations in the relative biological indicators. While in the general population the reference AgU value is 0.40±0.2 µg/l (9), in workers employed on investment casting performed with the oxy-acetylene flame, AgU ranged between 5 and 261 µg/24 h. The maximum urinary Ag excretion is reached 24-36 hours after exposure.

REFERENCES

1. C. Vitiello: Creficeria Moderna, Hoepli ed., Trento, 1983.
2. C. Minoia, M.C. Oppezzo, G. Micoli: Atti XXI Seminario. Spettrochimico AIM - V Riunione Ital. Francese di Spettroscopia Atomica, Frascati (Roma) 4-7/6/1985.
3. C. Minoia, E. Camurati, A. Mazzuccotelli, L. Pozzoli: Boll. Chim. Ig. 35, 453, 1984.

4. A. Mazzuccotelli, C. Minoia, G. Micoli, M. Colli, S. Angeleri, P. Colecchi, M. Esposito: Abstracts XXI Int. Congr. Occ. Health-Dublin (Ireland) 1984, p. 611.
5. C. Minoia, A. Cavalleri: Determinazione dei metalli in tracce nel laboratorio clinico e tossicologico. ed. La Goliardica Pavese, Pavia 1985.
6. S. Meret, R.I. Henkin: Clin. Chem. 7, 369, 1971.
7. S. Nomoto, F.W. Sunderman Jr., Clin. Chem. 16, 477, 1970.
8. N.S. Viera, J.W. Hansen: Clin. Chem. 27, 73, 1981.
9. C. Minoia, S. Angeleri, A. Mazzuccotelli, M. Colli, A. Cavalleri: Abstracts XXI Int. Congr. Occ. Health-Dublin (Ireland) 1984, p. 610.

44

Biological monitoring of workers occupationally exposed to cadmium fumes in São Paulo , Brazil

H.V. Della Rosa*, J.R. Gomes, F.R. Pivetta†, S.C. Jacob†, R. Mazon†† –**
*Univ. de S. Paulo (USP) – Fac. Cienc. Farmac. – S. de Toxicologia;
**USP – Fac. Saúde Pública; †INCQS – Fundaçao Oswaldo Cruz;
††Ass. Bras. Prev. Acid. (ABPA).

INTRODUCTION

Cadmium (Cd) and its compounds have been increasingly used in industrialized countries in the present century. In Brazil, Cadmium production and consumption have increased significantly.

From the point of view of toxicology, Cadmium acquired importance in the 1940's in Japan with the episode of Itai-Itai disease producing a great social, economical and political impact. Cases of Cadmium intoxication due to occupational exposure have been reported since 1858 (Niosh, 1976).

Although there is no information about the aproximate number of individuals occupationally exposed in Brazil, this number is surely significant, specially because of the increasing industrial consumption of Cadmium.

Since the toxic effects caused by Cadmium are multiple and very serious, they must be studied very carefully in order to obtain better knowledge of their mechanism of action, so that any biological change can be detected before clinical symptoms appear.

The kidney is the target organ for Cadmium in the organism and proteinuria is the most characteristic sign of a nephrotoxic effect. It was this fact that prompted us to evaluate the problem in Sao Paulo, Brazil.

There are many parameters available related to renal function to evaluate occupational exposure to Cadmium, neverthless, not all of them are biological indicators recommended for the early detection of biological changes. Among these indicators, Beta$_2$-microglobulin (β_2-M) and retinol binding protein (RBP) in urine are so far, considered the most sensitive tests for detecting an early tubular damage in workers occupationally exposed to Cadmium (Alessio et al., 1983).

The interest in evaluating RBP is due to the fact that this protein, compared to β_2-M, possesses more chemical stability in relation to urine pH variations in the bladder, which makes evaluation of tubular function with this test more accurate because the losses due to chemical degradation in vesical pH and body temperature are not significant (Bernard & Lauwerys, 1981). Therefore this protein can become an adequate biological indicator of occupational exposure to Cadmium.

MATERIAL AND METHODS

A total of 23 workers were examined from a metallurgical plant located in Sao Paulo, Brazil, which is developing a program for improving the working environment. These workers were divided into four groups, according to their exposure potential.

Group I - 3 gas Cadmium alloy melting furnace operators (Area I).

Group II - 3 workers in an adjacent area that was connoted to the furnace room. One of the workers, the one in charge, stayed in the room most of the time, while the other two moved around (Area II).

Group III - 8 workers in another adjacent area that was connected with the furnace room (Area III).

Group IV - 9 workers from the same plant but not occupationally exposed (control group).

All the workers in groups I, II and III had been exposed for at least 4 years.

Cadmium in air (Cd-A) was determined on samples collected on membrane filters using personal samplers by means of flame atomic absorption spectrophotometry as recommended by Niosh (1976).

Blood and urine samples were collected at the end of a shift and Cd-B and Cd-U were determined by graphite-furnace atomic absorption spectrophotometry (GF-AAS), according to the method described by Perkin-Elmer (1977). Urinary retinol binding protein (RBP-U) concentration was determined by radial immunodiffusion in LC-partigen-retinol binding protein plates from the Behring Institute. Pre-shift urine samples were previously dialized and concentrated using a Savant Speed Vac concentrator (300 times).
Results were analysed using mann Whitney's U-test for group differences.

RESULTS

Cd-A concentrations are shown in Table 1 and Cd-B, Cd-U and RBP-U concentrations found in exposed and non-exposed (control) workers are shown in Table 2.

Table 1: Cadmium concentrations in air (Cd-A) in the working environment.

AREA	SAMPLES	MEAN (Cd) mg/m^3	S.D.
I	7	0.388	0.233
II	3	0.078	0.020
III	3	0.043	0.024

Table 2: Cadmium in blood (Cd-B) and in urine (Cd-U) and urinary retinol binding protein (RBP-U) in samples from exposed and non-exposed workers.

GROUP	WORKERS	Cd-B µg/100 ml	Cd-U µg/l	RBP-U µg/l
I	1	5.40	329.4	264.3
	2	8.26	688.7	530.0
	3	9.08	206.1	177.1
	MEAN	7.58±1.93	408.06±250.73	323.8±183.81
II	4	1.79	49.6	38
	5	1.91	64.3	272.6
	6	0.55	10.4	39.5
	MEAN	1.42±0.75	41.43± 27.86	116.7±135.01
III	7	0.31	2.3	80.2
	8	0.46	4.5	84.0
	9	3.26	105.7	42.7
	10	0.28	10.9	62.2
	11	0.66	4.7	134.2
	12	0.47	13.2	148.8
	13	0.57	4.6	120.8
	14	0.46	7.5	50.2
	MEAN	0.81±1.00	19.17± 35.15	90.38± 39.79
IV (CONTROL)	15	BDL	0.7	23.3
	16	BDL	0.4	15.7
	17	BDL	1.9	14.8
	18	BDL	0.8	47.4
	19	BDL	2.1	86.7
	20	BDL	1.4	69.6
	21	BDL	1.4	34.1
	22	BDL	0.6	39.3
	23	BDL	2.0	BDL
	MEAN	-	1.25±0.65	41.36± 25.73

Detection limit for GF-AAS: Cd-B 0.005 µg/100 ml
Cd-U 0.070 µg/l

Cd-U and RBP-U concentrations were adjusted to specific gravity 1.024.

BDL = below detection limit.

The graphic representation of the data from Table 2 are shown in Figure 1.

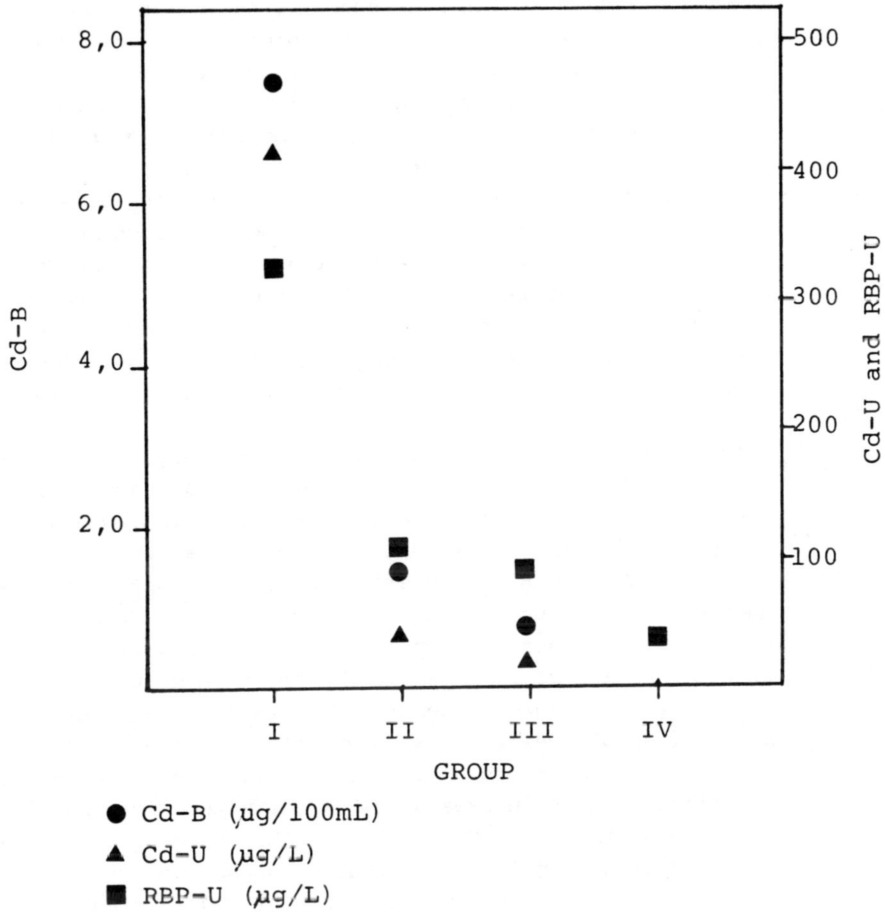

Fig. 1. – Mean (Cd-B) (Cd-U) and (RBP-U) concentrations in exposed and control workers.

DISCUSSION AND CONCLUSION

Cd-B concentrations were significantly different in the three working environments studied. In two of the three areas Cd-A levels were above the TWA of 0.05 mg/m^3 fixed by A.C.G.I.H.

(1984) for Cadmium fumes. In area I (furnace room) the observed value is compatible with the occurrence of renal dysfunction and respiratory alterations.

Cd-B values were also significantly different in all groups examined. All the workers in group I, as well as two (66.6%) in group II had Cd-B values above the biological limit value (BLV) of 1 μg/100 ml (W.H.O, 1980). Only one worker (12.5%) from group III had Cd-B value above the limit. This can be explained by the fact that the worker's job included frequent visits to the furnace room and higher exposure to Cd fumes.

Cd-B values of all workers in the control group were found to be below the detection limit of the method (0.05 μg/ml - Table 2). Urinary Cadmium concentrations were also found to be significantly different in all groups. The renal damage reflected in the increasing excretion of low molecular weight proteins like $Beta_2$-microglobulin (β_2-M), and RBP becomes more evident when Cd-U is above BLV of 10 μg/l (Lauwerys, 1983; W.H.O, 1980).

Among the 14 workers that had been exposed, in 7 the RBP exceeded the value of 110 μg/l (Peterson & Berggard, 1971), i.e., all in group I, one in group II (33.3%), and three in group III (37.5%) (Table 2). Four of these workers had Cd-U values above the BLV. In all workers in group I and in worker number 5 in group II the Cd-binding site tissues should be satured with consequent renal dysfunction (Lauwerys, 1983). The same thing may occur in workers 4 and 6 in group II. In group III, because of lower exposure, Cd-U values are always below 10 μg/l and should reflect the amount of Cd in the organism mainly in the kidneys (Alessio et al., 1983). Worker number 9, an exception, had an increased Cd-U value without any renal damage as shown by the low RBP-U excretion (Haguenoer & Furon, 1981).

Renal dysfunction in these conditions and exposures is to be expected according to Bernard et al., (1970) and Buchet et al., (1980).

Although the recorded clinical findings for these workers are incomplete and have been available only for the last four years, it was noticed that worker number one in group I, with more than 10 years of exposure, showed a renal lithiasis and underwent nephroctomy. Worker number 2 from the same group with 8 years of exposure still shows arterial hypertension.

From these data it can be concluded that:

- the Cd-B test can be used as a biological indicator of Cadmium absorption;
- the Cd-U test can be used not only as a biological indicator of absorbtion but also as a biological indicator of the

concentration thereshold at which renal damage might occur;
- the RBP test in urine can be used as a biological indicator of tubular dysfunction. The use of such a test in the periodical control of occupationally Cadmium-exposed workers could be of great value in preventive medicine.

REFERENCES

- Alessio L. et al.: Human biological monitoring of industrial chemical series, Commission of the European Communities, 1983, p. 23-44.
- American Conference on Governmental Industrial Hygienists Threshold limit values for chemical substances in the work environment adopted by A.C.G.I.H. for 1984-85, 116p.
- Bernard A. et al.: Eur. J. Clin. Invest., 9,11-12, 1979.
- Bernard A. & Lauwerys R.: Clin. Chem., 27, 1781-2, 1981.
- Buchet J.P. et al.: J. Occup. Med., 22,741-50, 1980.
- Perkin-Elmer Analytical Methods for atomic absorption spectrophotometry, Determination of Cadmium in blood, plasma and urine, 1977, BC-6.
- Haguenoer J.M. & Furon D.: Cadmium in toxicologie et hygiene industrielles 1981, V. 1, p. 213 - 51.
- Lauwerys R.R.: Cadmium in industrial chemical exposure, Guidelines for biological monitoring. Biom. Publ., 1983, p. 17-22.
- National Institute for Occupational Safety and Health: Criteria for a recommended standard for occupational exposure to Cadmium, 1976, p. 86.
- Organizacion Mundial de la Salud: Cadmio in limites de exposicion professional a los metales pesados que se recomiendam por razones de salud, 1980, p. 22-39.
- Peterson P.A. & Berggard I.: J. Biol. Chem., 246,25-33, 1971.

45

Study of kidney function of workers with chronic low level exposure to inorganic arsenic

V. Foà, A. Colombi, M. Maroni, F. Barbieri – Institute of Occupational Health, Clinica del Lavoro L. Devoto, University of Milan, Milano, Italy

I. Franchini, A. Mutti – Institute of Semeiotica Medica, University of Parma, Parma, Italy

E. De Rosa, G.B. Bartolucci – Institute of Occupational Health, University of Padova, Italy

Although kidney is the target organ for the toxic effects of some metals, possible renal toxicity of arsenic in humans has never been investigated. Explorative studies carried out in single subjects with chronic low-level exposure to inorganic arsenic had shown some cases of increased β_2 microglobulin and total protein excretion. This study was undertaken to evaluate sensitive biochemical and immunological markers of early glomerular and tubular damage more extensively in workers occupationally exposed to inorganic arsenic.

The study was designed as a cross-sectional investigation comparing workers exposed to inorganic arsenic with a matched control group.

Workers with prolonged occupational exposure to arsenic trioxide in glassware factories were investigated.

Arsenic trioxide in the concentration of 2% was used as an ingredient of glass-mix to obtain discoloration of the glass. Exposure to arsenic in this process occurred through inhalation of As-containing dust as well as skin contact; intensity of exposure varied with the work task of each subject. 17 glass-mix makers with remarkable As exposure were identified and admitted to the study. A comparable group of non-exposed subjects was used as controls.

Characteristics of subjects under study: arsenic exposed, number of subjects, 17; age 43 ±9.7 years (mean ±SD); smokers 47%; alcohol consumption, $<$ 40 g/day: 76%, $>$ 40 g/day:24%; lenght of exposure 14.5 ±8.0 years (mean ±SD); controls, number of subjects, 22; age 37 ±4.6; smokers 40%; alcohol consumption 68% and 32%.

Exposure to As was assessed by measuring urine As concentration in three spot urine samples collected after four days of exposure in the morning, at the beginning of the workshift and at the end of the workshift.

The results of urinary As concentrations are shown in Table 1.

As it is known, total arsenic concentration in urine is the sum of different fractions. Some of them - In As, MMA and DMA - are the terminal metabolites of inorganic arsenic in the body. Other fractions - not shown in Table - represent organically linked As which derives from dietary sources, particularly marine food.To evaluate occupational exposure, only the inorganic As metabolites have to be taken into consideration (Foà et al, 1984). The data shown in Table 1 indicate a significant increase in urinary As concentration in the glass workers when compared with a reference population. In fact, the sum of the forms deriving from inorganic arsenic is about 20 fold higher than the corresponding reference value. It must be reminded that the recommended biological threshold level for exposure to As is 100 µg/1 (Nordberg, et al. 1979).

As far as arsenic excretion at the different times of the day is concerned, no significant differences were seen among the urine samples collected in the morning, at the beginning or at the end of the workshift. The presence of high As concentrations in urine also in the morning and pre-shift samples, indicates that As exposure in these workers was remarkable and long-standing.

Renal function parameters investigated were: urinary albumin, retinol binding protein and β_2-microglobulin (determined by latex immunoassay), brush-border antigen (measured by an Elisa assay) as described by Mutti et al. (1985).

The brush-border antigen was isolated from human kidney tubules and then monoclonal antibodies were produced in the mouse. In addition to intense fixation to the brush-border, the monoclonal antibody was found to cross-react also with the vascular endotelium of the kidney arterioles (Mutti et al. 1985).

This marker was used because urinary excretion of a kidney

antigen may be a reliable indicator of shedding due to epithelial exfoliation or increased cellular turnover as a consequence of tubular damage induced by nephrotoxic substances.

No significant differences were observed within the exposed group for each of these parameters among the urine spot samples collected in the morning, at the beginning and at the end of the workshift.

Table 2 shows the comparison of renal function parameters between exposed and control subjects. Values are expressed as creatinine ratio and reported as arithmetic and geometric means.

The Kolmogoro-Smirnov test was used to check whether the distributions after results were normal before and after logarithmic transformation.

Since all the variables were not significantly different from a normal distribution, the t test was used, adopting the pooled or separate variance estimate method according to the test of homogeneity of the variance.

As it may be seen, no statistical differences were found for most variables. Only the retinol binding protein showed a borderline significance, being the values increased in the exposed group.

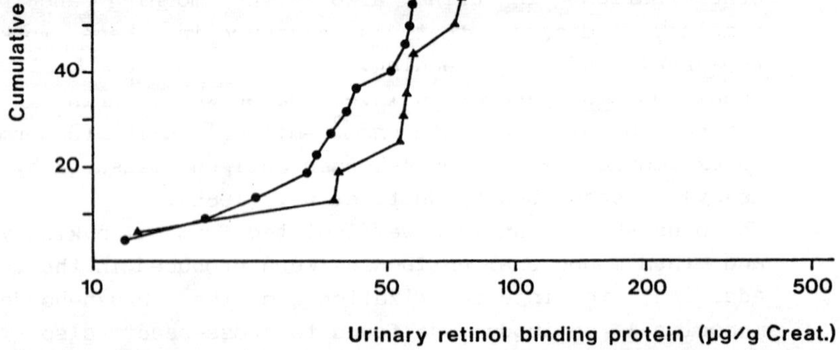

Fig. 1. – Cumulative frequency distribution of retinol binding protein in urine of control and exposed workers.

This difference is graphically shown in Figure 1, where the cumulative frequency distribution in the two groups is reported.

The existence of a significant difference between the two distributions was also tested with the Chi square method after partitioning the distribution in five frequency classes, but no significant difference was found with this method.

In conclusion, this study has not revealed any major differences in renal function parameters of subjects exposed to As around the recommended biological threshold level when they are compared with a matched reference group. RBP was the only variable of borderline significance and this deserves further investigation in more numerous groups of exposed subjects.

Previous preliminary observations in glass workers had shown some cases of moderate increase in β_2-micro-globulin and total-protein excretion. These findings have not been confirmed and extended population surveys would be desirable for a complete definition of such subtle effects.

This study was partly supported by grant EV3V.0845.1 (TT) from the Commission of the European Communities.

REFERENCES

Foà V., Colombi A., Maroni M., Buratti M., Calzaferri G.
The speciation of the chemical forms of arsenic in the biological monitoring of exposure to inorganic arsenic. Sci Tot. Environ 34, 241-259 (1984).

Mutti A., Lucertini S., Valcavi P., Neri T.M., Fornari M., Alinovi R., Franchini I.
Urinary excretion of brush-border antigen revealed by monoclonal antibody: early indicator of toxic neophropathy. Lancet, 26, 914-917 (1985).

Nordberg G.F., Pershagen G., Lauwerys R.
"Inorganic arsenic. Toxicological and epidemiological aspects." Report to the Commission of the European Communities, Dept. of Community Health and Environmental Medicine,Odense, University Printing Office, 1979, Denmark.

Table 1: Urine as concentration in glass workers and in a reference population.

GROUP	NUMBER OF SUBJECTS	SAMPLING TIME	URINE ARSENIC (µg/l) In As	MMA	DMA
exposed glass workers	(n = 15)	morning	26.6 ±17.8	15.6 ±13.1	46.7 ±33.9
	(n = 11)	beginning of workshift	25.3 ±8.5	18.5 ±11.5	67.6 ±36.4
	(n = 16)	end of workshift	22.8 ±13.3	18.1 ±11.6	67.7 ±44.6
reference * population	(n =148)	morning	1.9 ±1.2	1.9 ±1.4	2.1 ±1.5

In As = Inorganic Arsenic; MMA = monomethylarsonic acid; DMA = dimethylarsinic acid;
* Data from Foà et al 1984.

Table 2: Renal function parameters in control and arsenic exposed subjects.

PARAMETER		SUBJECTS		T-TEST
		CONTROL	EXPOSED	SIGNIFICANCE
Albumin	mean (SD)	6.0 (3.1)	10.4 (12.8)	ns
(mg/g creat.)	geometric mean	5.2	6.0	ns
Retinol binding	mean (SD)	60.0 (30.6)	125.6 (122.6)	0.052
protein (μg/g)	geometric mean	52.1	85.9	0.044
β_2-microglobulin	mean (SD)	57.1 (31.0)	109.5 (177.9)	ns
(μg/g)	geometric mean	45.7	57.7	ns
Brush-border	mean (SD)	1.3 (1.0)	2.1 (1.6)	ns
antigen (U/g)	geometric mean	1.0	1.2	ns

Urine collected at the beginning of the workshift

ns = $P > 0.05$

Report of session IV. Biochemical and cellular indices of renal changes induced by exogenus chemicals

Robert A. Goyer, M.D. – Deputy Director, National Institute of Environmental Health Sciences, Research Triangle Park, North Carolina, U.S.A.

This paper is a summary and commentary on the scientific presentations in Session IV, "Biochemical and Cellular Indices of Renal Changes Induced by Exogenous Chemicals." A brief look at the current status of treatment and diagnosis of chronic renal disease provides a strong argument for greater emphasis on early diagnosis and identification of etiology so that there can be greater opportunity for prevention. Advances made in treatment of end-stage renal disease (ESRD) are certainly noteworthy but dialysis and renal transplantation are not ideal solutions to this problem. In the United States, the number of beneficiaries of financial assistance (Medicare payments) rose from 18,412 in 1974 to over 70,000 in 1982, and renal transplants increased from 3,179 to 5,358 in the same period. The cost of this program went from $229 million in 1974 to over $1.8 billion in 1982. And the program is still growing. An estimated 7,800 renal transplants were performed in 1985 (HCFA, 1986). Although the cost of the program largely reflects improved care and survival of patients with ESRD rather than change in incidence, it also illustrates the lack of progress in prevention of occurrence or progression of the disease. A major impediment to early recognition is lack of clinical and laboratory tests with appropriate specificity and sensitivity. The most widely used clinical indicator of renal disease, serum creatinine, is not abnormal (between 1.5 and 2.0 mg/dl blood) until 50 percent or more of renal function is lost. Although chronic renal disease may be asymptomatic or only associated with non-specific symptoms, there is little opportunity for prevention in identifying cause at this stage of the disease. Opportunity for prevention or therapeutic intervention will be greatly assisted by the availability of biologic indicators of the early stages of renal pathology.

The purpose of this summary is to bring attention to or highlight various aspects of the presentations on biologic indicators of renal disease. Because of the brevity of this presentation, the comments are highly selective and not meant to be a complete digest of the detailed presentations.

Dr. Lauwerys reviewed a number of tests that are currently available and he presented an overview of the mechanisms of proteinuria. Although gross proteinuria has long been used as a simple clinical test, separation into a high molecular weight fraction (albumin) and low molecular weight protein provides a means to independently identify glomerular and tubular disease. That is, proteinuria without increase in low molecular weight proteinuria identifies glomerular disease. Alternatively, low molecular weight proteinuria without albuminuria suggests primary tubular disease. Further specificity may be achieved by measuring specific low molecular weight proteins such as beta-2-microglobulinuria or retinol-binding-protein. Such studies may be regarded as suitable for routine use today.

A number of new tests are emerging as having greater sensitivity and/or specificity but vary in their stage of development and validation. Dr. Price described the use of urinary enzymes as early indicators of renal change. The lysosomal enzyme n-acetyl-β-D-glucosaminidase (NAG) provides a sensitive indicator of renal tubular cell injury, particularly as occurs in acute drug or chemical toxicity. It has the advantage in that it is stable and may be used as a tool for primary screening. It may also be a good prognostic indicator with urine level falling rapidly with recovery (acute mercury toxicity) but when it remains elevated it may be a sign of poor prognosis (diabetic nephropathy). Transplant rejection is often preceded by an increase in NAG activity. Increase in NAG is more a sensitive measure of hypertensive nephropathy than gross proteinuria, so the method has promise of improving early detection of some common forms of renal disease that may progress to end-stage disease. Problems with enzymuria as a clinical test, however, are that enzymes are generally only stable over a narrow pH range (6.5-8.0); NAG is among the most stable.

Dr. Vesterberg reviewed progress in newer methods for fractionation of urine proteins. Basically, there are two approaches. Isoelectric focusing is based on differences in charge, sensitive to difference produced by a single amino acid. Polyacrylamide gel separates proteins by size so that when the two methods are combined in a two-dimensional system an overview of all the proteins in a urine sample may be obtained. The method, however, is labor intensive and not all proteins separated by this method have been positively identified, but changes in the urinary excretion of multimolecular forms of proteins may provide insight into mechanisms of proteinuria and renal toxicity at the molecular level, particularly when coupled with complementary analysis of serum proteins.

Perhaps the most sensitive and specific method for identifying proteins in urine is the use of monoclonal antibodies to human kidney antigens described by Dr. Mutti. Monoclonal antibodies are now available that react with brush border and peritubular capillaries, with brush border and glomeruli, and exclusively with brush border antigens. By using very sensitive immunochemical techniques, the use of these antibodies makes it possible to determine site specific renal disease. One of the most interesting findings to date noted by Dr. Mutti is the observation that both low and high molecular weight proteins may occur in the urine of persons exposed to nephrotoxins, and slight increases in albuminuria among subjects exposed to nephrotoxic agents can be correlated with increased levels of brush border antigens in the urine. This may be due to independent effects of the nephrotoxin but is also consistent with the hypothesis that a primary tubular lesion may lead to immune-mediated glomerular damage, a notion that has important implications in terms of pathogenesis of immunologically-mediated glomerular disease and the role of nephrotoxic agents (drugs and chemicals), as etiologic factors in the more common forms of glomerular nephropathies.

Among the poster presentations pertaining to indicators of renal disease are two summarized studies on occupational exposure to cadmium. Dr. Jarup and co-workers from the National Institute of Environmental Medicine, Stockholm, found that cumulative blood cadmium may be used to identify persons at risk for developing cadmium-induced renal damage as determined by urinary beta-2-microglobulin excretion. In another study on cadmium workers, Drs. Huang and Lin of the China National Center for Preventive Medicine found that increase in urinary beta-2-microglobulin (B-2-M) was not related to increase in serum B-2-M in workers chronically exposed to cadmium and that cadmium-induced urinary B-2-M was irreversible in most workers for several years even after removal from exposure.

And, finally, the discussion of these papers brought attention to the problem of assessing urine flow rate for the quantitation of various indicators of renal disease in "spot" urine samples. Common practice is to standardize the urine sample with respect to creatinine. The assumption is that creatinine is excreted at a roughly constant rate of one gram per twenty-four hours in adults. But, urine creatinine excretion is influenced by a number of factors, e.g., age, diet, lean body mass, exercise. Correlations between specific gravity and urine flow rate are good for urine having a specific gravity greater than 1.016 but not less than 1.016. Probably the most practical approach is to use creatinine and obtain a repeat sample when necessary. Discussants also suggested that search for better reference factors be continued and hydroxyproline and 3-methyl histidine were cited as examples of substances that have potential as markers.

REFERENCE

HCFA (Health Care Financing Administration). End-Stage Renal Disease Quarterly Report. March 31, 1986. United States Department of Health and Human Services, Baltimore, Maryland.

Part V

Biological indices for assessment of human genotoxicity induced by exogenous chemicals

Coordination:
G. Rubino (Torino, Italy)
G. Della Porta (Milano, Italy)

46

Dose indicators in monitoring of exposure to carcinogens in man

H. Vainio, K. Hemminki* – International Agency for Research on Cancer, Lyon, France and *Institute of Occupational Health, Helsinki, Finland

ABSTRACT

Until recently, the only means of assessing human exposure to carcinogens has been environmental monitoring - the measurement of known agents in, e.g., ambient air. Now, however, dose indicators are available. These include direct measurement of chemicals or their metabolites in body fluids or excreta; less direct methods to monitor biologically relevant doses, such as measuring protein or DNA adducts; and nonspecific dose indicators, such as measurement of mutagenic activity in body fluids. In situations in which a strong relationship exists between a dose indicator, e.g., a DNA adduct, and cancer, the indicator may be used as an early warning sign of the risk of cancer.

INTRODUCTION

Primary control of exposures to carcinogens must be by hygienic means, which are indicated by environmental monitoring and measurements in air, water and food. This approach, while still the most important for exposure assessment, has limitations for estimating the dose received by an individual. Biological monitoring provides time-integrated data on actual internal doses and enables an assessment of the body burden of substances absorbed in the body by various routes. Biological monitoring can be used to evaluate internal dose, the biologically relevant (target) dose and, in some instances, early reversible biological effects (Fig. 1).

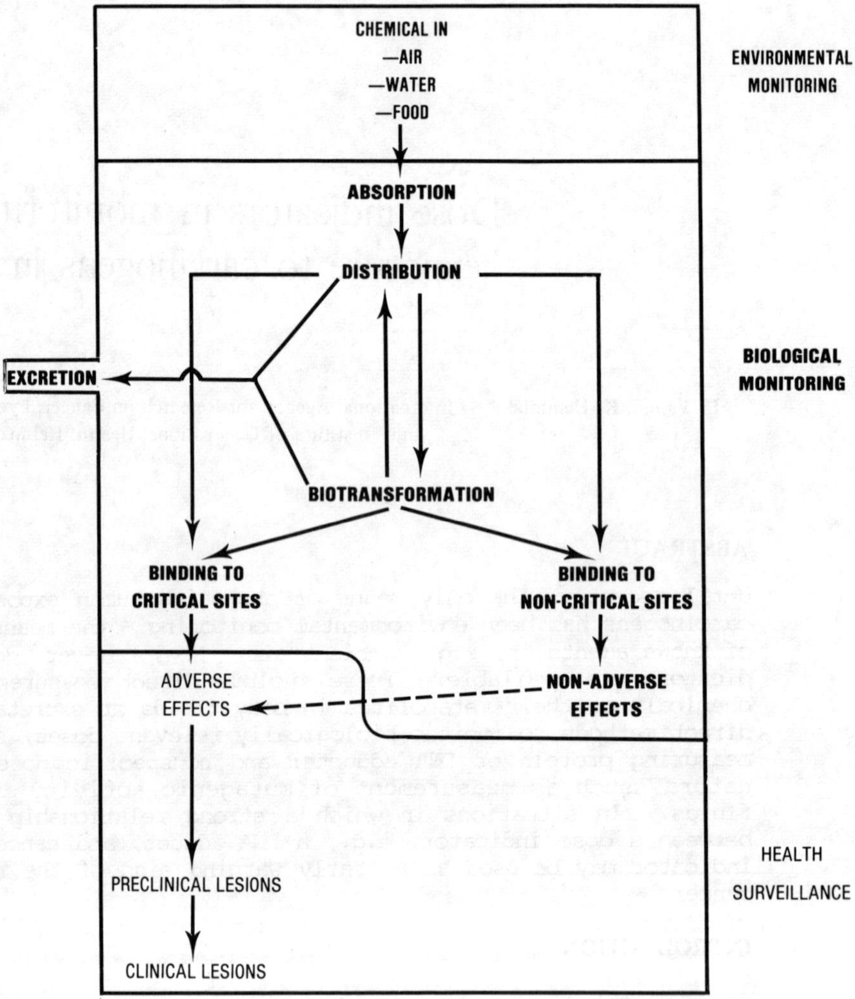

Fig. 1. – Pathway of a chemical from the enviroment to target molecules in the organism.

Experimental work carried out over the past three decades has indicated that many carcinogenic, mutagenic and cytotoxic chemicals are electrophiles, or are converted to electrophiles metabolically, and can thus react covalently with cellular nucleophiles, including DNA. Such alkylation products damage the structure and function of DNA, which may lead to mutations, malignant transformation or cellular death (Office of Science and Technology Policy, 1985).

We briefly review here some of the basic principles that underly the monitoring of internal dose and of biologically relevant dose in terms of DNA and protein adducts. The detection of early effects, such as clastogenic changes, is covered by other presentations in this volume.

MEASUREMENT OF INTERNAL DOSE

Chemicals and their metabolites

While knowledge about dose-effect and dose-response relationships is normally a prerequisite for the interpretation of the results of biological monitoring, these data are not usually available for carcinogens. Assessment of exposure to carcinogens involves simply estimation of the amount of the chemical which has entered the body. Such levels reflect the exposure of the individual (or population) and, in turn, determine the adverse effects that may be manifested in the same individual or in the population. Internal dose is a function of current exposure as well as of accumulated past exposure. It can be determined by measuring the substance itself, or its metabolite, in blood, urine, expired air, hair, adipose tissue, etc. Methods for determining chemicals and their metabolites in biological samples are usually specific, and allow individual interpretation of data (for further discussion, see Berlin et al., 1984).

Most of the biological monitoring tests presently available measure specifically a chemical or its metabolite in some body fluids or excreta (Berlin et al., 1984). The measurement of various classes of carcinogens and/or their metabolites has been summarized recently (van Sittert, 1984). Some of the organic agents that have been detected in urine of exposed persons are listed in Table 1.

Analysis of mutagenic activity in body fluids

Assay of excreta for genotoxic activity has been used widely as a nonspecific means of estimating exposure to carcinogens (Vainio et al., 1984) since Yamasaki and Ames (1977) first reported the mutagenicity of urine from cigarette smokers. The test has both advantages and disadvantages. Its advantages include the fact that is noninvasive, it can detect many chemicals at the same time, it is highly sensitive and can be done

Table 1: Levels of urinary metabolites of some carcinogens in occupationally exposed and nonexposed persons.

Chemical	Urinary metabolite	Urinary concentration in		Reference
		non-exposed	occupationally exposed	
Acrylonitrile	Acrylonitrile	< 5 ug/l	10.9 ug/l (0.17 ppm; 8 h) 105 ug/ml (3.7 ppm; 8 h)	Sakurai et al. (1978)
Benzene	Phenol	< 20 mg/l	45-50 mg/l (10 ppm; 8 h)	Lauwerys (1979)
Methyl chloride	S-Methylcysteine	0-90 umol CH_3SH/ mmol creatinine	50-700 umol CH_3SH/ mmol creatinine (30-90 ppm; 8 h)	van Doorn et al. (1980)
4,4-Methylene-bis(2-chloroaniline) (MOCA)	MOCA	< 40 ug/l	250 ug/l (< 0.02 mg/m^3)	Linch et al. (1971)
Vinyl chloride	Thiodiglycolic acid	< 0.05-1.3 mg/l	0.3-4.0 mg/l (0.14-7.0 ppm; 8h)	Müller et al. (1978); 1979)

[a]Modified from van Sittert (1984)

at reasonable cost. Its disadvantages include the fact that it can be used only for certain classes of carcinogens, and the results may be confounded by factors such as diet and smoking.

Measurement of bacterial mutation cannot be used to identify mutagenic agents, unless the active constituents are structurally identified. The presence of mutagens in body fluids has not been translated into a known risk to the individual; the excretion of mutagens may in fact be a detoxification process. For example, urinary mutagenicity assays conducted on rats did not distinguish carcinogenic compounds from their noncarcinogenic analogues (Malaveille et al., 1986). At a given exposure level, a lower level of mutagens in excreta could mean that a higher fraction of the dose is maintained in the body, and that the tissue dose may be higher than with a higher level of excreted mutagens. Therefore, a detailed knowledge of the toxicokinetics of the agents in question is critical to the proper interpretation of the results. In the present state of knowledge, the occurrence of genotoxic activity in the body fluid or excreta is an indicator of (short-term) exposure and absorption, not an indicator of effect.

MEASUREMENT OF BIOLOGICALLY EFFECTIVE DOSE – MACROMOLECULAR ADDUCTS

The basic assumption in measuring DNA adduct levels for estimation of exposure is that they correlate with the extent of exposure. While directly acting (lipophilic) alkylating agents usually react similarly with DNA of various tissues, metabolically activated agents display large differences between tissues, depending on their metabolic capacities to activate and inactivate chemicals and to remove adducts from DNA.

The assumption underlying measurement of DNA adducts for estimation of cancer risk is that the adduct levels in target tissues correlate with proneness to tumour formation. However, high levels of adducts may also accumulate in non-target tissues, indicating that formation of adducts is not the only condition for tumour formation. While no study has been done in which adduct levels and appearance of cancer have been assessed at the same time, in several studies adduct levels have been measured separately and found to correlate either with their carcinogenic potency or organ specificity (Lutz, 1979; Phillips et al., 1979; Hemminki, 1983; Phillips et al., 1984). It is not clear what effects the site of binding in DNA or the size and structure of the ligand have on carcinogenic potencies. Some of the nucleotide analogues incorporated into DNA may not be carcinogenic. For methylating and ethylating carcinogens, alkylation at exocyclic oxygen atoms appears to correlate with carcinogenic potency, but whether this property relates to the relative stability of these adducts in DNA or to their "procarcinogenic" nature remains to be established. As aflatoxin B_1 reacts principally at the N-7 of guanine, aromatic amines at C-8 and polycyclic aromatic hydrocarbons at N^2, binding at these sites and at oxygen atoms appears to lead to "procarcinogenic" lesions. It may be, however, that the presence of a large alkyl group at any DNA base perturbs the structure in such a way as to bring about cancer, given other necessary conditions. It should be emphasized that DNA adduct formation is a very early part of the complex process leading to the development of cancer, and further understanding of this process is necessary before the role of DNA adducts can be stated clearly (Office of Science and Technology Policy, 1985).

Methods for DNA adduct monitoring

Immunological techniques: Immunological techniques have been developed with sensitivities approaching those required to detect modified DNA in exposed human populations (Poirier, 1981). These are based on the use of antibodies (usually produced in rabbits) against modified DNA, polynucleotides, nucleotides or nucleosides. The assay method is a competitive immunoassay in which two identical haptens (e.g., modified DNAs) compete for the antibody-binding sites. In radioimmunoassays, the concentration of the radioactive hapten is kept constant, and various concentrations of the unknown sample (e.g., a DNA sample from blood cells) are used to generate an inhibition curve in order to quantify the level of modification in the sample. Enzyme-linked assays involve generation of coloured (enzyme-linked immunosorbent assay, ELISA) or radioactive (ultrasensitive enzymatic radioassay, USERIA) reaction products. Because the levels of the antigen-antibody reaction are enzymatically multiplied in ELISA and USERIA, they are usually, but not always, some orders of magnitude more sensitive than conventional radioimmunoassays. Antigens reported in the literature include alkyl derivatives of guanosine and thymidine and adducts of acetylaminofluorene and benzo[a]pyrene; other antibodies include those raised against adducts of aflatoxin B_1 and cisplatin. The affinities of the antibodies described vary between 10^6 and 10^{11} l/mol. With the best available antibodies, it is possible to detect one modification per 10^6-10^7 bases.

Some experience has been gained using samples derived from humans treated with cisplatin (Poirier et al., 1985) or exposed to polycyclic aromatic hydrocarbons (Shamsuddin et al., 1985; Everson et al., 1986). Although the technique is not yet applicable for monitoring of individuals exposed to carcinogens, there is every reason to believe that it could be so used in the future.

^{32}P-Postlabelling assay: Another highly sensitive assay, which entails the enzymatic incorporation of radiolabelled ^{32}P into DNA constituents, has been developed recently (Randerath et al., 1981; Gupta et al., 1982; Reddy et al., 1984). It enables the detection and quantification of DNA adducts, and can therefore be used directly on DNA from animals or cultured cells exposed to test compounds. Application of the method to a large number of carcinogens of diverse structures has been documented (Reddy et al., 1984).

The lower level of adduct detection is about one adduct in 5 x 10^7-10^8 nucleotides with the standard procedure; for many aromatic adducts, one adduct in about 10^9 nucleotides can be detected with ATP-deficient labelling. The technique entails simultaneous labelling of normal and adduct nucleotides, thereby enabling accurate quantification of DNA adduct levels. It is also possible to isolate the adducts first and then label them in the virtual absence of normal nucleotides. This technique affords an increase in sensitivity of detection to one adduct in about 10^{10} normal nucleotides.

Experience with human samples is limited to placentas of smokers and nonsmokers (Everson et al., 1986) and to white blood cells of foundry workers (Reddy, Randerath, Hemminki, unpublished). In the placental samples, an aromatic-type of adduct was identified, but its presence was related only weakly to the intensity of smoking. Furthermore, the postlabelling and the antibody method agreed qualitatively but not quantitatively (Everson et al., 1986).

Electrophore labelling: This technique entails derivatization of nucleic acid bases by fluoride-containing electrophores, which allow sensitive detection by gas chromatography and negative-ion chemical ionization mass spectrometry (Mohamed et al., 1984). Instrumentation is expensive and requires special training. The sensitivity of the method with model pyrimidine compounds appears excellent (femtogram range), but further work is needed on purines and in-vivo samples before the applicability of the method can be fully assessed.

Detection of protein binding

Exposure to carcinogens and mutagens can be estimated by measuring alkylated proteins in blood. Since there is a certain relationship between the alkylation of amino acids and the alkylation of nucleic acids, such results can also be used for estimation of the genotoxic risks of exposed workers (Ehrenberg et al., 1974; Ehrenberg & Hussain, 1981; Neumann, 1984). Ethylene oxide and methyl methanesulfonate were found to react directly with nucleophilic groups in experimental animals. In contrast, ethylene, propene, N-nitrosodimethylamine, and vinyl chloride reacted with haemoglobin only after their transformation into reactive metabolites (see Table 2).

Table 1: Haemoglobin adducts measured in in-vivo sample.[a]

Methylating agents
Methyl methanesulphonate
N-Nitrosodimethylamine
N-Methyl-N-nitrosourea
Dacarbazine
Dichlorvos
Aromatic amines
t-4-Dimethylaminostilbene
Benzidine
2-Acetylaminofluorene
Acetanilide
Paracetamol
4-Aminobiphenyl
Other compounds
Ethylene
Ethylene oxide
Propylene
Propylene oxide
Styrene
Styrene oxide
Vinyl chloride

[a]From Hemminki and Randerath (1986)

Alkylation of haemoglobin has also been measured in the blood of people exposed to ethylene oxide (Calleman et al., 1978). Although the concentrations of ethylene oxide in the work environment were as low as 5-10 ppm, it was possible to isolate alkylated residues of the amino acid histidine from the haemoglobin of exposed persons. Furthermore, hydroxyethyl residues have now been detected in the haemoglobin of unexposed individuals at levels which may imply a role of these alkylations in human cancer. Measurement of the alkylation of haemoglobin has not yet been adapted for routine measurements.

Haemoglobin is considered to have special advantages as a target of alkylation because it is abundant and it has a long half-life. Proteins with a shorter half-life, e.g., the serum proteins, are suitable for verifying exposures of short duration. Another advantage of certain serum proteins is that they are synthesized in the liver cells with the greatest capacity for metabolic activation of agents that alkylate indirectly.

USE OF DOSE INDICATORS

The determination of (pro)carcinogens or their (activated) metabolites and levels of macromolecular adducts could be used in monitoring, with at least three aims:

(1) Estimation of exposure to known carcinogenic and mutagenic agents, i.e., for occupational or environmental monitoring (Sorsa et al., 1982; Vainio et al., 1983). Data on individuals may relate to exposure patterns and can be used to suggest improvements in hygienic conditions. Alternatively, individual results may reflect differences in constitutional factors, which may indicate susceptibility to cancer. Populations that have more adducts in DNA than properly selected control populations may be at risk of developing cancer; theoretical calculations of the magnitude of such risk have been made (Ehrenberg & Hussein, 1981), but such extrapolations may or may not be directly applicable in the estimation of individual risk. Adduct determinations can provide biochemical markers for cancer epidemiology if large numbers of reliable measurements are available.

(2) Identification of potentially mutagenic and carcinogenic components of complex mixtures (Vainio & Sorsa, 1985). For example, studies are under way to identify the products of tobacco smoke and of exposures in foundries leading to the formation of DNA adducts.

(3) Design of individual cancer chemotherapy (Hemminki & Ludlum, 1984). Preliminary results suggest that patients responsive to cisplatin therapy also generate higher and more stable levels of cisplatin adducts in the DNA of their lymphocytes (Poirier et al., 1985).

When a strong relationship exists between an exposure indicator, e.g., a DNA adduct, and the end effect, cancer, the indicator can be used as an early warning sign of the effect. Although our knowledge about the predictiveness of DNA adducts is still very incomplete, the fact that all chemicals known to produce DNA adducts in vivo also have carcinogenic activity calls for prudence. Since all genotoxic agents studied so far have been shown to react with haemoglobin in vivo, wide applicability of haemoglobin adducts would be indicated (Neumann, 1984). The quantitative relationship between macromolecular adducts (with either protein or DNA) and cancer risk has still to be elucidated.

REFERENCES

Berlin, A., Draper, M., Hemminki, K. & Vainio, H., eds (1984) Monitoring Human Exposure to Carcinogenic and Mutagenic Agents (IARC Scientific Publications No. 59), Lyon, International Agency for Research on Cancer, 457 pp.

Calleman, C.J. Ehrenberg, L., Jansson, B., Osterman-Golkar, S., Segerbäck, D., Svensson, K. & Wachtmeister, C.A. (1978) Monitoring and risk assessment by means of alkyl groups in hemoglobin in persons occupationally exposed to ethylene oxide. J. environ. Pathol. Toxicol., 2, 427-442

van Doorn, R., Borm, P.T.A., Leijdekkers, C.M., Henderson, P.T., Reuves, J. & van Bergen, T.J. (1980) Detection and identification of S-methylcysteine in urine of workers exposed to methylchloride. Int. Arch. occup. environ. Health, 46, 99-109

Ehrenberg, L. & Hussain, S. (1981) Genetic toxicity of some important epoxides. Mutat. Res., 86, 1-113

Ehrenberg, L., Hiesche, K.D., Osterman-Golkar, S. & Wennberg, I. (1974) Evaluation of genetic risks of alkylating agents: Tissue doses in the mouse from air contaminated with ethylene oxide. Mutat. Res., 24, 83-103

Everson, R.B., Randerath, E., Santella, R.M., Cefalo, R.C., Avitts, T.A. & Randerath, K. (1986) Detection of smoking related covalent DNA adducts in human placenta. Science, 231, 54-57

Gupta, R.C., Reddy, M.V. & Randerath, K. (1982) ^{32}P-Post-labelling analysis of non-radioactive aromatic carcinogen-DNA adducts. Carcinogenesis, 3, 1081-1092

Hemminki, K. (1983) Nucleic acid adducts of chemical carcinogens and mutagens. Arch. Toxicol., 52, 249-285

Hemminki, K. & Ludlum, D.B. (1984) Covalent modification of DNA by antineoplastic agents. J. natl Cancer Inst., 73, 1021-1028

Hemminki, K. & Randerath, K. (1986) Detection of genetic interaction of chemicals by biochemical methods. Determination of DNA and protein adducts. In: Mechanisms of Cell Injury: Implications for Human Health, Berlin, Springer Verlag (in press)

Lauwerys, R. (1979) Human Biological Monitoring of Industrial Chemicals. 1. Benzene (Document EUR 6570), Luxembourg, Health and Safety Directorate, Commission of the European Communities

Linch, A.L., O'Connor, G.B., Barnes, J.R., Killian, A.S. & Neeld, W.E. (1971) Methylene-bis-ortho-chloroaniline (MOCA): Evaluation of hazards and exposure control. Am. ind. Hyg. Assoc. J., 32, 802-819

Lutz, W.K. (1979) In vivo covalent binding of organic chemicals to DNA as a quantitative indicator in the process of chemical carcinogenesis. Mutat. Res., 65, 289-356

Malaveille, C., Brun, G. & Bartsch, H. (1986) Mutagenicity of urine from rats treated with benzo(a)pyrene, pyrene, 2-acetylaminofluorene and 4-acetylaminofluorene in Salmonella typhimurium TA100 or TA98 strains. In: Ashby, J., de Serres, F.J., Shelby, D., Margolin, B.H., Ishidate, Jr. & Becking, G.C., eds, Progress in Mutation Research Volume - Evaluation of Short-Term Tests for Carcinogens: Report of the International Programme on Chemical Safety's Collaborative Study on In Vivo Assays. Proceedings of the 2nd ICPS Collaborative Study on in-vivo Tests for Carcinogens (in press)

Müller, G., Norpoth, K., Kusters, E., Herweg, K. & Versin, E. (1978) Determination of thiodiglycolic acid in urine specimens of vinylchloride exposed workers. Int. Arch. occup. environ. Health, 41, 199-205

Müller, G., Norpoth, K. & Wickramasinghe, R.H. (1979) An analytical method, using GC-MS, for the quantitative determination of urinary thiodiglycolic acid. Int. Arch. occup. environ. Health, 44, 185-191

Mohamed, G.B., Nazaret, A., Hayes, M.J., Giese, R.W. & Vouros, P. (1984) GC-MS characteristics of methylated perfluoroacyl derivatives of cytosine and 5-methyl cytosine. J. Chromatogr., 314, 211-217

Neumann, H.-G. (1984) Analysis of hemoglobin as a dose monitor for alkylating and arylating agents. Arch. Toxicol., 56, 1-6

Office of Science and Technology Policy (1985) Chemical carcinogens: a review of the science and its associated principles. Fed. Regist., Part 2

Phillips, D.H., Grover, P.L. & Sims, P. (1979) A quantitative determination of the covalent binding of a series of polycyclic hydrocarbons to DNA in mouse skin. Int. J. Cancer, 23, 201-208

Phillips, D.H., Reddy, M.V. & Randerath, K. (1984) ^{32}P-Postlabelling analysis of DNA adducts formed in the livers of animals treated with safrole, estragole, and other naturally-occurring alkenylbenzenes. II. Newborn male B6C3F$_1$ mice. Carcinogenesis, 5, 1623-1628

Poirier, M.C. (1981) Antibodies to carcinogens-DNA adducts. J. natl Cancer Inst., 67, 515-519

Poirier, M.C., Reed, E., Zwelling, L.A., Ozols, R.F., Litterst, C.L. & Yuspa, S.H. (1985) Polyclonal antibodies to quantitate cis-diammine-dichloroplatinum (II)-DNA adducts in cancer patients and animal models. Environ. Health Perspect., 62, 89-94

Randerath, K., Reddy, M.V. & Gupta, R.C. (1981) ^{32}P-Postlabelling test for DNA damage. Proc. natl Acad. Sci. (USA), 78, 6126-6129

Reddy, M.V., Gupta, R.C., Randerath, E. & Randerath, K. (1984) ^{32}P-Postlabelling test for covalent DNA binding of chemicals in vivo: application to a variety of aromatic carcinogens and methylating agents. Carcinogenesis, 5, 231-243

Sakurai, H., Onodera, M., Utsunomiya, T., Minakuchi, H., Iwai, H. & Matsumura, H. (1978) Health effects of acrylonitrile in acrylic factories. Br. J. ind. Med., 35, 219-222

Shamsuddin, A.K.M., Sinopili, N.T., Hemminki, K., Boesch, R.R. & Harris, C.C. (1985) Detection of benzo(a)pyrene: DNA adducts in human white blood cells. Cancer Res., 45, 66-68

van Sittert J.J. (1984) Biomonitoring of chemicals and their metabolites. In: Berlin, A., Draper, M., Hemminki, K. & Vainio, H., eds, Monitoring Human Exposure to Carcinogenic and Mutagenic Agents (IARC Scientific Publications No. 59), Lyon, International Agency for Research on Cancer, pp. 159-172

Sorsa, M., Hemminki, K. & Vainio, H. (1982) Biological monitoring of exposure to chemical mutagens in the occupational environment. Teratog. Carcinog. Mutagenesis, 2, 137-150

Vainio, H. & Sorsa, M. (1985) Application of short-term tests in monitoring occupational exposure to complex mixtures. In: Waters, M.D., Sandhu, S.S., Lewtas, J., Claxton, L., Strauss, G. & Nesnow, S., eds, Short-term Bioassays in the Analysis of Complex Environmental Mixtures IV, New York, Plenum Publishing Corporation, pp. 291-302

Vainio, H., Sorsa, M. & Hemminki, K. (1983) Biological monitoring in surveillance of exposure to genotoxicants. Am. J. ind. Med., 4, 87-103

Vainio, H., Sorsa, M. & Falck, K. (1984) Bacterial urinary assay in monitoring exposure to mutagens and carcinogens. In: Berlin, A., Draper, M., Hemminki, K. & Vainio, H., eds, Monitoring Human Exposure to Carcinogenic and Mutagenic Agents (IARC Scientific Publications No. 59), Lyon, International Agency for Research on Cancer, pp. 247-258

Yamasaki, E. & Ames, B.N. (1977) Concentration of mutagens from urine by adsorption with the nonpolar resin XAD-2: cigarette smokers have mutagenic urine. Proc. natl Acad. Sci. USA, 74, 3555-3559

47

Quantitation of carcinogen-DNA adducts with monoclonal antibodies

Regina M. Santella, Xaio Yen Yang – Institute of Cancer Research and School of Public Health, Division of Environmental Sciences, Columbia University, New York, 10032, U.S.A.

ABSTRACT

Sensitive methods for the quantitation of carcinogen-DNA adducts are now available and can be used to screen human populations exposed to environmental carcinogens. We have developed monoclonal antibodies to several carcinogen-DNA adducts and characterized them as to sensitivity and specificity by enzyme linked immunosorbent assay (ELISA). Highly sensitive competitive ELISA with color or fluorescence detection can quantitate one adduct measured per 10^8 nucleotides. Adducts are currently being measured in lung, placenta and pheripheral white blood cell DNA of humans exposed to a variety of environmental carcinogens.

INTRODUCTION

Over the past few years immunological methods have been developed for the sensitive detection of damaged DNA. Highly specific antibodies have been generated to a number of altered DNA bases including carcinogen-DNA adducts as well as UV damaged nucleosides. For a review, see (1,2). These antibodies have been utilized in highly sensitive assays to quantitate low-levels of damaged DNA, providing information about the formation and repair of adducts. In addition, immunohistochemical studies have provided information on the distribution of adducts in specific cell types and localization in tissue sections.

Antibodies can be generated by two methods. The modified DNA can be electrostatically complexed to methylated bovine serum albumin (mBSA) or methylated keyhole limpet hemocyanin (mKLH) before mixing with adjuvant for immunization. Alternatively, the ribose form of specific monoadduct derivatives can be covalently coupled to protein through oxidation of the ribose group by periodate (3). Methods are also available for the coupling of deoxyribose nucleosides to protein (4). Table 1 summarizes

Table 1: Polyclonal and monoclonal antibodies against carcinogen-DNA adducts.

Antigen	Ref
O^6Me dGuo	(5,6)
O^2Me dThd	(1)
O^6Et dGuo	(7,8,9)
O^4Et dThd	(5,8)
O^6Bu dGuo	(5,8)
C8-acetylaminofluorene dGuo	(9,10)
acetylaminofluorene-DNA	(11)
Benzo[a]pyene diolepoxide-I-DNA	(12,13,14)
cis platinum-DNA	(15,16,17)
Aminopyrene-DNA	(18)
8 methoxypsoralen-DNA	(19,20)
Aflatoxin B_1-DNA	(21,22)
OsO_4-oxidized poly(dT) (thymine glycol)	(23)
UV-DNA (thymidine dimer)	(24)

currently available polyclonal and monoclonal antisera using as antigens either DNA or monoadducts. Antibodies to a monoadduct are preferable when individual adducts are to be quantitated. These adducts can be isolated from enzymatic digests of DNA by HPLC or other chromatographic separation. One advantage of this approach is that large amounts of DNA can be digested and the adduct isolated for quantitation. Sensitivites and specifities can be increased by this procedure.

Antibodies to modified DNA have been useful for quantitating adduct levels in a number of animal and tissue culture studies. In addition, they have been essential in studies monitoring humans for exposure to environmental carcinogens. The carcinogen does not have to be radiolabeled and adducts can be quantitated with sensitivities in the femtomole (10^{-15}) range. One approach in these monitoring studies has been to measure the biologically effective dose of the carcinogen, that is the amount of material that has been absorbed, metabolized and bound to critical cellular targets (25). For chemical carcinogenesis, DNA binding is believed to be the critical event in initiation of tumor induction (26). It is hoped that studies providing information about the levels of carcinogen DNA adducts in the target tissue or a surrogate will provide more relevant data in human exposure studies than has been previously available. This should lead to more effective procedures for assessing quantitatively the risks to humans of various environmental chemicals. However the tissue samples most readily accessible for study are often remote from the actual target tissue. Therefore, information on the relationship between the endpoint measured in the target and sampled tissue is needed. For example, in many studies carcinogen adduct levels are measured in placental or peripheral white blood cell DNA while the target may be lung, liver, etc.

Although DNA is believed to be the critical target in chemical carcinogenesis, most carcinogens also bind to RNA and protein (26,27). Therefore, the use of carcinogen binding to hemoglobin has been suggested as a surrogate for DNA binding in human dose monitoring studies (28,29,30). This approach has several advantages, including the fact that large quantities of hemoglobin are available in small amounts of blood, there is no repair system for modified hemoglobin and chronic low levels of exposure may be detectable due to accumulation over the lifetime of the erythrocyte (about four months in humans). However, for validation of this approach, the relationship of adduct levels on globin with those on DNA at the target tissue must be determined for each chemical.

MONOCLONAL ANTIBODIES TO BENZO[A]PYRENE DIOL EPOXIDE I MODIFIED DNA AND PROTEIN

We have isolated and characterized a total of five stable clones producing antibody specific for DNA modified by 7,8-dihydroxy-9,10-epoxy-7,8,9,10-tetrahydrobenzo[a]pyrene (BPDE-I) (13,31). One clone (8E11) was obtained from the spleen cells of a BPDE-I-G-BSA immunized animal while four clones (5D11, 5D2, 1D7, and 4C2) were from the spleen cells of several BPDE-I-DNA immunized animals. Competitive ELISAs were used to determine the sensitivity and specificity of the antibodies. Ninty six well micro plates were coated with 5 ng of BPDE-I-DNA. The antibody was mixed with various concentrations of the competitors before adding to the wells. The amount of antibody bound to the well was determined with a second antibody, goat anti mouse IgG coupled to alkaline phosphatase. The substrate for the enzyme, p-nitophenylphosphate, was used to quantitate amount of enzyme present and therefore amount of antibody bound to the plate. The data is expressed as amount of competitor that results in 50% inhibition, lower values indicating higher antibody binding. Antibody 8E11 has a 50% inhibition value with BPDE-I-DNA of 350 fmole. It has better reactivity against the monoadduct BPDE-I-dG (50% inhibition at 145 fmole) than with DNA, as is typical of antibodies generated against monoadducts. In addition, it shows good cross-reactivity with BPDE-I tetraols (50% at 250 fmole), the hydrolysis products of BPDE-I. This suggests that the antibody recognizes mainly determinants present on the BP ring. Antibody 5D11 has better sensitivity for BPDE-I-DNA (50% inhibition at 19 fmole) than for the monoadduct BPDE-I-dG (50% inhibition at 21000 fmole) and no cross-reactivity with BPDE-I-tetraols. Neither acetylaminofluorene or 1-aminopyrene modified DNA (AAF-DNA or 1-AP-DNA) cross-reacted with any of the antibodies when tested at high concentrations. Both of these carcinogens produce adducts at the C-8 position of guanine instead of the N-2 position to which BPDE-I binds. In contrast 5D11 does react with DNAs modified by other trans diol epoxides. When tested with DNAs modified by benz[a]anthracene trans-3,4-diol anti 1,2-epoxide, benz[a]anthracene, trans-8,9-diol anti 10,11-epoxide or chrysene trans-1,2-diol anti-3,4-epoxide 50% inhibition values of 16,26 and 162 ng respectively, were obtained. This can be compared to 1.1 ng for

DNA modified by BPDE-I to an extent of 0.57%. Since the antibody can recognize N-2 guanine adducts that are formed from trans diol epoxides with similar stereochemistry to that of BPDE-I, sample antigenicity will result from antibody binding, with variable affinity, to any trans diol epoxide-adducts which may be present.

Quantitation of adducts on human samples is also carried out by competitive ELISA. It is possible to test a maximum of 50 ug of DNA in a micro well. Recently, the assay has been made more sensitive with the development of 96 well plate fluorescent readers. All steps in the ELISA are carried out as in the standard color assay except methyl umbelliferyl phosphate is utilized as the substrate for alkaline phosphatase. This compound becomes fluorescent only after phosphate hydrolysis.

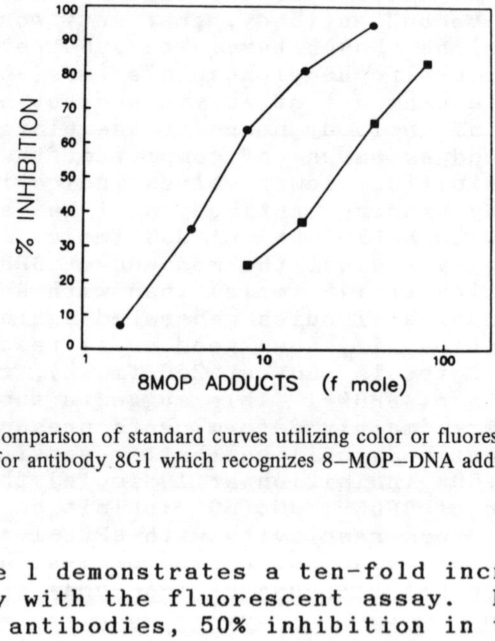

Fig. 1. – Comparison of standard curves utilizing color or fluorescence endpoints for antibody 8G1 which recognizes 8–MOP–DNA adducts.

Figure 1 demonstrates a ten-fold increase in sensitivity with the fluorescent assay. For most monoclonal antibodies, 50% inhibition in the 2-20 fmole range are now obtainable. With this sensitivity and 50 ug of DNA per well, it is possible to quantitate levels of adducts in the range of one per 10^7 to 10^8 bases. Others have used radiolabeled substrates or biotin avidin complexes to also increase sensitivities (32,33). Several recent studies have begun to utilize this approach to monitor humans for exposure to BP. Samples have included lung tissue and/or white blood cell DNA from lung cancer patients, white blood cell DNA of roofers, foundry workers and volunteer laboratory workers (34,35,36). We have also examined adduct levels in the placental DNA of women smokers and nonsmokers (37). The mean value for smokers was 1.65 fmole adduct/ug DNA and for nonsmokers 0.96 fmole/ug. This result is not statistically significant and may reflect other exposures to this ubiquitous carcinogen, including passive smoking, dietary and environmental (such as kerosene heaters) exposure. Assays on the same samples by the postlabeling technique (discussed below) carried out by Dr. Kurt Randerath suggest another unidentified adduct may be more specific for exposure to cigarette smoke (37).

We have recently developed several monoclonal antibodies specific for BPDE-I modified protein (38). These antibodies were to be used to quantitate levels of BP bound to human hemoglobin as an alternate marker of exposure to BP. However the reactivity of the antibodies for the BP chromophore when it was bound to protein was about ten fold lower than when it was free in solution. This is probably due to the burying of the chromophore in hydrophobic regions of the protein making it less accessible for antibody binding. Previous studies have demonstrated that tetraols can be released from modified protein with acid treatement. These tetraols can then be quantitated in an ELISA with high sensitivites since large amounts of globin can be treated to release tetraols. We are currently investigating the applicability of this method for human hemoglobin samples.

MONOCLONAL ANTIBODIES TO 8–METOXYPSORALEN PLUS UVA MODIFIED DNA

8-Methoxypsoralen (8-MOP) plus ultraviolet A light (320-400 nm) is used clinically in the treatment of psoriasis (39) and more recently extracorporeally as an experimental cytoreductive treatment for the leukemic phase of cutaneous T-cell lymphoma. 8-MOP forms two monoaddition adducts on thymine as well as a cross-linked derivative (40). Several case reports in the literature have associated treatment with 8-MOP plus UVA with cutaneous carcinoma (41), and leukemia (42). Patients undergoing this treatment provide an ideal population for validating adduct level monitoring since a defined quantifiable exposure has occurred. We have therefore developed a panel of monoclonal antibodies which specifically recognize 8-MOP modified DNA to quantitate adduct levels in patients (19). A sensitive fluorescence ELISA was developed for quantitating DNA adduct levels in lymphocytes from patients undergoing extracorporeal photophoresis. In addition serum levels of free 8-MOP were measured by an HPLC technique (43). There was a seven-fold difference in 8-MOP serum levels (range 13-87 ng/ml) which probably reflects differences in absorption of the drug. The difference in adduct levels was much smaller (5.5-7.5 adducts/10^7 nucleosides) and may indicate that UV dose is a limiting factor in adduct formation. Additional samples from these patients as well as patients undergoing PUVA treatment for psoriasis are currently being assayed.

IMMUNOHISTOCHEMICAL STUDIES

A number of studies have recently begun to take advantage of the specificity and sensitivity of antibodies to determine at the cellular level carcinogen-DNA adducts. By immunoperoxidase techniques detection levels of 5 O^6EtdGuo/10^6 bases (about 5×10^4 O^6EtdGuo per diploid genome) and 0.4 AAF-dGuo/10^6 bases (44) were reported in treated rats. Two studies have looked at AAF adducts by immunofluorescence (46,47). A detection limit of about 2000 adducts per cell was reported for fibroblasts treated in culture with N-Aco-AAF (46) while an animal feeding study indicated 10^5 adducts per cell could be seen in liver sections (47). Similar sensitivities were reported by this group for detection of benzo[a]pyrene adducts in mouse keratinocyte nuclei (49). Antibodies specific for UV damaged DNA were used to study damage and repair of nuclear DNA in nude mouse and human skin *in vivo* by immunofluorescence and immunoperoxidase (48).

Highly sensitive computer assisted techniques are now available for quantitation. Fluorescence images can be amplified by an image intensifier and fed into an image analysis system with a high sensitivity camera. Using highly purified monoclonal antibodies to O^6EtdGuo coupled directly to a fluorescent probe approximately 700 O^6-Et d Guo residues per diploid genome could be detected (50). This corresponds to about 7 adducts per 10^8 nucleosides or 700 adducts per cell.

We have carried out immunofluorescence studies on human keratinocytes treated with 8-MOP and UVA. After treatment with 250 ng of 8-MOP/ml and 12 J/cm^2 specific nuclear staining can be detected (Figure 2).

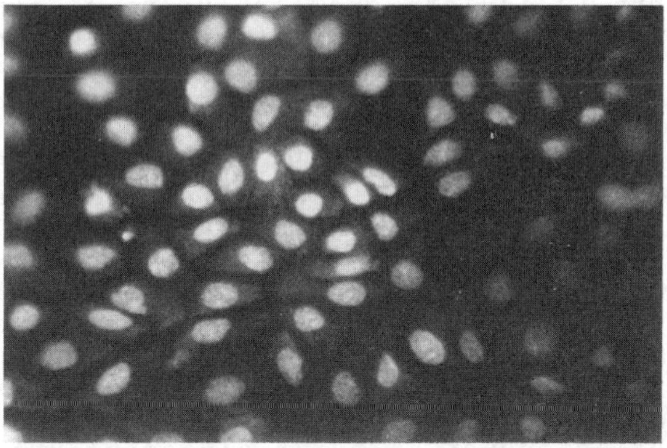

Fig. 2. – Immunohistochemical localization of 8–MOP–DNA adducts in human keratinocytes treated with 250 ng/ml 8–MOP and 12 J/cm^2.

DISCUSSION

A large number of polyclonal and monoclonal antibodies are now available for detecting and quantitating carcinogen DNA adducts. These antibodies can be utilized in immuno-histochemical studies to localize carcinogen adducts in various tissues or cell types. In addition, they are an essential tool for quantitating adduct levels in exposed populations by RIA or ELISA. This approach, termed "molecular cancer epidemiology", can help identify at a molecular level specific exogenous agents or host factors that play a role in human cancer causation. A limited number of studies have already begun to apply these techniques to human monitoring. Several studies (discussed above) have used polyclonal and monoclonal antibodies to BPDE-I-DNA to quantitate adducts in lung, placental and white blood cell DNA of smokers and nonsmokers as well as several occupational groups (34 37). Antibodies against cis platinum modified DNA have been used to quantitate adduct levels in the white blood cells of patients undergoing chemotherapy with this agent (51). Another group has used antibodies to the various mono and cross-linked cisplatinum adducts to quantitate their levels in hydrolysates of DNA from cells treated in culture as well as a limited number of human samples (52). Alkylation of DNA was measured in oesophageal tissue samples from cancer patients in a high risk region of China (53). In this study the DNA was digested and the O^6-alkylguanine adducts isolated by HPLC before quantitation by RIA. Antibodies to aflatoxin have be used to make affinity columns to isolate from the urine of exposed populations the excised aflatoxin-guanine adduct as well as metabolites for quantitation (54).

In addition to the immunologic approaches to quantitating adduct levels on DNA a ^{32}P post-labeling assay has been developed (55,56). For this procedure, samples of DNA are enzymatically digested to 3'monophosphates. With ^{32}P labeled ATP a phosphate group is attached to the 5' position. Extensive chromatographic seperation removes the labeled normal nucleotide diphosphates and allows visualization of the modified diphosphates. A recent improvement in the assay is reported (57) to give sensitivites of one adduct in 10^{10} nucleosides. This procedure was used to detect an adduct in human placental DNA related to exposure to cigarette smoke (37). Fluorescence techniques have been used to quantiate adducts of those carcinogens that fluoresce. BPDE-I-DNA adducts have been measured in the lymphocytes of workers by synchronous fluorescence (35,58). Excised aflatoxin-guanine adducts have been quantitated in the urine of an African population exposed to high levels of this compound (59) by HPLC with fluorescence detection. The number of studies monitoring DNA adducts in human populations is increasing. When these laboratory techniques are combined with epidemiologic studies, they will provide a useful basis for assessing quantitatively human risk from environmental exposure. In addition, they should provide valuable information on human cancer causation.

ACKNOWLEDGMENTS

The authors wish to thank Drs. F. Gasparro and R. Edelson for 8-MOP samples, V. DeLeo for human keratinocytes, C.D. Lin and N. Dharmaraja for excellent technical asistance; and S. Allen for secretarial assistance. This work was supported by grants from the Council for Tobacco Research U.S.A., Inc., #1483A; NIH CA 21111, and SCOR ES 03881.

REFERENCES

1. Strickland, P.T., and Boyle, J.M. Immunoassay of carcinogen-modified DNA, In: Progress in Nucleic Acid Research & Molecular Biology, (ed) Cohn, W. E., Academic Press, New York, 31, pp. 1-58 (1984).
2. Poirier, M.C. Antibodies to carcinogen-DNA adducts, J. Natl. Cancer Inst. 67, 515-519 (1981).
3. Erlanger, B.R., and Beiser, S.M. Antibodies specific for ribonucleosides and ribonucleotides and their reaction with DNA, PNAS USA 52, 68-74 (1964).
4. Erlanger, B.F. Principles and methods for the preparation of drug protein conjugates for immunological studies, Pharmacl. Rev 25, 271-280 (1973).
5. Mullcr, R., and Rajewsky, M.F. Antibodies specific for DNA components structurally modified by chemical carcinogens, J. Cancer Res. Clin. Oncol. 102, 99-113 (1981).
6. Wild, C.P., Smart, G., Saffhill, R., and Boyle, J.M. Radioimmunoassay of 06 methyldeoxyguanosine in DNA of cells alkylated in vitro and in vivo, Carcinogenesis 4, 1605-1609 (1983).
7. Muller, R., and Rajewsky, M.F. Immunological quantification by high-affinity antibodies of 06-ethyldeoxyguanosine in DNA exposed to N-ethyl-N-nitrosourea, Cancer Res. 40, 887-896 (1980).
8. Rajewsky, M.F., Muller, R., Adamkiewicz, J., and Drosdziok, W. Carcinogenesis: Fundamental Mechanisms and Environmental Effects, Reidel. Press, pp. 207-218 (1980).
9. Van der Laken, C.J., Hagenaars, A.M., Hermsen, G., Kriek, E., Kuipers, A.J., Nagel, J., Scherer, E., and Welling, M. Measurement of 06-ethyl-deoxyguanosine and N-(deoxyguanosine-8-yl)-N-acetyl-2- aminofluorene in DNA by high-sensitive enzyme immunoassays, Carcinogenesis 3, 569-572 (1982).
10. Poirier, M.C., Yuspa, S.H., Weinstein, I.B., and Blobstein, S. Detection of carcinogen-DNA adducts by radioimmunoassay, Nature (London) 270, 186-188 (1977).
11. Sage, E., Fuchs, R.P., and Leng, M. Reactivity of the antibodies to DNA modified by the carcinogen N-acetoxy-N-acetyl-2-aminofluorene, Biochemistry 18, 1328 (1979).
12. Poirier, M.C., Santella, R., Weinstein, I.B., Grunberger, D., and Yuspa, S.H. Quantitation of benzo[a]pyrene-deoxyguanosine adducts by radioimmunoassay, Cancer Res. 40, 412-416 (1980).
13. Santella, R.M., Lin, C.D., Cleveland, W.L., and

Weinstein, I.B. Monoclonal antibodies to DNA modified by a benzo[a]pyrene diol epoxide, Carcinogenesis 5, 373-377 (1984).
14. Slor, H., Mizusawa, N., Nechart, T., Kakefuda, R., Day, R.S., and Bustin, M. Immunochemical visualization of binding of the chemical carcinogen benzo[a]pyrene diol epoxide to the genome, Cancer Res. 41, 3111-3117 (1981).
15. Malfoy, B., Hartmann, B., Macquet, J.P., and Leng, M. Immunochemical studies of DNA modified by cis dichlorodiammine platinum (II), Cancer Res. 41, 4127-4131 (1981).
16. Poirier, M.C., Lippard, S., Zwelling, L.A., Ushay, M., Kerrigan, D., Santella, R.M., Grunberger, D., and Yuspa, S.H. Antibodies elicited against cis - diamminedichloroplatinum(II)-modified DNA are specific for cis- diamminedichloroplatinum(II)-DNA adducts formed in vivo and in vitro, Proc. Natl. Acad. Sci. USA 79, 6443-6447 (1982).
17. Fichtinger-Schepman, A., Baan, R., Luiten-Schuite, A., VanDijk, M., and Lohman, P.H.M. Immunochemical Quantitation of Adducts Induced in DNA by cis-Diamminedichloroplatinum(II) and analysis of Adduct Related DNA-Unwinding, Chem-Biol. Interactions 55, 275-288 (1985).
18. Hsieh, L.L., Jeffrey, A.M., and Santella, R.M. Monoclonal antibodies to 1-aminopyrene-DNA, Carcinogenesis 6, 1289-1293 (1985).
19. Santella, R.M., Dharmaraja, N., Gasparro, F.P., and Edelson, R.L. Monoclonal Antibodies to DNA modified by 8-methoxypsoralen and ultraviolet A light, Nucleic Acids Res. 13, 2533-2544 (1985).
20. Zarebska, Z., Jarbabek-Chorzelska, M., Chorzelski, T., and Zablonska, S. Immune serum against anti DNA-8-methoxypsoralen photoadduct, Z. Naturforsch 39, 136-140 (1984).
21. Haugen, A., Groopman, J.D., Hau, I.C., Goodrich, G.R., Wogan, G.W., and Harris, C.C. Monoclonal antibody to aflatoxin B1-modified DNA detected by enzyme immunoassay, PNAS USA 78, 4124-4127 (1981).
22. Hertzog, P.J., Smith, J.R.L., and Garner, R.C. Production of monoclonal antibodies to guanine imidazole ring opened aflatoxin B1 DNA, the persistent DNA adduct in vivo, Carcinogenesis 3, 825-828 (1982).
23. Leadon, S.A., and Hanawalt, P.C. Monoclonal antibody to DNA containing thymine glycol, Mutation Res. 112, 191-200 (1983).
24. Strickland, P.T., and Boyle, J.M. Characterisation of two monoclonal antibodies

specific for dimerised and non-dimerised adjacent thymidines in single stranded DNA, Photochem. Photobiol. 34, 595-601 (1981).

25. Perera, F.P., and Weinstein, I.B. Molecular epidemiology and carcinogen-DNA adduct detection: New approaches to studies of human cancer causation, J. Chron. Dis. 35, 581-600 (1982).

26. Miller, E.C. Some current perspectives on chemical carcinogenesis in humans and experimental animals: Presidential Address, Cancer Res. 38, 1479-1496 (1978).

27. Pereira, M.A., and Chang, L.W. Binding of chemical carcinogens and mutagens to rat hemoglobin, Chem-Biol. Interactions 33, 301-305 (1981).

28. Calleman, C.J. In vivo dosimetry by means of alkylated hemoglobin-a tool in the design of tests for genotoxic effects, In: Banbury Report: Indicators of Genotoxic Exposure, (eds) Bridges, B.A., Butterworth, B.E., and Weinstein, I.B., CSH Lab., New York, pp. 157-167 (1982).

29. Calleman, C.J., Ehrenberg, L., Jansson, B., Osterman-Golker, S., Segerback, D., Svensson, K., and Wachtneister, C.A. Monitoring and risk assessment by means of alkyl groups in hemoglobin in persons occupationally exposed to ethylene oxide, J. Environ. Path. & Toxic. 2, 427-442 (1978).

30. Ehrenberg, L., Hiesche, K.D., Osterman-Golkar, S., and Wennberg, I. Evolution of genetic risks of alkylating agents: Tissue doses in the mouse from air contaminated with ethylene oxide, Mutation Res. 24, 82-203 (1974).

31. Santella, R.M., Hseih, L.L., Lin, C.D., Viet, S., and Weinstein, I.B. Quantitation of exposure to benzo[a]pyrene with monoclonal antibodies, Env. Health Perspect. 95-100 (1985).

32. Shamsuddin, A.M., and Harris, C.C. Improved enzyme immunoassays using biotin-avidin-enzyme complex, Arch. Pathol. Lab. Med. 107, 514 517 (1983).

33. Leipold, B., and Remy, W. Use of avidin-biotin-peroxidase complex for measurement of UV lesions in human DNA by micro ELISA, J. Immun. Meth. 66, 227-234 (1984).

34. Perera, F.P., Poirier, M.C., Yuspa, S.H., Nakayama, J., Jaretzki, A., Curnen, M.M., Knowles, D.M., and Weinstein, I.B. A pilot project in molecular cancer epidemiology: determination of benzo[a]pyrene-DNA adducts in animal and human tissues by immunoassays, Carcinogenesis 3, 1405-1410 (1982).

35. Harris, C.C., Vahakangas, K., Newman, J.M., Trivers, G.E., Shamsuddin, A., Sinopoli, N., Mann, D.L., and Wright, W.E. Detection of benzo[a]pyrene diol epoxide-DNA adducts in peripheral blood lymphocytes and antibodies to the adducts in serum from coke oven workers, Proc. Natl. Acad. Sci. USA 82, 6672-6676 (1985).
36. Shamsuddin, A.K.M., Sinopoli, N.T., Hemminki, K., Boesch, R.B., and Harris, C.C. Detection of benzo[a]pyrene: DNA adducts in human white blood cells, Cancer Res. 45, 66-68 (1985).
37. Everson, R.B., Randerath, E., Santella, R.M., Cefalo, R.C., and Avitts, T.A. Detection of smoking-related covalenty DNA adducts in human placenta, Science 231, 54-57 (1986).
38. Santella, R.M., Lin, C.D, and Dharmraja. N. Monoclonal Antibodies to a Benzo[a]pyrene diolepoxide modified protein, Carcinogenesis 7, 441-444 (1986).
39. Parrish, J.A., Fitzpatrick, T.B., Pathak, M.A., and Tannenbaum, L. Photochemotherapy of psoriasis with oral methoxysalen and long wave ultraviolet light, NE Jr. Med. 291, 1207-1211 (1974).
40. Song, P.S., and Tapley, J.K. Photochemistry and photobiology of psoralens, Photobiol. 29, 1177-1197 (1979).
41. Eskelinen, A., Halme, K., and Lassus, A. Risk of cutaneous carcinoma in psoriatic patients treated with PUVA, Photoderm. 2, 10-14 (1985).
42. Freeman, K., and Warin, A.P. Acute myelomonocytic leukemia developing in a patient with psoriasis treated with oral 8-methoxypsoralen and longwave ultraviolet light, Clinical & Exper. Derm. 10, 144-146 (1985).
43. Ljunggren, B., Carter, M., Albert, J., and Reid, T. Plasma levels of 8-methoxypsoralen determined by high pressure liquid chromotography in psoriatic patients injesting drug from two monofactors., J. Invest. Derm. 74, 59-62 (1980).
44. Menkveld, G.J., VanDerLaken, C.J., Hermsen, T., Kriek, E., Scherer, E., and Engelse, L.D. Immunohistochemical localization of O6-ethyldeoxyguanosine and deoxyguanosin-8-yl(acetyl) aminofluorene in liver sections of rats treated with diethylnitrosamine, ethylnitrosourea, or N-acetylaminofluorene, Carcinogenesis 6, 263-270 (1985).
45. Heyting, C., VanDerLaken, C.J., VanRaamsdonk, W., and Pool, C.W. Immunohistochemical detection of O6-ethyldeoxyguanosine in the rat brain after in vivo applications of N-ethyl-N-nitrosourea,

Cancer Res. 43, 2935-2941 (1983).
46. Muysken-Schoen, M.A., Baan, R.A., and Lohman, P. H.M. Detection of DNA adducts in N-acetoxy-2-acetylaminofluorene- treated human fibroblasts by means of immunofluorescence microscopy and quantitative immunoautoradiography, Carcinogenesis 6, 999-1004 (1985).
47. Huitfeldt, H.S., Spangler, E.F., Hunt, J.M., and Poirier, M.C. Immunohistochemical localization of DNA adducts in rat liver tissue and phenotypically altered foci during oral administration of 2-acetylaminofluorene, Carcinogenesis 7, 123-129 (1986).
48. Eggset, G., Volden, G., and Krokan, H. U.v.-induced DNA damage and its repair in human skin in vivo studied by sensitive immunohistochemical methods, Carcinogenesis 4, 745-750 (1983).
49. Poirier, M.C., Stanley, J.R., Beckwith, J.B., Weinstein, I.B., and Yuspa, S.H. Indirect immunofluorescent localization of benzo[a]pyene adducted to nucleic acids in cultured mouse keratinocyte nuclei, Carcinogenesis 3, 345-348 (1982).
50. Adamkiewicz, J., Eberle, G., Huh, N., Nehls, P., and Rajewsky, M.F. Quantitation and visualization of alkyl deoxynucleosides in the DNA of mammalian cells by monoclonal antibodies, Environ. Health Persp. 62, 49-55 (1985).
51. Poirier, M.C., Reed, E., Zwelling, L.A., Ozols, R.F., and Yuspa, S.H. The use of polyclonal antibodies to quantitate cis-platinum drug DNA adducts in cancer patients, Env. Health Prosp. 62, 89-94 (1985).
52. Plooy, A., Fichtinger-Schepman, A., Schutte, H. H., vanDijk, M., and Lohman, P.H.M. The quantitative detection of various Pt-DNA-adducts in Chinese hamster ovary cells treated with cisplatin: application of immunochemical techniques, Carcinogenesis 6, 561-566 (1985).
53. Umbenhauer, D., Wild, C.P., Montesano, R., Saffhill, R., Boyle, J.M., Huh, N., Kirstein, U., and Rajewsky, M.F. O-6 Methyldeoxy-guanosine in oesophageal DNA among persons at high risk of oesophageal cancer, Intl J. Cancer 36, 661-665 (1985).
54. Groopman, J.D., Donahue, j.P.R., Zhu, J., Chen, J., and Wogan, G.N. Aflatoxin metabolism in humans: Detection of metabolites and nucleic acid adducts in urine by affinity chromatography, Proc. Natl. Acad. Sci. USA 82, 6492-6496 (1985).
55. Gupta, R.C., Reddy, M.V., and Randerath, K. [32P]-postlabeling analysis of nonradioactive

aromatic caracinogen-DNA adducts, Carcinogenesis 3, 1081-1092 (1982).
56. Randerath, K., Randerath, E., Agrawal, H.P., and Reddy, M.V. Biochemical (postlabelling) Methods for Analysis of Carcinogenic-DNA Adducts, In: Monitoring Human Exposure to Carcinogenic and Mutagenic Agents, (eds) Berlin, A., Draper, M., Hemminki, K., and Vainio, H., IARC, Lyon France, pp. 217-232 (1984).
57. Gupta, R.C. Enhanced Sensitivity of 32P-Postlabeling Analysis of Aromatic Carcinogen: DNA Adducts, Cancer Res. 45, 5656-5662 (1985).
58. Vahakangas, K., Haugen, A., and Harris, C.C. An applied synchronous fluorescence spectrophotometric assay to study benzo[a]pyrene-diolepoxide-DNA adducts, Carcinogenesis 6, 1109-1116 (1985).
59. Autrup, H., Bradley, K.A., Shamsuddin, A.K.M., Wakhisi, J., and Wasunna, Q. Detection of putative adduct with fluorecence characteristics identical to 2,3-dehydro-2-(7'-guanyl)-3-hydroxyaflatoxin B1 in human urine collected in Murang's District, Kenya, Carcinogenesis 4, 1193-1195 (1983).

48

Cytogenetic methods for assessing human exposure to genotoxic chemicals

A. Forni – Institute of Occupational Health, University of Milan, Italy

INTRODUCTION

Assessment of genotoxicity in man by cytogenetic methods is applied in order to detect exposure or early effects of exposure to known or potential mutagenic/carcinogenic chemicals. Ideally, genomic effects at target organs should be studied, but this is seldom practicable. Peripheral blood lymphocytes are the cell system most extensively used in cytogenetic studies, due to ease of blood sampling, the possibility of obtaining large numbers of scorable cells in short-term cultures and the demonstrated sensitivity of this cell system in detecting visible chromosome damage induced by a well known mutagen, such as ionizing radiations. Compared to radiations, however, evaluating genotoxicity of chemicals is more difficult for several reasons, such as different mechanisms of action of different substances, individual differences in metabolism, positive or negative interference by several agents, no knowledge of the concentration of the active metabolite/s at the target tissues (Wolff, 1982). Despite these problems and those concerning interpretation of the results, which will be discussed later, cytogenetic monitoring is increasingly performed mainly in occupationally exposed groups (see review by Ashby and Richardson, 1985).

GENERAL ISSUES IN CYTOGENETIC MONITORING

Three cytogenetic endpoints in cultured lymphocytes are presently considered: structural chromosome aberrations (CA), sister chromatid exchanges (SCE) and micronuclei (MN). Each method has its own potentialities and limitations, which will be separately discussed. However, several issues in cytogenetic monitoring of groups at risk are common to the three tests.

Selection of study subjects and of controls is the first important step in designing cytogenetic studies. In preliminary investigations, groups of subjects with heavy or long-term exposures should be preferentially chosen, in order to detect the sensitivity of the endpoint. Exposure data should be carefully collected and exposure, when feasible, should be measured. The groups should be as homogenous as possible in regard to exposure. The group size should be in the order of tens (see in Archer et al., 1981), generally larger if exposure is lower, in order to obtain statistically significant results.

An equal number of controls matched for age, sex and life habits (e.g., smoking), with no history of occupational exposure/s, should be studied. Individuals with important exposures to other known chromosome-damaging agents (e.g., therapeutic or heavy diagnostic X-radiation, recent viral diseases or vaccinations) should not be considered for study or as controls. Prospective studies, with subjects serving as their own controls might be desirable.

Standardization of techniques, scoring and reporting is essential. Lymphocyte cultures should be standardized for type of material (whole blood versus purified lymphocytes), culture medium, culture supplement/s, culture time and so on, at least for the whole period of study. Control cultures should be performed concurrently, in order to avoid possible intralaboratory differences due to change of methods. The cytogenetic endpoint should be scored on a suitable number of cells, which is related to the background frequency (100-200 metaphases for CA, 25-50 metaphases for SCE, 2000-3000 cells for MN), and criteria for scoring and reporting should be standardized (see Forni, 1984, for CA; Archer et al., 1981, for SCE; Högstedt, 1984, for MN).

The results obtained in the study group/s should undergo statistical analysis in comparison to those of the control group by a suitable test, as discussed, f.i., by Archer et al. (1981). Sophisticated statistical methods might be applied in order to take into account several variables. If reliable exposure data is available, the possibility of dose-response relationships should be investigated. In large studies, splitting the whole group in smaller, but sufficiently large, subgroups with different exposure levels might serve to detect an effect, which otherwise might go undetected due to dilution.

STRUCTURAL CHROMOSOME ABERRATIONS

Acquired chromosome aberrations (CA) are the first cytogenetic endpoint investigated in assessing exposure to mutagens since the sixties, and the most extensively studied. The test has been largely discussed on several occasions (see, e.g., Archer et al., 1981; Gebhart, 1981; Forni, 1984).

From the methodological point of view, it is generally agreed that 2-day lymphocyte cultures are the most suitable, since they enable to score cells at first mitosis in vitro and so to evaluate the in vivo situation. Longer culture times may influence the aberration yield, due to loss of heavily damaged cells after 1st mitosis and, on the other hand, due to increase of culture-borne aberrations.

The aberrations scorable at direct observation are of the chromatid-type, if only one chromatid is involved, or "chromosome-type", if both chromatids are equally involved (see classification in Buckton and Evans, 1973; ISCN, 1985). In evaluating the in vivo effect of ionizing radiations and of some radiomimetic chemicals, only chromosome-type aberrations are considered. Most chemicals, however, are S-dependent, i.e. reveal an unrepaired damage during the S-phase of the next cell cycle, giving rise to a chromatid-type aberration at 1st division and to a "derived" chromosome-type aberration at 2nd division. Therefore, when evaluating the effect of exposure to these agents, both types of aberrations should be considered. One should be aware, however, that chromatid-type aberrations can be also induced in culture by technical factors, which might partially account both for inter- and for intralaboratory variations of results.

Chromatid- and chromosome-type aberrations, due to their different significance, should be reported separately. Statistical analysis of results should be done both for total abnormal metaphases and for single type of aberration or groups of aberrations (see for details Forni, 1984).

Structural chromosome aberrations are a specific indicator of clastogenic effect, but are aspecific for physical, chemical and biological clastogens. Therefore, the possible factors of confounding have to

be taken into account, when choosing the subjects for study and interpreting the results.
Persistence of induced chromosome aberrations for years and decades after cessation of exposure has been demonstrated both in irradiated subjects (Buckton et al., 1962) and in individuals heavily exposed to benzene (see Forni, 1979, for review). If, on one hand, this represents an advantage, since it enables detection of a past damage and makes CA, within certain limits, an indicator of cumulative exposure, on the other hand this renders the test unsuitable to evaluate the possible effect of present low-level exposure/s in subjects with past heavier exposures.
No sufficient knowledge so far exists regarding dose-response relationships and the sensitivity of the test for low-level exposures. Evans et al. (1979) could demonstrate a correlation between cumulative low-level exposure and "unstable" chromosome-type aberrations, independent from effect of age, by pooling an enormous number of data obtained in a long-term follow-up of a group of subjects occupationally exposed to low levels of ionizing radiations. For exposure to chemicals, the problem might be even greater, due to difficulties in measuring exposure.
Finally, a limitation in using this method routinely on a large scale is the rather high cost, due to the fact that scoring is difficult and time-consuming and that large number of metaphases are to be scored, thus requiring several hours of well-trained personnel.

SISTER CHROMATID EXCHANGES (SCE)

SCE in lymphocytes is another cytogenetic endpoint which in the last 10 years has become increasingly popular in monitoring groups at risk of exposure to mutagens-carcinogens (see Archer et al., 1981; Sorsa, 1984; Ashby and Richardson, 1985, for reviews). Proposed in 1975 by Perry and Evans as a very sensitive in vitro method, with possible in vivo application to man, as a simpler and more sensitive substitute for chromosome aberration testing, this test has revealed over the years its advantages and weaknesses, and the initial enthusiasm has somewhat decreased (Wolff, 1982).
The test consists in culturing cells, e.g., lymphocytes, for two cell cycles in the presence of bromodeoxyuridine (BUdR), which, when incorporated into DNA, induces at 2nd metaphase a differential stainability of sister chromatids, and thus enables to detect exchanges between the two. The mechanism of SCE formation and the biological significance of this endpoint

are still unknown. BUdR itself is responsible, at least in part, for "spontaneous" SCE frequency. From the experience available, SCE are a sensitive indicator of exposure to S-dependent chemicals, while they are not significantly increased by ionizing radiations and some radiomimetic agents.

From a technical point of view, concentration of BUdR in relation to cycling cells (Carrano et al., 1980), time between blood sampling and initiation of culture (Nemenzo et al., 1985), culture time (Parkes et al., 1985) etc. seem to influence the mean baseline number of SCE/cell. The mean background frequency in control subjects is rather high, and variable in different reports (5-14 SCEs/metaphase). There is also large variation of SCE/cell (from 0 to > 20) in the same culture, suggesting the existence of different lymphocyte subpopulations with different "spontaneous" susceptibility to SCE induction.

The high sensitivity of the test and its aspecificity for agents are responsible for interference from other exposure/s or life habits. Smoking is reported by most authors to increase SCE, but recently also cessation of smoking has been reported to positively affect this endpoint (Wulf et al., 1985).

The effect of age and sex is debated. In a recent report, aging resulted to be positively correlated to SCE frequency, and this might be due either to a cumulative damage or to selection with age of lymphocytes with high susceptibility to SCE induction (Sarto et al., 1985).

In monitoring in vivo exposure to mutagens, SCE induction is fast, but seems to be less persistent than CA, which makes the test more suitable for evaluating the effect of current or relatively recent exposure.

From a review comparing SCE and CA in mutagenicity testing, about 30% of the agents adequately tested by both methods gave qualitatively discordant results in either direction, thus suggesting that the mechanisms of induction of these endpoints are different (Gebhart, 1981). Also in human monitoring, despite a rather general agreement of positive or negative results between CA and SCE, there is evidence that, at least for exposure to some chemicals (e.g., benzene) or in certain situations (e.g., past exposure to vinyl chloride), one or the other test is more suitable. The two methods might, in a way, be complementary (Anderson, 1981; Gebhart, 1981).

MICRONUCLEI

First described in human bone marrow cells of subjects receiving chemotherapy, as an acute short-las-sting effect, MN represent an endpoint extensively exploited as an in vivo short-term test of genotoxicity in experimental animals (see review in Heddle, 1983).

MN are small, additional nuclei resulting from exclusion from the main nucleus of a daughter cell at the end of mitosis, of chromosome fragments or of whole lagging chromosomes. The test is therefore an indirect indicator both of chromosome breakage and of partial impairment of the mitotic spindle, and can detect the effect both of clastogens and of mitotic spindle poisons.

In human monitoring, the determination of MN frequency has been applied to investigate genotoxicity at target tissues, e.g.,oral mucosa epithelial cells in groups at risk for oral cancer (Stich et al., 1982) and, even more recently, on lymphocytes (Högstedt et al., 1983), with the claim of simplicity of scoring. From the few studies available, culture time seems to influence this endpoint (see Högstedt, 1984 for methodological problems). Due to very low background frequencies, a large number of cells must be scored. Recently, an effect of age has been demonstrated by Norman et al. (1985). According to these authors, an increased frequency of MN seems to be predictive of chromosome-type aberrations; therefore, the test might be used to screen and select persons for CA studies. However, due to recent application and scantiness of data, the test must still be validated.

INTERPRETATION AND HEALTH SIGNIFICANCE OF CYTOGENETIC ENDPOINTS

From the data reported and from a comparison of the three cytogenetic methods considered, it is obvious that no ideal test exists, since all are aspecific for agents and confounded by several factors. Therefore, one should be very careful in interpreting the results and attributing positive data to a certain exposure, unless factors of confounding can be ruled out. The use of matched controls can partially solve these problems.

An issue on which there is general agreement is that interpretation of results of cytogenetic studies should be done mainly on a group basis (see Anderson, 1981, and several papers in Berlin et al., 1984).

One of the most debated issues concerning these tests is the health significance of positive findings (see in Archer et al., 1981; Wolff, 1982; several papers in Berlin et al., 1984). It is generally agreed that groups with significantly increased rates of CA might be at an increased risk for cancer. No evidence so far exists for a relationship between CA in lymphocytes and later cancer occurrence at the individual level, except for a few cases of benzene leukaemia where a chromosomal effect, with abnormal clone formation, could be demonstrated also in the target tissue, the bone marrow (see Forni, 1979, for review). For SCE, which do represent an effect on DNA without genomic changes, the relationship with possible mutational events in somatic or germinal cells is even more poorly understood, and based on the capability of most carcinogens to induce SCE in vitro or in vivo. To try to clarify these important points, epidemiological follow-up of groups cytogenetically studied for reproductive effects and cancer risk is urgently needed. These uncertainties raise also ethical problems: one concerns the information of the subjects studied about the results of the cytogenetic investigations; the other is whether cytogenetic testing should be considered as biological monitoring or rather surveillance of selected occupationally or environmentally exposed groups for detection of early adverse effects.

REFERENCES

Anderson D. (1981) Mutagenicity testing. Rev. Environ. Health 3, 369-433.

Archer P.G., Bender M., Bloom A.D. et al. (1981) Report of Panel 1: Guidelines for cytogenetic studies in mutagen-exposed human populations. In: Guidelines for studies of human populations exposed to mutagenic and reproductive hazards, Bloom A.D. ed., March of Dimes Birth Defects Foundation, White Plans, N.Y., pp. 1-35.

Ashby J. and Richardson C.R. (1985) Tabulation and assessment of 113 human surveillance cytogenetic studies conducted between 1965 and 1984. Mutat. Res. 154, 111-113.

Berlin A., Draper H., Hemminki K. and Vainio H. eds. (1984) Monitoring human exposure to carcinogenic and mutagenic agents. IARC Scientific Publications No. 59. International Agency for Research on Cancer, Lyon.

Buckton K.E. and Evans H.J. eds. (1973) Methods for the analysis of human chromosome aberrations. WHO, Geneva.

Buckton K.E., Jacobs P.A., Court Brown W.M. and Doll R. (1962) A study of the chromosome damage persisting after X-ray therapy for ankylosing spondylitis. Lancet 2, 676-682.

Carrano A.V., Minkler J.L., Stetka D.G. and Moore D.H. (1980) Variation in the baseline sister chromatid exchange frequency in human lymphocytes. Environ. Mutagen. 2, 325-337.

Evans H.J., Buckton K.E., Hamilton G.E. and Carothers A. (1979) Radiation-induced chromosome aberrations

in nuclear-dockyard workers. Nature 277, 531-534.
Forni A. (1979) Chromosome changes and benzene exposure. A review. Rev. Environ. Health 3, 5-17.
Forni A. (1984) Chromosomal aberrations in monitoring exposure to mutagens-carcinogens. In Berlin A. et al. eds., pp. 325-337.
Gebhart E. (1981) Sister chromatid exchange (SCE) and structural chromosome aberration in mutagenicity testing. Human Genet. 58, 235-254.
Heddle J.A., Hite M., Kirkhart B. et al. (1983) The induction of micronuclei as a measure of genotoxicity. Mutat. Res. 123, 61-118.
Högstedt B. (1984) Micronuclei in lymphocytes with preserved cytoplasm: a method for assessment of cytogenetic damage in man. Mutat. Res. 130, 63-72.
Högstedt B., Åkesson B., Axell K. et al. (1983) Increased frequency of lymphocyte micronuclei in workers producing reinforced polyester resin with low exposure to styrene. Scand. J. Work Environ. Health 9, 241-246.
ISCN (1985) An International System for Human Cytogenetic Nomenclature. Harnden D.G. and Klinger H.P. eds. Karger, Basel.
Nemenzo J.H., Richmond G.W. and Hine C.H. (1985) Unstable SCE rates of blood lymphocytes of workers exposed to ethylene oxide. J. Occup. Med. 27, 718-718.
Norman A., Bass D. and Roe D. (1985) Screening human populations for chromosome aberrations. Mutat. Res. 143, 155-160.
Parkes D.J.G., Scott D. and Stewart A. (1985) Changes in spontaneous SCE frequencies as a function of sampling time in lymphocytes from normal donors and cancer patients. Mutat. Res. 147, 113-122.
Perry P. and Evans H.J. (1975) Cytological detection of mutagen-carcinogen exposure by sister chromatid exchange. Nature 258, 121-125.
Sarto F., Faccioli M.G., Cominato I and Levis A.G. (1985) Aging and smoking increase the frequency of sister-chromatid exchanges (SCE) in man. Mutat. Res. 144, 183-187.
Sorsa M. (1984) Monitoring of sister chromatid exchange and micronuclei as biological endpoints. In Berlin A. et al. eds., pp. 339-349.
Stich H.F., Curtis J.R. and Parida B.B. (1982) Application of the micronucleus test to exfoliated cells of high cancer risk groups: tobacco chewers. Int. J. Cancer 30, 553-559.
Wolff S. (1982) Difficulties in assessing the human health effects of mutagenic carcinogens by cytogenetic analyses. Cytogenet. Cell Genet. 33, 7-13.
Wulf H.C., Husum B. and Neibuhr E. (1985) Cessation of smoking enhances sister chromatid exchanges in lymphocytes. Hereditas 102, 195-198.

49

Monitoring human exposure to genotoxic chemicals

J. Ashby – Imperial Chemical Industries PLC, Alderley Park, Cheshire, SK10 4TJ U.K.

INTRODUCTION

There is a need for methods to determine the extent of human exposure to carcinogens and genotoxic chemicals, both in the environment and the workplace. The stimulus is provided by the fact that about 30 environments have been associated with the chemical induction of cancer in man. Cancer epidemiology has the advantage of being direct, ie, it measures an increase in the incidence of cancer mortality. But of course, such studies only monitor carcinogenesis and can only lead to a corrective response after the event; further, the available techniques are mostly insensitive. As a consequence, a range of methods has been developed over the past 20 years aimed either at quantifying the extent of exposure to known genotoxins or at monitoring exposed individuals for evidence that a carcinogen or a genotoxic agent has interacted with and/or modified somatic tissue. The present article attempts to compare several of the techniques and to relate them to some which are currently under development. Thus, this review starts with an appraisal of cytogenetic methods employing peripheral blood lymphocytes and concludes with consideration of techniques employing the recognition of chemical-DNA adducts via the use of monoclonal antibodies.

OBJECTIVES OF HUMAN SURVEILLANCE STUDIES

An acute dilemma is apparent: Surveillance studies are almost exclusively initiated in situations where people are exposed to a known animal or human carcinogen, but the data from the available surveillance techniques cannot be used to predict an increased risk of cancer for the exposed individual. Even in the case of cytogenetic studies, where discernible changes can be recognised in the structure of chromosomes, it is considered inappropriate to communicate individual results to those exposed. Such uncertainty regarding the personal significance of surveillance data will necessarily become more marked when considering the significance of the detection of one chemical lesion per cell or of discerning the level of such lesions which should be considered not to represent any hazard, even to a group of individuals. This rather negative note is sounded to encourage animal studies, or human epidemiological studies in which several parameters are compared and whose aim would be to discern the relationship between the parameters used in current human surveillance studies, and the eventual induction of heritable mutagenic effects or cancer. If such studies are not done, or if a poor correlation is found, then effort should be devoted to the development of techniques which will act as an early warning to the eventual development of such events. It is accepted that cancer may not be the only consequence of a chemical interacting with genetic material, but it is the only one which can be used at present to determine the sensitivity and relevance of individual surveillance techniques. Therefore, in this article, it has not been assumed that a technique is valuable just because it is 'sensitive' to the agent under study. The particular danger is that effort may be focused solely on increasing the sensitivity of a particular technique, irrespective of the phenotypic significance to the individual of the parameter being studied. Thus, the validity of particular techniques is judged herein by the extent to which they provide a true measure of the damage induced by a genotoxic chemical to genetic material - some techniques, may, as suggested below, provide a quite misleading picture.

CLASSES OF SURVEILLANCE TECHNIQUE

The probable relevance of human surveillance data to the eventual induction of cancer in an exposed individual may be estimated, very approximately, using the matrix shown below.

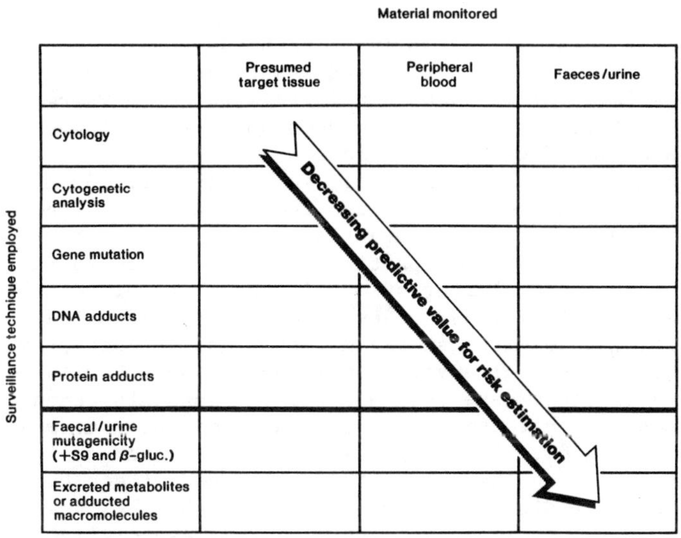

This matrix cannot be used precisely, but the concepts contained within it may contribute to an understanding of the strengths and weaknesses of a given surveillance method in a particular environment. Thus, in situations where the occurrence of a particular human tumour type may be anticipated, the most relevant surveillance data (beyond hygiene) would probably be target tissue cytology. Such situations are relatively rare, but may be illustrated by exfoliative bladder cell studies (see Maltoni, herein) or cytogenetic studies using buccal mucosal cells (see Sarto et al, herein). The development of cellular gene expression techniques, employing oncogene probes, may produce highly specific and early markers of neoplasia in the future; this could enhance the importance of cytological surveillance methods. In contrast, the least definitive surveillance data are probably estimates of the metabolites of a chemical in the urine of exposed people, or determination of the mutagenicity of urine when determined in vitro in the presence of deconjugating enzymes. Such meausurements are only meaningful if the parameter is known to be reflective of a similar, but unmonitored, modification to somatic cell DNA. This will usually be unlikely, and is discussed later.

Most of the surveillance techniques in use at present are contained within the centre of the matrix, ie, those involving DNA or protein studies in the cells of peripheral blood. These methods can be graded in terms of their probable relevance to cancer induction, and they are discussed in this order below.

PERIPHERAL BLOOD CYTOGENETIC STUDIES

This is a well established technique whose general performance has recently been reviewed (Ashby and Richardson, 1985, Galloway et al 1986, Forni, this volume). There is growing evidence that specific chromosomal rearrangements may be associated with one or more of the critical stages of chemically induced cancer; in particular, the modification of normal gene-expression and the disturbance of cellular homeostasis and differentiation. However, it is not yet possible to assign particular significance to the occurrence of chromosomal aberrations in the white blood cells of an exposed individual. Nevertheless, the technique is directly associated with genetic material and as such provides useful data. Several environments containing established human carcinogens have been monitored using this technique, and positive effects were recorded. Nonetheless, many small studies have been reported where the database is insufficient to conclude that a chemically-induced effect has been produced in the individuals sampled. This highlights a problem common to all surveillance techniques, but especially for the cytogenetic methods, because most data exist for them, that an adequate knowledge of intra- and inter-individual control variability is an absolute pre-requisite for a successful study. This subject has been discussed in detail elsewhere (eg, Ashby and Richardson 1985; Galloway et al 1986; Sinha et al 1984, 1986; Forni, this volume) and the reader is directed to these papers, particularly to that of Galloway et al (1986) which presents critical data on the cytogenetic penalties of smoking and age.

The sensitivity of this technique is subject to technical aspects of the study design, and probably to the particular chemical under study. Comparative data exist for people exposed to ethylene oxide (EO) where Sarto et al (1984) and Galloway et al (1986) were able to detect chromosomal aberrations elicited by exposure of workers to between 5-10 ppm. This sensitivity is similar to that of the protein adduct technique for EO (see later). Although more marked effects were observed in people exposed to higher dose-levels of EO, the trend to improved industrial hygiene reduces the likelihood of encountering such populations again. The question of whether new techniques should be developed to monitor exposure to very low levels of EO focuses the central dilemma of surveillance studies. Thus, if EO is employed on a plant under approved conditions (ie taking account of current control limit values, etc), and if a chromosomal study of those individuals revealed the absence of cytogenetic damage, then should the sensitivity of the surveillance technique be increased just to confirm that a very low level of exposure probably still exists? Control limit values are established based on the most recent human or animal carcinogenicity data, and such controlled environments are assumed not to present a carcinogenic hazard. Therefore, a negative cytogenetic study study under such circumstances could be viewed as confirmatory of the validity of the control limit value. Such a situation occurred when a cytogenetic study of the lymphocytes of vinyl chloride (VC) exposed workers was carried out in parallel with a a reduction in the TLV of VC; previously positive effects become negative (Anderson et al 1980). Continued operation of that plant under such conditions of exposure is regarded as justifiable, but it is probable that monoclonal antibody techniques could be devised to detect some evidence of VC interaction with somatic tissue in those workers - the question remains, are such endeavours justified?

In studies where chromosomal analysis (CA) has been combined with an assessment of sister chromatid exchanges (SCE) similar qualitative results have been observed (reviewed in Ashby and Richardson, 1985). Moreover, due to the superior statistical design of most SCE studies, and to the relative ease of assessment of SCE, increased sensitivity of SCE over CA studies seems possible (Galloway et al 1986). However, Galloway et al (1986) observed that SCE was not a good predictor of CA within the same individual, and while this does not devalue the use of the SCE technique as a general surrogate for CA, it raises an interesting alert regarding the assumption that all surveillance techniques are measuring essentially the same phenomenon, but to different levels of sensitivity. The fact that the induction of SCE may be more related to DNA damage than to the induction of chromosomal aberrations means that chemical lesions on DNA may likewise not be predictive of CA in the same person. As aberrations are not regarded as automatically predictive of human carcinogenic risk, then the concept of the matrix shown previously becomes apparent.

It is worth drawing on the extensive database derived over the past 20 years from the 120 human cytogenetic studies conducted. The fact that some of the important study design factors and data assessment criteria are only now becoming apparent suggests that newer surveillance techniques should not be employed before adequate control trials have been completed. In addition to the control variables recently defined in great detail by one group of workers (Soper etal 1984, Stolley et al 1984 and Galloway et al 1986), two further recent matters illustrate these concerns. The first is the demonstration that sodium selenite, which is clastogenic to rat lymphocytes cultured in vitro and to rat bone marrow cells in vivo, is non-clastogenic to the peripheral lymphocytes of the same animals. This example provides an instance of where a 'marker cell' is less sensitive than the presumed target cell (Newton and Lilly 1986). These authors cite the report by Norppa et al (1980) in which a negative clastogenic reponse was seen in the peripheral lymphocytes of humans exposed to therapeutic dose levels of sodium selenite.

The second report of possible general significance is that by Galloway et al (1986) where significant variability was observed between different cytogeneticists reading the same slide. This subjective difference was rationalized by the authors and demonstrated not to have affected their results, but such apparently small matters will become of increasing importance as the level of exposure to genotoxins is reduced and the sensitivity of surveillance techniques is increased.

PERIPHERAL BLOOD LYMPHOCYTE MUTATION ASSAYS

Albertini (1985) has described an assay which determines the induction of gene mutations by the assessment of 6-thioguanine resistant T-lymphocytes in human blood. Recently, the method has been adapted for use with flow cytometry techniques (Amneus and Eriksson, 1986), and Nicklas et al (1986) have described new sensitive gene probes to quantify the expression of these mutations. These techniques do not yet appear to have been employed in a human surveillance study. The importance of the method is that it is the only one available at present to measure somatic cell gene mutations. It is pertinent to observe that the flow cytometric method was developed due to problems encountered with the use of the original autoradiographic protocol. This suggests that sufficient time should pass between the development of a new technique and its routine use in humans.

ADDUCT FORMATION ASSAYS

Three main categories of surveillance assay have been designed to recognize and quantitate the products of the (presumed) covalent interaction of a chemical, or its metabolites, with either DNA or proteins. The first of these, the DNA post-labelling technique developed by Randerath (Gupta et al 1982), is the most closely related in concept to the cytogenetic techniques discussed above. This is because a complete spectrum of adducts with DNA can be recognized without any being chemically characterized. Consequently, this technique can be employed in mixed chemical environments, much as Sawsen (this volume) recorded a clastogenic response in individuals exposed to Pb/Hg/TNT without knowing which was the active agent. The post-labelling assay (Gupta et al 1982) involves isolation of DNA, its hydrolysis to single nucleosides, their radiolabelling _via_ ^{32}P phosphorylation and the final production of a 2-dimensional autoradiographic nucleoside spectrum. The assay is still undergoing development (Santella, this volume) but already a sensitivity of 1 adducted base in 10^{10} (ie 1 adduct per cellular genome) is claimed. As a means of determining the net cellular exposure of an individual to total electrophiles in a given environment, this technique seems to be unrivalled. Further development will be required before it can be employed routinely, and it is likely that it will remain best suited for specific research uses - ie, it is unlikely to become a basic screening test in the way that cytogenetic analysis now is. In the immediate future this system may find application in experimental carcinogenicity research, since its performance in the recent IPCS study of animal genotoxicity tests was encouraging (Ashby et al 1986). Such experimental studies may also help to resolve the major current problem with this system, namely, its ability to detect levels of adducts which are so low as to be of debatable phenotypic significance.

Assays which quantitate defined adducts on DNA or protein are quite different in concept to those discussed so far and are open to greater interpretative problems when applied to the study of exposed human populations. Both types of assay require the initial preparation of pure and chemically defined adducts which are then searched for in the somatic tissue of exposed individuals. This may require substantial chemical and biochemical preparations and involve the investigator having to select the DNA or protein adducts most worthy of study. This has not presented too great a problem to date because well-studied genotoxins have usually been employed. Thus, the monoclonal antibody recognition of intact DNA-adducts has been concentrated on direct-acting agents such as diaminodichloro platinum, or the ultimate carcinogenic form of benzo[a]pyrene, a diol epoxide. Likewise, the vast majority of the haemoglobin (protein) adduct studies have involved the use of direct acting agents such as ethylene oxide (EO) or simple methylating agents such as methyl methanesulphonate (MMS). This has probably led to these assays appearing to be of greater relevance to cancer induction than may actually be the case. It is suggested that while showing great promise as future human surveillance techniques, each should be subject to critical appraisal at this stage of their development. Sentinel concerns for each assay are discussed next.

MONOCLONAL ANTIBODY RECOGNITION OF SPECIFIC DNA ADDUCTS

The greatest problem facing this class of assay is that most genotoxic chemicals (perhaps aided by one or more electrophilic metabolites) produce a variety of lesions on DNA. Some of these lesions will be rapidly repaired while others may remain intact. It remains to be confirmed, unequivocally, which type of lesion leads to the gene mutations/chromosomal aberrations that presage carcinogenicity. This means that both the selection of adducts for study and the sampling time could affect dramatically the outcome of a human surveillance study. In particular, there is a very real prospect that decisions taken in the early stages of such a human surveillance study could lead to the identification of a subgroup of individuals who give a maximum response for the assay parameters selected, but whom, in fact, may be at least risk of developing cancer. This prospect is less likely to present in a cytological/ cytogenetic study. If this point sounds to be too negative one has only to remember the two decades of cancer research that were diverted due to the erroneous assumption that the readily detected K-region epoxide of benzo[a]pyrene was the ultimate carcinogenic species, or to the revelation that the N-7 methylation of guanosine, while easy to monitor due to its chemical stability, was, in fact, only a benign marker for the chemically unstable, but mutagenic O-6 methylguanosine lesion.

CHEMICAL–PROTEIN ADDUCT ASSAYS

Ehrenberg and Osterman-Golkar in Sweden and Farmer in England have pioneered the concept that electrophilic chemicals capable of interacting covalently with DNA will also interact in a quantitatively similar manner with protein. Thus, by isolating protein in the form of haemoglobin from exposed individuals, the derived constituent amino acids can be assayed for the presence of pre-synthesised amino acid-chemical adducts. A fundamental assumption of this technique is that the covalent interaction of a chemical, or one of its metabolites, with protein is quantitatively reflective of its (unmonitored) covalent interaction with cellular DNA. This assumption is difficult to validate in general due to the fact that the majority of published studies have involved exposure to reactive (direct acting) agents such as ethylene oxide (EO). In the EO case, and for the related species propylene oxide, the levels of hydroxyalkyl adducts of histidine, cysteine, etc, have provided useful surrogates for chromosomal analysis of the peripheral lymphocytes of the same individuals. For natural electrophiles and their simple precursors (eg ethene as a metabolic precursor of EO) these techniques show great promise; however, two concerns are apparent. The first is that in the case of simple electrophiles such as EO and MMS a background level of adducts exists, and this may confound the interpretation of low-level effects. Second, the selective interaction of an electrophile with protein may reduce quantitatively the extent to which it can alkylate DNA. Thus, in the case of certain chemicals, the greater the extent of interaction with protein, the less will be the extent of interaction with DNA. Such concerns are most likely to apply to those agents least likely to react with the critical centres of DNA, such as the O-6 atom of guanine. Thus, in a multi-chemical study conducted in rats by Pereira and Chang (1981) it was observed that the greatest extent of protein adduct formation was observed for the very weak animal carcinogen MMS (3264 pmol/gHb) while the potent human carcinogens aflatoxin B_1 and benzidine gave levels of adducts 2 orders of magnitude lower (47.3 and 22.5 pmol/gHb, respectively).

The danger is, therefore, that protein adduct surveillance techniques may sometimes provide a reverse indicator of individual cancer hazard while still providing a gross measure of exposure to electrophiles.

A final example of the subtle problems which may be encountered when employing protein adduct assays in human studies concerns experiences with 3 aromatic amines, 4-aminobiphenyl (4AB), benzidine (BZD) and aniline. These closely related chemicals present the extremes of possible human hazard: 4AB and BZD are established human bladder and rodent carcinogens while aniline is generally regarded as non-carcinogenic to both species. All 3 compounds are oxidised to an N-hydroxy derivative which is capable of reacting with haemoglobin to form a chemically unstable derivative. This reaction is characterised clinically by the production of methaemoglobinaemia (cyanosis) in exposed individuals. In the case of 4AB and BZD the same N-hydroxylated metabolites are <u>also</u> conjugated as glucuronides. In the bladder these conjugates are acted upon by the enzyme β-glucuronidase, and this leads to the formation of an electrophilic species associated with the eventual induction of bladder cancer by these agents. Thus, the haemoglobin adducts formed for these chemicals may also provide a reverse measure of the actual extent of reaction with DNA. In partial support of this suggestion are the data of Pereira and Chang (1981) who have reported an identical level of haemoglobin adducts for BZD and aniline in rats (22.5 and 24.8pmol/gHb, respectively). This observation must surely be considered a totally unacceptable resolution of these two chemicals. Such concerns influence the interpretation of the similar studies reported for 4AB in rats (Green et al 1984) and humans (Skipper et al 1986). Finally, Lewalter and Korallus (1985) have demonstrated that the level of aniline-haemoglobin adducts formed in exposed individuals is directly related to their acetylator status. Thus, fast acetylators primarily excrete aniline as acetanilide in their urine while slow acetylators produce high levels of aniline-haemoglobin adducts.

The above papers illustrate that when moving away from simple alkylating agents to complex organic molecules, several metabolites may be encountered, each of which may compete for interactions with macromolecules. In such situations it may be difficult to detect a single marker reaction which relates both directly and quantitatively to the cancer-critical reaction of the compound with nuclear DNA. Even the underlying assumption that the chemical reactivity of an agent to DNA and proteins is the same may not hold true. For example, the fact that the 4AB-haemoglobin product can be broken down to yield free 4AB is inconsistent with the expected properties of the presumed cancer critical 4AB-DNA adduct.

Although this section on adducts appears to be very negative, such caution befits endeavours which might be perceived as providing risk-estimation data on the interaction of an individual with an established human carcinogen.

FAECAL AND URINE MUTAGENICITY ASSAYS

Venitt (1984) has reviewed the use of faecal mutagenicity assays and Clonfero et al (1986) have recently discussed the use of urine mutagenicity assays on individuals exposed to polycyclic aromatic hydrocarbons. These assays, especially the latter, provide a convenient method of sampling humans exposed to genotoxins. A conceptually curious aspect of both methodologies is that excreted conjugates are broken down with the aid of faecalase or β-glucuronidase, respectively, and then activated with induced rat liver S9 mix. This seems to be throwing nature into reverse gear, and while this may be found acceptable in general screening programs, it precludes the use of the derived data for individual risk assessment purposes. For example, were acetanilide to be mutagenic to bacteria, then application of the urine mutagenicity technique to the aforementioned human surveillance study of aniline exposed people, reported by Lewalter and Korallus (1985), would have yielded high mutagenic activity in urine for individuals with low haemoglobin adduct levels, a curious confrontation of two methods of human surveillance! These assays share an advantage with cytogenetic techniques, of producing a response in the absence of exact knowledge of the causative agent. Despite this, they are placed in the lower right hand side of the matrix shown earlier, ie, they are probably the least definitive of the available human surveillance techniques.

CHEMICAL ANALYSIS OF URINE

The assessment of urine for chemicals or their metabolites may provide a useful analytical biological monitoring approach in some situtations. Thus, in this volume, Sanguinetti et al, Luotamo et al and Karacic et al describe the use of (and the problems encountered with) the measurement of phenol as a marker for exposure to benzene. However, these techniques, while providing a useful measure of individual (idiosyncratic) response to a genotoxin, are not optimal for estimating atmospheric concentrations of a chemical. For example, Karacic et al (this volume) monitored phenol in the urine of workers in the shoe industry in order to determine that they were probably exposed to 10 ppm of benzene: this could have been established more easily and accurately by standard atmospheric monitoring techniques.

SUMMARY

Developing technology offers many opportunities to monitor quantitatively the acute consequence of exposing humans to potentially carcinogenic/mutagenic agents. It is suggested that if these techniques are to be employed with confidence certain basic questions should be considered.

a) Is the technique to be employed to estimate exposure/absorption to genotoxins or to predict individual hazard?

b) Is the technique sufficiently well developed and validated to be employed with confidence in a human population?

c) Is there access to a sufficiently large concurrent control population and do adequate historical control data exist for the technique.

d) Have sufficient experimental animal studies been conducted using the technique to enable data to be interpreted with confidence in terms of possible human risk?

e) Is it possible to anticipate the many consequences in terms of risk management which could ensue from the study?

If such questions are not answered the unacceptable situation could present of investigations commencing as risk assessment studies and ending merely as contributions to the control database for the assay, this due to a failure to be able to interpret the derived data definitively. In this context the value of adequate preliminary animal studies cannot be over emphasized. As an illustration of this concern, Bailey et al (1981) demonstrated, as part of the development of their cysteine alkylation assay, that dimethylnitrosamine administered to rats at a dose-level of 12.5 mg/kg gave no increase in urinary S-methylcysteine; but that dose-level is known to be carcinogenic to the rat. It would have been easy to interpret 12.5 mg/kg as a threshold carcinogenic dose-level for exposure of rats to DMN, but that would have been incorrect. Such an interpretation translated to the human situation could prove to be disastrous. Equally, the unthinking use of a monoclonal antibody capable of detecting one aniline-DNA adduct per cellular genome may not necessarily be the most useful tool in the armament of a clinician responsible for the industrial hygiene of an aniline plant.

ACKNOWLEDGEMENT

I am grateful to Bruce Woollen of these laboratories for making available to me his recent literature search of protein/DNA adduct papers, and to Peter Blain, Bruce Woollen and Mike Stonard for very helpful discussions.

REFERENCES

Albertini R J (1985). Somatic gene mutations in vivo as indicated by the 6-thioguanine resistant T-lymphocytes in human blood, Mutation Res, 150, 411-422.

Amneus H and Eriksson L (1986). The frequency of 6-thioguanine-resistant human peripheral blood lymphocytes as determined by flow cytometry and by clonal propogation, Mutation Res, 173, 61-66.

Anderson D, Richardson C R, Weight T M, Purchase I F H and Adams W G F (1980). Chromosomal analysis in vinyl chloride exposed workers. Results from analysis 18 and 42 months after an initial sampling, Mutation Res, 79, 151-162.

Ashby J and Richardson C R (1985). Tabulation and assessment of 113 human surveillance cytogenetic studies conducted between 1965 and 1984, Mustation Res, 154, 111-133.

Bailey E, Connors T A, Farmer P B, Gorf S M and Rickard J (1981). Methylation of Cysteine in haemoglobin following exposure to methylating agents, Cancer Res, 41, 2514-2517.

Clonfero E, Zordan M, Cottica D, Venier P, Pozzoli L, Cardin E L, Sarto F and Levis A G (1986). Mutagenic activity and PAH levels in urine of humans exposed to therapeutical coal tar, Carcinogenesis, 7, 819-823.

Galloway S M, Berry P K, Nichols W W, Wolman S R, Soper K A, Stolley P D and Archer P (1986). Chromosome aberrations in individuals occupationally exposed to ethylene oxide, and in a large control population, Mutation Res, 170, 55-74.

Green L C, Skipper P L, Turesky R J, Bryant M S and Tanneubaum S R (1984). In vivo dosimetry of 4-animobiphenyl in rats *via* a cysteine adduct in haemoglobin, Cancer Res, 44, 4254-4259.

Gupta R C, Reddy M V and Randerath K (1982), ^{32}P-Postlabelling analysis of non-radioactive aromatic carcinogen-DNA adducts, Carcinogenesis, 3, 1081-1092.

Lewalter J and Korallus V (1984). Blood protein conjugates and acetylation of aromatic amines. New findings on biological monitoring. Int Arch Envir Health, 56, 179-196.

Newton M F and Lilly L J (1986). Tissue-specific clastogenic effects of chromium and selenium salts *in vivo*, Mutation Res, 169, 61-19.

Nicklas J A, O'Neill J P and Albertini R J (1986). Use of T-cell receptor gene probes to quantify in vivo hprt mutations in human T-lymphocytes, Mutation Res, 173, 67-72.

Norppa H, Westermark T, Laasonen M, Knuutila K and Knuutila S (1980). Chromosomal effects of sodium selenite in vivo. Aberrations and sister chromatid exchanges in human hymphocytes, Hereditas 93, 93-96.

Periera M A and Chang L W (1981). Binding of chemical carcinogens and mutagens to rat hemoglobin, Chem Biol Interactions, 33, 301-305.

Sarto F, Cominato I, Pinton A M, Brovedani P G, Faccioli C M, Bianchi V and Levis A G (1984). Cytogenetic damage in workers exposed to ethylene oxide, Mut Res, 138, 185-195.

Sinha A K, Linscombe V A, Gallapudi B B, Jersey G C and Flake R E (1986). Cytogenetic variability of lymphocytes from phenotypically normal men, J Tox Envir Hlth, 17, 327-345.

Sinha A K, Linscombe V A, Gollapudi B B, McClintock M L, Flake R E and Bodner K M (1984). The incidence of spontaneous cytogenetic aberrations in lymphocytes from normal humans, Canadian Journal of Genetics and Cytology, 26, 528-531.

Skipper P L, Bryant M S, Tannenbaum S R and Groopman J D (1986). Analytical methods for assessing exposure to 4-aminobiphenyl based on protein adduct formation, J Occup Med (in press).

Soper K A, Stolley P D, Galloway S M, Smith J G, Nichols W W and Wolman S R (1984). Sister chromatid exchange (SCE) report on control subjects in a study of occupationally exposed workers, Mutation Res 129, 77-88.

Stolley P D, Soper K A, Galloway S M, Nichols W W, Norman S A and Wolman S A (1984). Sister chromatid exchanges in association with occupational exposure to ethylene oxide, Mutation Res 129, 89-102.

Venitt S (1982). Mutagens in human faeces, are they relevant to cancer of the large bowel? Mut Res, 98, 265-286.

50

Factors influencing sister chromatid exchanges (SCE) in man

F. Sarto, M.C. Faccioli, L. Mustari, P.G. Brovedani, E. Clonfero, A.G. Levis* – Istituto di Medicina del Lavoro dell'Università. Via Facciolati N. 71 – 35127 Padova, Italy
*Dipartimento di Biologia dell'Università. Via Loredan N. 10 – 35131 Padova, Italy

INTRODUCTION

Among the genotoxicity assays, the use of cytogenetic tests, such as the detection of sister chromatid exchanges (SCE), in studies aimed at monitoring animal and human exposure to genotoxins is a topic of current interest. A great deal of enthusiasm was initially raised by the rapidity of SCE test and its greater sensitivity, in some cases, as compared to the chromosome aberration test. This probably explains the scarce consideration given to factors such as the variability due to the different methods used to detect SCE and the individual confounding factors present in the populations studied. In fact the evaluation of significant differences in SCE frequencies between persons either with breakage syndromes in particular diseases or exposed to genotoxic agents and their respective controls, relies heavily on the weight which must be given to the factors which can influence SCE.
The aim of this study was to evaluate the variability of the SCE test due to the following factors: a) Methodological; b) Readers c) Subjects examined (age, sex, life-style). An attempt to reduce or control the factors known to influence SCE frequencies was also made.

MATERIALS AND METHODS

Cytogenetic methods. In all experiments we used the same standard cytogenetic technique: 2 replicated 4 ml cultures with fetal calf serum and PHA (Difco) were added to 200 ul of whole blood and 100 ul of BRdU (final concentration: 100 μM). Lymphocytes were cultured in the dark for 73-75 hours and the metaphases blocked during the last 210 min. with 100 μl of colchicine (final concentration: 0.2 μg/ml). The FPG staining method and other technical details are reported elsewhere /2/.
Subjects examinated. Subjects were students (age 16-20 years) or health workers not occupationally exposed to genotoxic agents and clinically healthy. Subjects who had had viral hepatitis or suffered from recent colds, influenza, allergies, or who had been recently vaccinated were not considered. Passive smoking, defined as the daily contact in confined quarters with a subject smoking more than 10 cigarettes/day, was established for 31 non-smokers. The examined subjects, aged from 16 to 70 years, were selected into 11 age classes of 5 years each; each class included 8 subjects: 2 male smokers, 2 male non-smokers. 2 female smokers, 2 female non-smokers. Other details of the population are reported elsewhere /2/.
Statistical analysis. The t statistic for independent samples and for paired data and the analysis of variance were used. To evaluate the difference in two different groups inside the same analysis of variance the linear contrast method was used. Multiple regression was done by a step-up method, and the variate giving the greatest additional contribution to the sum of the squares due to regression was selected. To evaluate the effects of smoking and aging on SCE at individual level, the Carrano and Moore /1/ method for the analysis of persons with high SCE frequency cells (HFCs) was used.

RESULTS

The methodological factors which influence the SCE frequency in cultured human lymphocytes are: the number of cells in culture, the concentration of PHA, the concentration of BRdU, the type of culture medium, the culture medium aging, the temperature during the culturing, the pH of the culture, the time of cell harvesting, the staining methods and the number of metaphases scored; among these factors the major source of variability is however the BRdU concentration.
From preliminary experiments (data not shown), we found that there is a significant increase in SCE (t paired = 3.86; p =

0.013) in cultures prepared with an old medium (not expired!) compared to replicated cultures prepared with a different batch of fresh medium. We also compared the SCE frequency in the same samples, stained with 2 different methods: fluorochrome plus giemsa and alkaline giemsa; we did not find a significant difference (t paired = 1.35; p = 0.23) when slides were read by the same reader. With a high standardization of the methodological factors, using the same batch fresh medium and the same staining procedure, we did not find any significant difference in SCE (t = 1.3; p = 0.89) in replicated cultures analysed by the same reader. Thus for each subject we used 2 or 3 replicated cultures processed together.

First, we evaluated the inter-reader variability on random metaphases: 3 readers scored 4 different samples; in Table 1 the results of the statistical analysis are summarized (F = analysis of variance. c.v. = coefficient of variation, d.f. = degrees of freedom).

Table 1

Subject (code)	readers	F	p=	d.f.	c.v.%
034	A-C-D	0.468	0.879	2,104	4.35
035	A-C-D	0.694	0.759	2,95	5.99
036	A-C-D	0.634	0.789	2,88	4.84
037	A-C-D	0.399	0.915	2,91	4.48

Second, we evaluated the inter-reader variability on the same metaphases: the same 3 readers scored the same metaphases (Table 2).

Table 2

Subject (code)	readers	F	p=	d.f.	c.v.%
034	A-C-D	0.060	1.000	2,96	1.7
035	A-C-D	0.100	0.990	2,93	2.1
036	A-C-D	0.001	1.000	2,93	0.0
037	A-C-D	0.140	0.998	2,87	2.6

Third, in repeated samples we measured separately the variability in SCE due to individual variation over time and the intra-reader variability: Two readers analysed SCE in repeated cultures from 5 subjects during a period of time from 97 to 154 days. The same reader reanalysed the first sample every time a new repeated culture was analysed. Table 3 shows A - the evaluation of the variation, by using the analysis of variance and the coefficient of variation, in repeated samples and B - the evaluation of the weight of the intra-reader variation on the variability of repeated samplings, by using the linear contrast method.

Table 3

		A				B			
Subject	reader	F	p=	d.f.	c.v.%	F	p=	d.f.	c.v.%
118	C	1.07	0.37	4,229	5.6	0.11	0.99	2,117	1.7
116	C	2.54	0.06	3,180	8.9	0.70	0.76	3,221	4.6
120	C	0.20	0.99	2,138	2.4	--	--	--	--
121	D	0.11	0.99	2,147	1.7	0.45	0.90	3,219	3.0
128	D	2.89	0.06	2,147	8.1	--	--	--	--

To study the confounding factors due to the subjects examined, 6 readers, who had at least a one year training period, analysed SCE in 30-55 well spread metaphases for each subject in coded slides. 3,248 metaphases were examined in total. In Table 4 mean frequencies of SCE per cell in the paired groups of the examined population are reported.

Besides age and smoking, the influence of other variables on SCE was evaluated by means of multiple regression.

In this study conducted on 88 subjects, the percentage of variance in SCE explained by aging was 18.35 (F test = 18.32; $p < 0.001$), by smoking 13.67 (F test = 15.84; $p < 0.001$), by readers 6.60 (F test = 1.48; $p = 0.19$), by sex 1.32 (F test = 1.49; $p = 0.22$) and by passive smoking 1.32 (F = 1.32; $p = 0.23$). Other factors, such as the consumption of alcoholic beverages and coffee, the use of antiinflammatory and antinevralgic drugs and diagnostic exposure to X-rays during the last 5 years, were not found to influence the SCE frequencies ($F = 0.01$, $p = 0.86$). By means of multiple regression a significant correlation was found among SCE, age and number of cigarettes smoked per day ($R = 0.56$; $p < 0.001$).

Table 4

compared groups	SCE m ± D.S.	t statistic	p=
smokers (44)*	12.0±1.9		
non-smokers (44)	10.5±1.6	3.92	0.001
males (44)	11.5±2.0		
females (44)	11.0±1.7	1.15	0.27
male smokers (22)	12.3±2.1		
male non-smokers (22)	10.6±1.6	3.00	0.001
female smokers (22)	11.6±1.7		
female non-smokers (22)	10.4±1.6	2.52	0.01
male smokers (22)	12.3±2.1		
female smokers (22)	11.6±1.7	1.18	0.22
male non-smokers (22)	10.6±1.6		
female non-smokers (22)	10.4±1.6	0.52	0.6
16-40 y. old subjects (40)	10.3±1.5		
46-70 y. old subjects (40)	12.1±1.9	4.61	0.001
non-smokers 16-40 y. old sub. (20)	9.8±1.1		
non-smokers 46-70 y. old sub. (20)	11.2±1.6	3.33	0.001

* number of subjects in brackets.

In Figure 1 age is plotted against the standard deviations of the SCE averages for the 44 non smokers; the coefficient of correlation for the regression is statistically significant. This result indicates that the frequency distribution of SCE frequencies is more skewed from the mean in old than in young subjects.

The results of the comparison between the frequency distribution of the HFCs according to Carrano and Moore /1/ are as follows: in smokers we had 15 subjects with HFCs (within 95% confidence limits), in non smokers we had 5 subjects with HFCs (χ^2 = 5.24; p = 0.022). In the 46-70 year-old subjects we had 22 persons with HFCs, in the 16-40 year-old subjects we had 7 persons with HFCs (χ^2 = 10.6; $p < 0.001$).

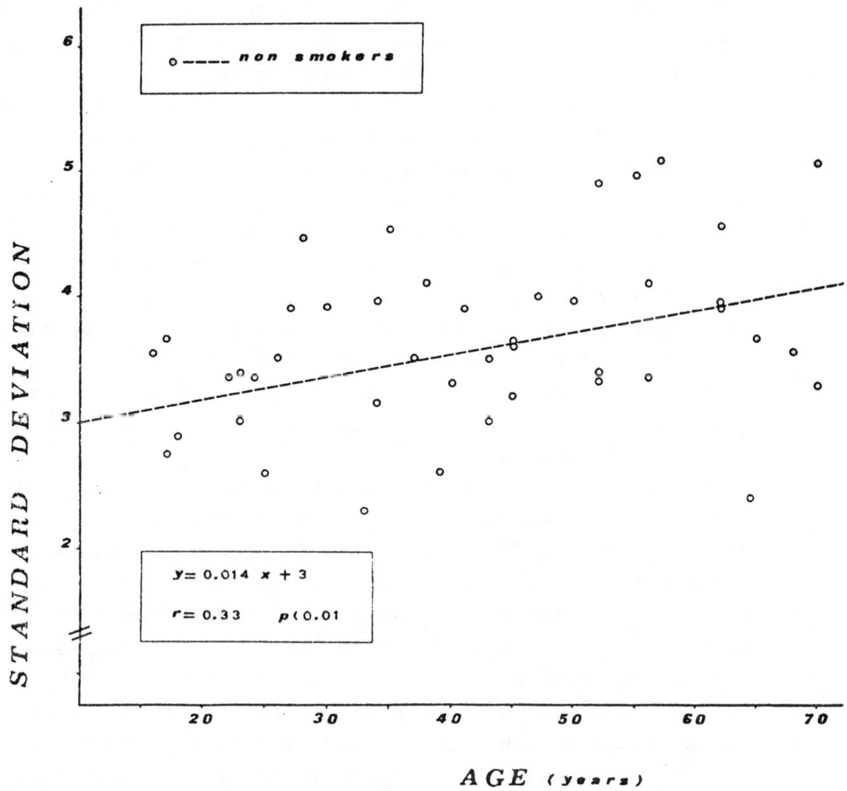

Fig. 1.

CONCLUSIONS

Many factors, which relate to the preparation techniques, if not strictly controlled and standardized, introduce errors in the estimated frequency of SCE. It is important to use a medium from the same batch even for prolonged experiments; it is also better not to change the staining method even though this factor does not significantly influence the SCE frequency. Using higly standarized methods and with correctly trained readers, we can conclude:

1) The intra-reader variability is very low on the same metaphases (c.v. from 1.7 to 2.6%), the intra-reader variability is higher, even though not statistically significant, on random metaphases. This could be due: a)

to differences in scoring because the pattern of the readers is not stable over time, i.e., even the same reader (not only different readers) may analyse different metaphases in the same slides, b) to differences in sampling because there are differences in SCE frequency among slides when different readers were not reading the same slides. In our study, in which 88 subjects were analysed, the factor "readers" explained 6.6% of the variance in SCE. In the study by Soper et al. /3/, made on 479 control subjects, the same factor explained 19.3% of the variance, but 29 readers from 3 laboratories partecipated in this study.

2) In cultures repeated over months the variability was greater but still not significant (c.v. from 1.7 to 8.9%); the largest source of variability seems to be represented by factors relating to subjects examined and secondly by methodological factors and readers.

3) The study on the confounding factors influencing SCE in man, planned experimentally, clearly demonstrates that aging, besides smoking, increases the SCE frequency in man. Smoking enhances SCE independently of age and sex; we were not able to establish whether there was a difference between the types of tobacco smoked because we only had 4 subjects who smoked high tar-content cigarettes.

In our study sex was not a significant confounding factor but in the study by Soper et al. sex explained 3.9% of the variance (the SCE frequency being higher in females). This could be due to the fact that the females in the present study did not take oral contraceptives which may induce SCE /1/. Practically speaking, in occupational cytogenetic monitoring, a high standardization of the techniques and a control of the intra - and inter - reader variation is necessary; furthermore infections, drugs allergies and life-style are difficult factors to control, however matching of the exposed/control subjects should be performed at least on the basis of at age, smoking habits and sex.

ACKNOWLEDGEMENTS

Supported by grants from the National Research Council of Italy (C.N.R., P.F. "Oncologia" and "Medicina Preventiva e Riabilitativa").

REFERENCES

/1/ Carrano A. and Moore D. The rationale and methodology for quantifying SCE in humans. in: Mutagenicity, Heddle J. ed., A.P. New York, 1982, pp. 267-304.

/2/ Sarto F. et al. Aging and smoking increase the frequency of sister chromatid exchanges in man. Mutation Research, 144, 183-187, 1985.

/3/ Soper K. et al. Sister chromatid exchanges report on control subjects in a study of occupationally exposed workers. Mutation Research, 129, 77-88, 1985.

51

Monitoring of urinary genotoxicity by workers exposed to cracking-products of mineral oils

Z.W. Myslak, H.M. Bolt – Institute of Occupational Health, University of Dortmund, Ardeystr. 67, D-4600 Dortmund, Federal Republic of Germany

ABSTRACT

Genotoxic activities in ambient air samples and in urine of workers occupationally exposed to mineral oils in the glass industry and of non-exposed controls were investigated using the SOS chromotest. Both groups (exposed and non-exposed) included smokers and non-smokers. Direct genotoxicity was found in the eluates of air filter samples. Genotoxicity in urine was not significantly higher in the group of workers exposed to mineral oils when compared with controls. But irrespective of occupational exposure was genotoxicity significantly higher in smokers versus non-smokers. Therefore, the direct genotoxic potency of oil mist in the glass industry is not reflected in higher urinary excretions of genotoxic metabolites in exposed workers, when compared with non-exposed control subjects.

INTRODUCTION

The use of cooling- and lubricate-agents in many industrial operations results in formation of mineral oil mists. Their toxicological profile can be modified by additives or by thermal degradation products of mineral oils (Stalder, 1984). A carcinogenic potency of mineral oils and/or their decomposition products has been discussed for many years. However, the opinion that human cancer might be caused by inhaled mineral oils, is not supported by epidemiological studies. (Seidenstücker, 1983).

During the manufacture process in a glass foundry the lubricants are exposed to temperatures ranging up to 900°C. Biological properties of mineral oil degradation products are not known. The aim of our study was to investigate the possible genotoxic activity of oil mist which occurs in the glass industry.

MATERIAL AND METHODS

Genotoxicity in air samples and urine of workers occupationally exposed to mineral oils in a glass foundry was investigated using the SOS chromotest. The lubricants used in this industrial setting were: 1) "Kleen mold 202" (hydrocarbons, Ca- and Mg-salts, fats, oils, graphite), 2) "WTG-1" (paraffin oil), 3) "Acmos 426" (varnish/graphite base); 4) "46 lack" (varnish/graphite base). The exact composition of these commercial material was unknown.
The lubricants are usually exposed to temperatures between 400 and 900°C which results in massive formation of oil mist.

PREPARATION OF AIR SAMPLES

Ambient air at workplaces was collected during 8 hours using a "Personal Air Sampler P-4000" (DuPont). The sampling filters were then eluted with 4 ml dichloromethane, and the eluate was evaporated. This procedure was repeated on the next day. Then 3 ml of dichloromethane were passed through the filter, the dichloromethane was evaporated and the residue dissolved in 1 ml dimethylsulfoxide (DMSO). The eluate, after being diluted to 6% DMSO with saline, was tested by the SOS chromotest procedure, without and with activation with added "S-9 mix".

PREPARATION OF URINE SAMPLES

Urine samples were collected from 29 mal persons which were occupationally exposed to mineral oils in a glass foundry (exposure time between 1 to 37 years, mean 16 years). Out of these, 13 were smokers and 16 non-smokers. Also, urine was obtained from 39 controls (11 female and 28 male persons), 18 smokers and 21 non-smokers. Native urine was centrifugated, neutralised (with NaOH) to pH 7 and tested in the SOS chromotest (see below) without external activation.

DETERMINATION OF GENOTOXICITY

The general principles of the SOS chromotest have been described by Quillardet et al. (Quillardet et al. 1982, 1985; Ohta et al. 1984). The test was performed according to a modification of the original procedure usinge a commercial kit using microtiter plates (Orgenics Ltd., Israel). Serial dilutions of the testing material were distributed in separate wells and were incubated with bacterial cultures at 37°C for 3.5 h.

Production of ß-galactosidase (indicating the induction of SOS DNA-repair) and of alkaline phosphate (indicating viability of bacteria) was determined using the substrates o-nitro-phenyl galactosidase and p-nitrophenyl phosphate, respectively. The activity of the SOS-repair system was expressed as "SOS-IF" (induction factor).

$$\text{SOS-IF} = \frac{R\ (C)}{R\ (0)}, \text{ where}$$

R (C) = ratio (R) at concentration c, R (0) = ratio (R) at concentration 0 (background).

STATISTICAL ANALYSIS

Student's t test was used to quantitate the statistical significance of group differences.

RESULTS

Figure 1 presents the results obtained with the eluates from filters on which oil mist was collected at the workplace.

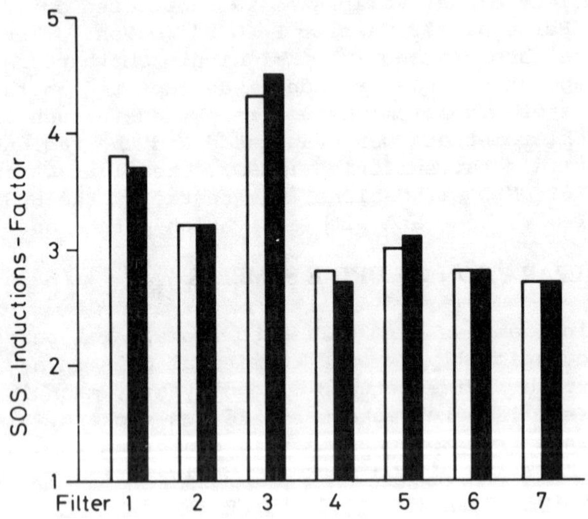

Fig. 1. – Genotoxicity (given as SOS–IF) detected in eluates of filters from the air samples (see text) with (black bars) and without (white bars) application of metabolic activation.

The SOS chromotest was already positive without S-9 mix, and external activation with S-9 mix did not result in higher levels of genotoxicity. This leads to the conclusion that the investigated oil mists contained some directly acting genotoxic compounds.

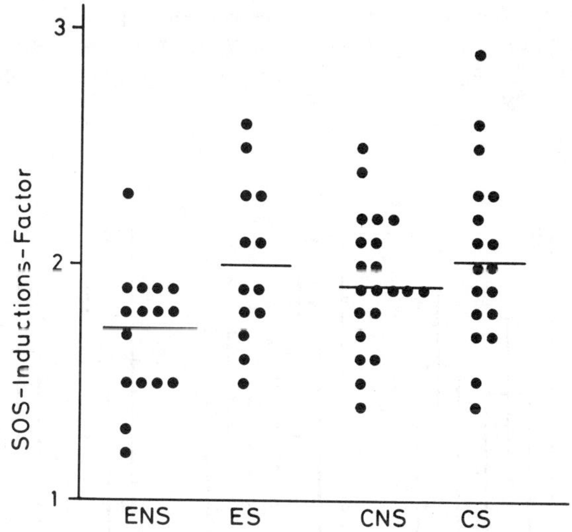

Fig. 2. – Genotoxicity (given as SOS–IF) detected in urine of workers exposed to mineral oils in a glass foundry (E), and controls (C), in each group smokers (S) and non-smokers (NS). Mean value of each group is shown by a bar.

The genotoxicity of individual urine samples is given in Figure 2.

There is some SOS chromotest-positive activity in the urine of all persons, including controls and non-smokers (background: SOS-IF = 1).

Therefore, urine may either physiologically contain traces of genotoxic substances, or this may be explained by particular test conditions. No correlation of this basic urinary activity with age or sex was observed.

No significant difference in urine genotoxicity was found between the group of exposed workers and the group of controls (p = 0,075; Figure 3 a). When smokers were compared with non-smokers, irrespective of their occupational exposure (Figure 3 b), the genotoxicity was consistently higher in the groups of smokers (p = 0,01).

Furthermore, the genotoxicity was compared between the group of exposed smokers versus control smokers ($p = 0,375$), of exposed non-smokers versus control non-smokers ($p = 0,02$), exposed smokers versus exposed non-smokers ($p = 0,01$) and control smokers versus control non-smokers ($p = 0,15$) (Figure 3 c). It appears (Fig. 3) that only smoking and not occupational exposure to mineral oils enhances the urinary genotoxicity in our study.

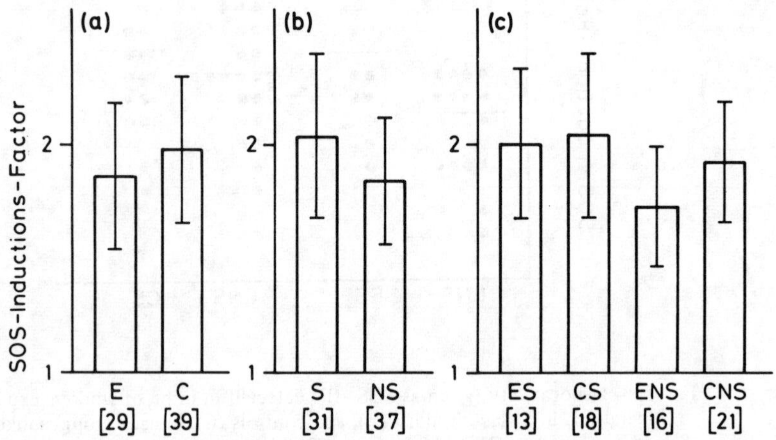

Fig. 3. – Genotoxicity (mean ± S.D.) detected in urine of workers exposed to mineral oils in a glass foundry and in controls (see text). Number of workers in each group is indicated in brackets. E – exposed workers, C – controls, S – smokers, NS – non-smokers.

DISCUSSION AND CONCLUSIONS

Our study corresponds with reports of others (Brockmeier and Norpoth, 1981; De Flora et al., 1985; Laires et al., 1982) that mineral oils used in industry may have genotoxic properties.

The results demonstrate, that oil mists, formed in glass foundries by lubricating the hot parts of machines (at temperatures ranging from 400 to 900°C) possess some directely genotoxic potency.

This corresponds with the negative results of genotoxicity testing of urine in exposed workers. From our results it can not be concluded that the directly genotoxic substances of the oil mists, do not produce genotoxic effects in the exposed workers. But we can draw the conclusion, that the apparent genotoxic potency of oil mists in the glass industry cause no changes in the level of genotoxic metabolites in urine of exposed workers, as compared with non-exposed subjects.

REFERENCES

Brockmeier, U., and Norpoth, K.: Ames-Test Untersuchungen zur Frage des Vorkommens atembarer Mutagene an verschiedenen Arbeitsplätzen. Verh. Dtsch. Ges. Arbeitsmedizin. 21, Gentner, Stuttgart 1981.

De Flora, S., De Renzi, G.P., Camoirano, A., Astengo, M., Basso, C., Zanacchi, P., and Bennicelli, C.: Genotoxicity assay of oil dispergants in bacteria (mutation, differential lethality, SOS-DNA-repair) and yeast (mitotic crossing-over). Mutat.Res. 158, 19-30 (1985).

Laires, Al., Borba, H., Rueff, J., Gomes, M.J., and Halpern, M.: Urinary mutagenicity in occupational exposure to mineral oils and iron oxide particles. Carcinogenesis 3, 1077-1079 (1982).

Ohta, T., Nakamura, N., Moriya, M., Shirasu, Y., and Kata, T.: The SOS function inducing activity of chemical mutagens in Escherichia coli. Mutat.Res. 139, 101-109 (1984).

Quillardet, Ph., Huisman, O., D'Ari, R., and Hofnung, M.: SOS chromotest, a direct assay of induction on an SOS function in Escherichia coli K-12 to measure genotoxicity. Proc.Natl.Acad. Sci. USA 79, 5971-5975 (1982).

Quillardet, Ph., and Hofnung, M.: The SOS chromotest, a colorimetric bacterial assay for genotoxins: procedures. Mutat.Res. 147, 65-78 (1985).

Seidenstücker, R.: Biologische Wirkung von Ölaerosolen. B.A.U.-Forschungsbericht 330 (1983).

Stalder, K.: Aerosole aus Kuhl-Schmiermitteln als arbeitshygienisches Problem. Arbeitsmed.Sozialmed.Präventivmed. 11, 265-268 (1984).

52

Mutagenic activity and polycyclic aromatic hydrocarbon (PAH) levels in urine of humans

E. Clonfero*, D. Cottica, P. Venier***, L. Pozzoli**, M. Zordan***, F. Sarto*, A.G. Levis***** – *Institute of Occupational Health, Padua, Italy. ** Occupational Health Clinic, Pavia, Italy. ***Department of Biology, Padua, Italy

INTRODUCTION

The possibility of monitoring the exposure of human populations to mutagenic/carcinogenic agents employing short term tests on biological fluids has been receiving growing attention in the last few years. The reports of increased urinary mutagenicity in humans variously exposed to genotoxic chemicals(1)support the interest in biological monitoring studies. Human exposure to the highest levels of PAH,a mixture of known genotoxic agents,occurs in the industrial processes involving the production or use of coal tar pitch(2). Urinary mutagenicity was detected in coke,foundry and anode plant workers,as well as in subjects therapeutically exposed to coal tar(4-6). Analytical determinations of urinary PAH are reported only for workers in the aluminium industry(7,9). In two recent studies we reported both the urinary PAH levels and the mutagenicity of subjects occupationally exposed to coal tar pitch volatiles(7)and therapeutically exposed to pharmaceutical coal tar(8), The levels of exposure to PAH occurring in the topical treatment of psoriatic patients are several fold higher than those encountered in the occupational exposure to these genotoxins. The psoriatic patients thus prove to be an interesting model for the study of the mutagenic activity and urinary excretion of PAH metabolites

resulting from human exposure to coal tar.
In the present paper the excretion of PAH and the mutagenicity of urinary extracts from workers occupationally exposed to coal tar pitch volatiles and from psoriatic patients treated cutaneously with pharmaceutical coal tar are described.

MATERIALS AND METHODS

Subjects analysed: The occupational exposure to PAH was monitored in 23 subjects employed in an anode producing plant.Of these,12 were exposed to PAH(7 smokers and 5 non smokers)and 11 were controls(6 smokers and 5 non smokers).In addition,3 non-smoking male psoriatic patients,presenting cutaneous lesions involving at least 30% of their body surface,were monitored during the therapy which consisted of a daily cutaneous application of crude coal tar for 3 consecutive days followed by exposure to u.v. rays. None of the studied subjects was either ill or took drugs during the period of urine collection.
Urine collection: 24 hours urinary samples from the 23 aluminium workers were collected in polyethylene containers at the end of the week working shift and immediately stored at -30 °C. For the psoriatic patients,each micturation was collected separately starting from 6 h after the first application of coal tar up to 36-48 h after the last application.
Mutagenicity assays: Urinary samples of 400 ml were used for mutagenicity testing after extraction and concentration on XAD-2 resin(for details see ref.7,8). Various doses of urinary extracts were tested in the Ames plate incorporation assay using strains TA98 and TA100 of S.typhimurium in presence of 50 μl/plate of S-9 rat liver homogenate(Aroclor induced)both in presence or absence of 200 U/plate of β glucuronidase. All mutagenicity data were corrected for the creatinine content of the urinary sample and are thus expressed as revertants/g of creat..
Analytical detemination of PAH:After concentration of the urinary samples on C-18 modified silica,they were reduced with hydriodic acid and subsequently the instrumental analysis of PAH was performed by h.r.g.c./m.s.(for details see ref.7,8). In reporting the urinary PAH concentrations,the recovery rate of the method was not taken into account.

RESULTS

In tab.1 the polycyclic aromatic hydrocarbon and urinary mutagenicity levels of the different groups analyzed are shown.

Table 1: Polycyclic aromatic hydrocarbons and mutagenic activity in human urine.

Group	N	Total PAH µg/g creatinine $\bar{x}$	(S.D.)	Mutagenic Activity[a] $\bar{x}$	(S.D.)
Control non-smokers	5	1.6	(0.7)	279	(312)
Control smokers	6	1.9	(1.3)	5,126	(2,613)
Non smoking aluminium workers	5	6.2	(4.3)	329	(241)
Smoking aluminium workers	7	7.1	(7.5)	5,000	(4,617)
Non smoking psoriatic patients[b]	3	217.3	(166.5)	62,462	(35,043)

a, induced revertants/g of creatinine on S.typhimurium TA100 or TA98 in presence of S-9 mix

b, peak value following coal tar applications

Among the occupationally exposed subjects, the increase in the urinary mutagenicity is detectable only in the occupationally exposed smokers. The non smoking aluminium workers show a slight increase in the urinary excretion of PAH but there is not a corresponding increase in the urinary mutagenicity value. The therapeutically exposed psoriatic patients, on the other hand, exhibit a pronounced increase both in the urinary PAH excretion and mutagenicity levels. Tab.2 summarizes the data regarding the urinary excretion of PAH for the different exposed groups considered. These data suggest that the increase of urinary PAH is largely ascribable to the excretion of low molecular weight compounds(Phenanthrene+Anthracene+Fluoranthene = 46-86% of the Total PAH), while the urinary concentration of the compounds with reported in vitro mutagenic activity(Benz(a)Pyrene and Benz(a)Anthracene) is only slightly increased by exposure to coal tar. Finally, fig.1 shows the regression line correlating Total urinary PAH to the mutagenic activity of the urinary extracts of psoriatic patients treated with therapeutical coal tar. A significant correlation between these two parameters was found($r=0.788, P<0.005$).

Table 2: Levels of benz (a) anthracene (BaA), benz (a) pyrene (BaP) and percentage contribution of low molecular compounds to the total PAH excreted in human urine.

Group	N	BaA µg/g creatinine $\bar{x}$ (S.D.)	BaP $\bar{x}$ (S.D.)	% low m.w. compounds
Control non-smokers	5	0.13 (0.10)	0.03 (0.02)	57.4
Control smokers	6	0.21 (0.19)	0.07 (0.06)	48.8
Non smoking aluminium workers	5	0.48 (0.82)	0.17 (0.22)	60.6
Smoking aluminium workers	7	0.19 (0.16)	0.08 (0.07)	86.6
Non smoking psoriatic patients[a]	3	15.30 (8.80)	0.72 (0.95)	81.0

a, peak value following coal tar applications

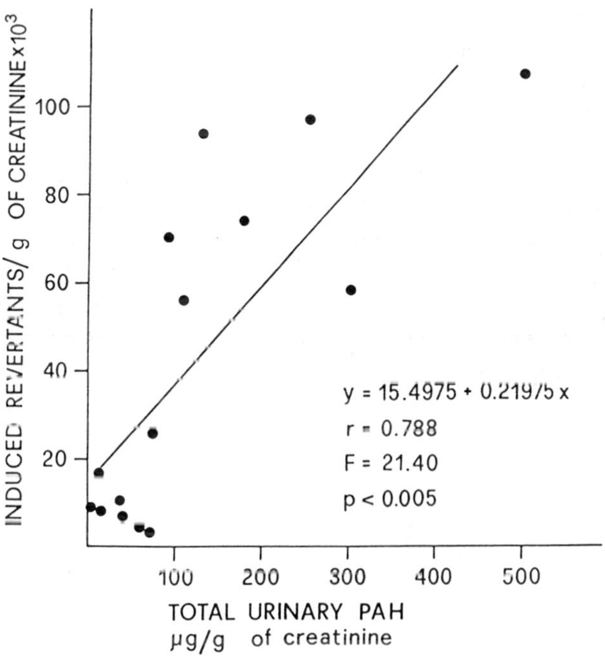

Fig. 1. – Regression line for total PAH levels and mutagenicity (plate test on Salmonella typhimurium strain TA 100+S9+β–glucuronidase) of urinary extracts of psoriatic patients treated with coal tar.

DISCUSSION

This paper presents urinary mutagenicity data to evaluate the occupational exposure of workers in the primary aluminium industry to coal tar pitch volatiles and the therapeutical exposure of subjects with psoriasis to crude coal tar. The mutagenicity found in the urine of aluminium industry workers can only be ascribed to smoking habits as the exposed non-smokers had mutagenicity values similar to those of the non-smoker controls. The increase in PAH levels detected in the urine of non-smokers following occupational exposure may be below the thresh old levels detectable by the mutagenicity assay used. This view is sustained by the finding of clear mutagenic effects in the urinary extracts of psoriatic patients treated with coal tar. In fact the excretion of PAH in the urine of these patients following treatment with coal tar is about twenty times higher than the corresponding levels detected in the group of non-smoker aluminium workers. The major contribution to the total PAH excreted in urine is given by the low m.w. compounds in both occupational and therapeutical exposure to PAH, while BaA and BaP excretion levels remain substantially unaltered, probably due to their prevalent fecal excretion (10). A significant correlation was found between the levels of PAH excreted in the urine of psoriatic patients following coal tar applications and the mutagenicity of their urinary extracts.

In conclusion it seems that notwithstanding the low levels of urinary PAH detectable following most occupational exposures to these genotoxins, the techniques available for determining the concentration of urinary mutagenic metabolites probably require improvements in order to compensate for the low sensitivity encountered in the mutagenicity assay procedures in use.

ACKNOWLEDGEMENTS

Supported by C.N.R., P.F. "Oncologia" and Centro di Alta Specializzazione in Cancerogenesi Ambientale of Veneto Region.

REFERENCES

1) Vainio,H.,Sorsa,M. and Falck,K.(1984)"Bacterial urinary assay in monitoring exposure to mutagens and carcinogens". In Berlin,A.,Draper,M., Hemminki,K. and Vainio,H.(Eds),Monitoring Human Exposure to Carcinogenic and Mutagenic Agents. IARC Scientific Publication n.53,IARC,Lyon pp.247-258
2) Linsted,G. and Sollemberg.J.(1982)"Polycyclic aromatic hydrocarbons in the occupational environment".Scand.J.Work Environ.Health,8,1-19
3) Schimberg,R.W.,Skytta,E. and Falck.K.(1981)"Iron foundry workers exposure to mutagenic polycyclic aromatic hydrocarbons". Staub.Reinh.Luft.,41,221-224
4) Kriebel,D.,Commoner,B.,Bollinger,D.,Brodson,A.,Gold,J. and Henry,J.(1983)"Detection of occupational exposure to genotoxic agents with a urinary mutagen assay".Mutat.Res.,108, 67-79
5) Heussner,J.C.,Ward.J.B. and Legator,M.S.(1985)"Genetic monitoring of aluminium workers exposed to coal tar pitch volatiles".Mutat.Res.,155,143-155
6) Wheeler,L.A.,Saperstein,M.D. and Lowe,N.J.(1981)"Mutagenicity of urine from psoriatic patients undergoing treatment with coal tar and ultraviolet light".J.Invest.Dermatol.,77, 181-185
7) Venier,P.,Clonfero,E.,Cottica,D.,Gava,C.,Zordan,M.,Pozzoli, L. and Levis,A.G.(1985)"Mutagenic activity and polycyclic aromatic hydrocarbon levels in urine of workers exposed to coal tar pitch volatiles in an anode plant". Carcinogenesis,6,749-752
8) Clonfero,E.,Zordan,M.,Cottica,D.,Venier,P.,Pozzoli L.,Cardin,E.L.,Sarto,F. and Levis,A.G.(1986)"Mutagenic activity and polycyclic aromatic hydrocarbon levels in urine of humans exposed to therapeutical coal tar".Carcinogenesis, 7,819-823
9) Becher,G.,Haugen,A., and Bjø rseth,A.(1984)"Multimethod determination of occupational exposure to polycyclic aromatic hydrocarbons in an aluminium plant".Carcinogenesis,5, 647 651
10) Camus.A.M.,Aitio,A.,Sabadie,N.,Wahrendorf,J. and Bartsch, H.(1984)"Metabolism and urinary excretion of mutagenic metabolites of benzo(a)pyrene in C57 and DBA mice strains". Carcinogenesis.5,35-39

53

Instability of lung retained welding fumes : a slow active Ni and Cr compartment

R.M. Stern 1), J. Lipka*, M.B. Madsen 2), S. Mørup 2), E. Thomsen 3), C.J.W. Koch 4)

ABSTRACT

Magnetopneumographic (MPG) measurements of welding fumes, as retained in the lungs of welders in vivo, or as found in low temperature ashed lungs from deceased welders in vitro, suggest that these samples contain material that more closely resembles magnetite than it does freshly produced welding fumes. Mössbauer studies of welding fume samples indicate that the magnetic iron complex is primarily in the form of disordered and/or substituted spinel (magnetite), which is apparently less substituted in autopsy specimens than in freshly formed fumes, as also found by x-ray diffractometry. Chemical analysis demonstrates the presence of soluble and insoluble Cr, Cr(VI) and Ni fractions in stainless

---

1) The Danish Welding Institute, DK-2605 Brøndby

2) Laboratory of Applied Physics II, Technical University of Denmark, DK-2800 Lyngby

*) J. Lipka, permanent adress: Dep. of Nuclear Physics and Technology, Slovak Technical University, Mlynska dolina, 81219 Bratislava, Czechoslovakia.

3) Chemiconsult, Røgerup, DK-4050 Skibby

4) Chemistry Dept., Royal Veterinary and Agricultural University, DK-1871 Frederiksberg

steel welding fumes.
It is suggested that metallic aerosols such as stainless steel welding fumes are not stable upon retention in the lungs, and represent a continual source of toxic elements such as Ni and Cr: a slow active compartment. These lung burdens which may contain as much as 1500 mg magnetic dust, as determined by MPG measurements, represent an initial deposit of similar amounts of Ni and Cr, and could contribute to elevated levels in blood and urine until depleted of trace elements, independent of current exposures. MPG measurements identify the small proportion (10-20%) of individuals with large body burdens of (magnetic) dust and are therefore a potential compliment to biological monitoring procedures.

INTRODUCTION

Magnetopneumographic (MPG) measurements of the induced remanent field of the thorax of individuals exposed to magnetic dusts have been used to determine the thorax dust burden of welders, asbestos miners, and coal miners (e.g. Cohen et al. 1981, Kalliomäki 1978; for a review see Lippman 1986). These studies are based on the assumption that magnetic characteristics of the inhaled particles are known, and are stable after retention in the lung and after incorporation in lung tissue, since interpretation of the lung burden in terms of the remanent magnetic field strength requires knowledge of the specific remanent magnetic moment of the retained dust. The few systematic MPG studies reported in the literature suggest that for cohorts with more than minimal experience there is only little if any correlation between retained dust and extent of exposure, but that some small fraction (10-20%) of any given cross-sectionally studied cohort will exhibit a signal indicative of a significant lung burden, i.e. >300 mg dust.

A number of recent studies of blood and urinary Cr and Ni as biological indicators of occupational exposure show that for stainless steel welders, average levels among exposed cohorts are elevated, although individual values are not necessarily indicative of intensity of exposure to welding fumes (e.g. Mutti et al. 1974, Tola et al. 1971, Welinder et al. 1983, Sjögren 1985, Sjögren et al. 1983, Granjean et al. 1980, Åkesson and Skerfving 1985; for a review see Aitio and Järvisalo 1986). In particular, the blood/urine ratio of these trace metals appears to be sensitive to the nature of the aerosol, some highly exposed individuals show relatively low levels, and some individuals exhibit elevated levels in the absence of recent exposure. Careful studies of Cr levels among stainless steel welders over a work-week, and after exposure has stopped, indicate that cross shift variations in urine Cr levels reflect immediate exposures which are characterized by a rapid clearance time (12 hrs), while background levels suggest the presence of long term clearance associated with a slow active compartment, some fraction of which has essentially an infinite half life. The presence of a slow and a fast compartment, together with significant uncertainties in determination of actual Cr content and distribution with respect to valence states and solubility fractions of welding exposures (Stern et al. 1984, Thomsen et al. 1986) introduces considerable uncertainty in the interpretation of biological monitoring in these occupational cohorts, with respect to exposure and internal dose.

There would appear to be a need to combine measurements of biological indicies of exposure with a technique which could determine existing body burden in the absence of recent exposure. Since stainless steel welding fumes are observed to be magnetic, it is suggestive that MPG measurements of stainless steel welders could be used to identify individuals with significant lung burden and provide information as to the nature of the retained material. Recent in vitro studies have suggested significant matrix and collection protocol dependent differences in the bioavailability and toxicity of Ni and Cr metabolites from various stainless steel welding fumes (Hansen and Stern 1984, Hansen and Stern 1985), and recent clinical studies have shown detectable cytological changes in the peripheral lymphocytes from stainless steel welders (Koshi et al. 1984, Koshi and Yagami 1984). If the magnetic and chemical characteristics and in vivo stability of welding fumes were known, then the MPG measurements could be used to give information on the toxic constituents associated with the lung burden, in particular with respect to the magnitude of the slow Ni and Cr compartment. It might then be possible to define and distinguish between risks due to external and internal sources of toxic substances, and to improve the utility of biological monitoring in general.

MATERIALS AND METHODS

Welding fumes, grinding dusts and standards

Welding fumes from a wide range of technologies and applications representative of current industrial uses are produced in a robot and collected on paper filters, from which they are scraped and stored in plastic bottles. The most common processes are Metal Inert/Active Gas (MIG/MAG) and Manual Metal Arc (MMA) welding of Mild Steel (MS) and Stainless Steel (SS). Metallic dusts are collected in a number of work places. Commercial samples of magnetite and maghemite (Tiede(Essingen) 605.1, 605.2) are used as standards for calibration of the MPG instrument.

Autopsy material

Low temperature ashed lung sections from several MS welders have been obtained. They are placed in similar containers as the standards for measurement in the MPG instrument.

The MPG instrument

A pilot model MPG instrument has been constructed which incorporates a pulsed DC remanent field system capable of characterizing magnetic dusts in terms of their remanent field strength and their coercive force (V_{hc}) together with an AC susceptibility bridge for determination of total magnetic metal content of the thorax (Drenck and Stern 1986, Stern and Drenck 1986). In the DC system, a uniform external magnetizing field is created by the discharge of a large capacitor bank through a pair of coils centered around the thorax of a subject placed on a movable non-magnetic bed. The maximum magnetizing field strength is 80 mT (pulse length 0.25 s) at 120 volts on the capacitor bank. The remanent field strength is measured by means of a flux-gate magnetometer centered at the magnetizing coil, operated in the gradient mode (40 cm base line), and placed so that the lower detector is 10 cm from the upper surface of the chest. A scan can be performed by means of lateral and transverse motion of the bed: the positional dependence of the magnetic signal is recorded continuously. In the AC system, a fluxgate detector is placed in the center of a set of compensating coils which are designed so that in the absence of a magnetic object in the energizing coils the field at the detector is identically balanced to zero. Measurements are made with a phase sensitive detector at the frequency of the energizing current (80 Hz). The ultimate sensitivity of the DC system is 1 mg magnetite distributed throughout the lungs; the AC system will detect the presence of 20-40 mg magnetic iron regardless of susceptibility, observed as a reduction of the magnitude of the signal of the unexposed thorax, which is normally diamagnetic due to the presence of water. The addition of ferrimagnetic material will decrease the net diamagnetic signal until, at a lung burden of approximately 200 mg magnetite, the net signal becomes ferromagnetic.

MPG measurements of welders

A series of experienced welders of mild and stainless steel, trainees, instructors, and clerical personnel (as controls) at the Danish Welding Institute were measured using the MPG instrument. For those individuals with significant remanent field values (e.g. >4 nT), a positional scan of the remanent field strength over the thorax was made on a 5 cm grid, and V_{Hc} and MPG(AC) was measured as well.

Chemical analysis

Fumes, dusts, standards and ashed lung samples are analyzed for metal content by means of atomic absorption spectrometry (AAS), and occasionally with proton induced x-ray spectral analysis (PIXE).

X-ray diffraction analysis

Some samples of fumes and ashed lungs are analyzed using an x-ray powder diffractometer.

Mössbauer spectroscopy

Samples of fumes, standards and ashed lungs are placed in standardized plexiglass holders containing 50 mg each, and analyzed with a conventional constant acceleration Mössbauer spectrometer with a source ^{57}Co in Rh. Velocities and isomer shifts are given relative to the centroid of the spectrum of α-Fe at room temperature.

RESULTS AND DISCUSSION

In Table 1 MPG results for each dust are expressed in terms of the effective content (weight %) of magnetite standard, and the coercive force is expressed as the magnitude of the demagnetizing pulse necessary to reduce the remanent field strength to zero after a standardized magnetizing pulse. The magnetic characteristics of selected welding fume samples, some occupational dust samples and standards are summarized here.

Table 1a: Magnetic parameters and AAS–results for welding fumes and other dusts.

Fume type	Relative Fe_3O_4 content (%)		AC/DC sensitivity	V_{hc} (V)	Chemistry (AAS) (iron content expressed as Fe_3O_4)
	DC	AC		(+/-2 V)	
MMA/MS	10-67	23-72	1.0-2.45	33-55	23-56
MMA/SS	4-42	5.2-68	0.8-4.4	19-49	6-14
MAG/MS	113-144	79-107	0.7-0.8	39-45	84-100
MAG/SS	60-67	61-87	0.9-1.5	29-49	52-72
MS/GD	20-200	1.00-200	1.4-2.8	13.5-25	(100% Fe)
SS/GD	50-64	18	0.28-0.36	25-34	(70% Fe)
Crocidolite	8	-	-	63-65	-
Sand	3	6	2.0	0	-
Coal fly ash	10.2	6	0.45	47	-

Table 1b: Magnetic parameters of standards.

Substance	DC Remanent Field (nT/g)	Susceptibility (AC dipole moment) (pT/g)	V_{hc} (V) (+/-2 V)
Magnetite	75 (Rigid sample) 160 (Col suspension)	250 400	52 0
Maghemite	197 (Rigid sample)	150	45
α-iron	4.04 (Rigid sample)	175	35
Water	0	125 pT/25 L	0

* 15 cm sample detector distance, 120 V (0.08 T pulse)

Magnetic parameters from autopsy samples from MS welders and clinical studies of SS welders with high lung burdens are given in Table 2.

Table 2a: Magnetic parameters and PIXE–results of low temperature ashed lungs in vitro.

Occupational history	Dust in sample	AC Signal	DC Signal	V_{hc} (V) (+/-2 V)	Chemistry (>0.5%) (PIXE)
Welder (MMA/MS)	5.5 g	63 pT	46 nT	49	P,Ca,Ti,Fe
Welder (MMA/MS)	12 g	1250 pT	1000 nT	44	P,S,Ca,Fe
Iron miner	1.9 g	500 pT	1000 nT	62	P,Ti,Fe

Table 2b: Magnetic parameters of stainless steel welders in vivo.

Occupation	Inferred lung burden (magnetite)	AC Shift from unexposed thorax	DC Signal	V_{hc} (V) (+/-2 V)	Possible dust type
SS welder 5 years MIG/SS, MMA/SS, TIG/SS	(30 mg)	15 pT	3 nT	54	Magnetite
SS welder 10 years TIG/SS, MIG/SS	(30 mg)	10 pT	3 nT	30	MMA/SS MIG/SS GD/SS
SS welder 13 years	(150 mg)	50 pT	16 nT	18	MMA/SS

The welding fumes, grinding dusts and standard samples are shown to have a wide range of magnetic characteristics. The magnetic characteristics of the low temperature ashed autopsy samples of MS welders' lungs more closely resemble those of the pure magnetic oxides than they do the majority of MMA/MS welding fumes tested. Similarly, the coercive force of the thorax burden of at least one of the SS welders does not resemble that found for most of the welding fumes, but is close to that of the standards.

The results of chemical analysis of selected welding fumes are given in Table 3.

Table 3: Content (weight %) and solubility of some metals from various welding fumes (for SS–fumes Fe is given as magnetite).

	Prompt Water Soluble (%)	Insoluble in water (%)	Total (%)
MMA/SS			
Cr(III)+Cr	0	0.2 - 2.1	0.2 - 2.1
Cr(VI)	2.2 - 4.3	0.03 - 0.42	2.2 - 4.3
Total Ni	0.01 - 0.31	0.27 - 1.6	0.38 - 1.9
Fe as Fe_3O_4	0	3.2 - 15	3.2 - 15
MIG/SS			
Cr(III)+Cr	0	3.56 - 13.78	3.56 - 13.78
Cr(VI)	0.005 - 1.50*	0.01 - 0.42	0.02 - 2.90
Total Ni	0.05 - 0.25	3.5 - 6.3**	3.5 - 6.5
Fe as Fe_3O_4	0	50 - 51	50 - 51
MMA/Ni			
Ni	<0.003	1.4	1.4
Cr	<0.002	0.014	0.02
Ba	6	34	40
Fe	0	1 - 3	1 - 3
MIG/Ni			
Ni	0.02 - 0.06 (Ni/NiO = 2:1 - 1:10)	53 - 60***	53 - 60
Cr	<0.004	0.04	0.04
Fe	0	0.1 - 5	0.1 - 5

* 4.5% soluble Cr(VI) if collected in an impinger.

** up to 4% Ni soluble in serum.

*** up to 40% soluble if collected in a serum filled impinger.

The chemical analysis shows that for MMA/SS fumes all of the water soluble Cr appears as Cr(VI), while the Ni content is essentially insoluble in water, and for MIG/SS that the Ni and Cr content is relatively insoluble; the combined Ni and Cr content is approximately 50% of that of magnetite.

The Mössbauer spectra of selected samples of welding fumes and lung samples are shown in Figure 1 and 2, respectively.

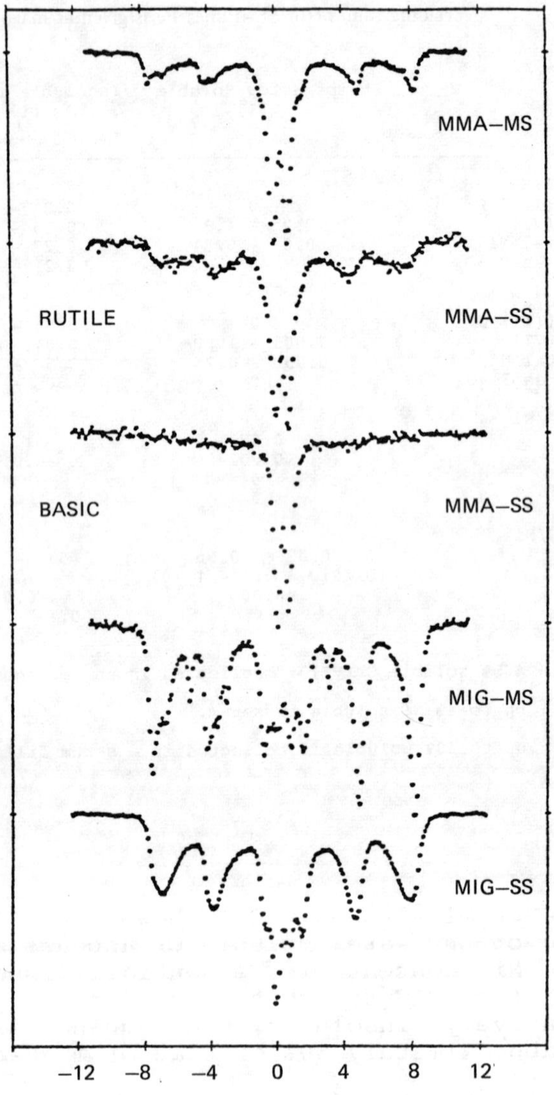

Fig. 1. – Mösbauer spectra of welding fumes obtained at room temperature.

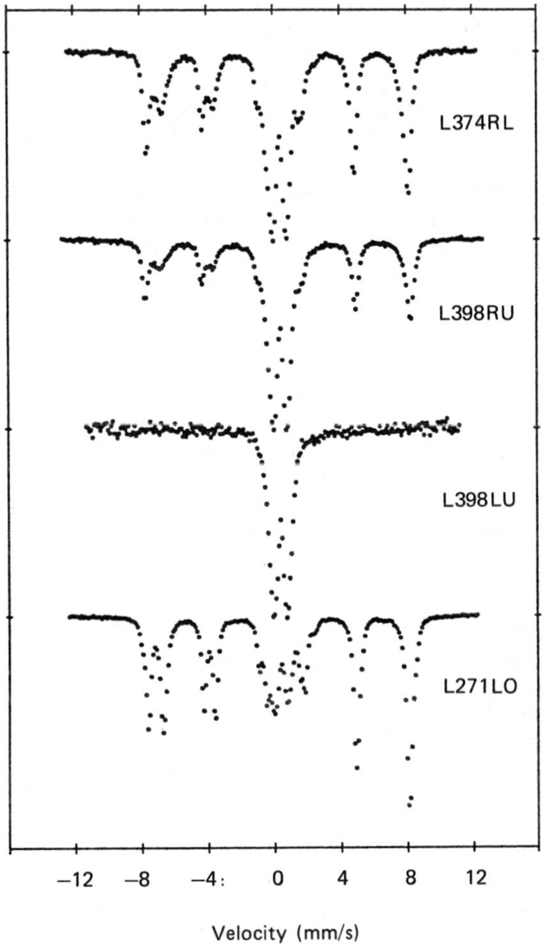

Fig. 2. – Mössbauer spectra of various lung samples obtained at room tempera temperature,

In the samples from the MMA welding process only iron containing oxides are found, whereas in the samples produced by MIG welding also metallic phases are found. The MIG/MS and MIG/SS welding fumes contain approximately 7% α-iron and 12% γ-iron, respectively.

The magnetically split (sixline) components in the spectra of the lung samples display the characteristic two-sextet pattern of magnetite. However, the relative areas of the two sextets show that some of the iron in the magnetite is substituted with other elements. The corresponding components in the welding fumes show larger variation in the relative areas, and severe line-broadning, presumably due to a higher degree of substitution than found in the lung samples.

A paramagnetic doublet with relative area from 20-100% of the total area is found in the Mössbauer spectra of the welding fumes and lung samples . The quadrupole splitting, Q, the isomer shift, I, and the relative area of this doublet are given in Table 4.

Table 4a: Room temperature Mössbauer parameters of paramagnetic doublet in spectra of welding fumes.

Sample	Q (mms^{-1})	I (mms^{-1})	Relative area (%)
MMA/MS	0.85	0.21	48
MMA/SS rutile	0.79	0.29	54
MMA/SS basic	0.67	0.32	100
MIG/SS	0.74	0.42	19

Table 4b: Room temperature Mössbauer parameters of paramagnetic doublet in spectra of low temperature ashed lung samples.

Sample	Q (mms^{-1})	I (mms^{-1})	Relative Area (%)	Occupational history
L 271 LO	0.84	0.37	21	Iron miner
L 374	0.90	0.39	29	Welder
L 374 RL	0.88	0.39	36	
L 398 LU	0.85	0.38	97	Welder
L 398 RM	0.83	0.38	55	

The isomer shifts of the paramagnetic doublets from the welding fumes and from the lung samples are different. This indicates that the component originates from different phases in the two types of samples.

The x-ray diffraction patterns of the welding fumes show the presence of two spinel phases, except for the MMA/SS(basic), which shows only one. Furthermore, the presence of alkali fluorides in some samples and metallic iron in all MIG samples are observed. Diffractographs show the presence of one spinel phase in most of the lung samples, however sample 398LU does not contain any detectable spinel phase. The Mössbauer spectra of this sample contain only a paramagnetic doublet, suggesting that the similiar doublets of the other lung samples are not due to the presence of spinel phases either. The Mössbauer parameters of this doublet are not in accordance with either heme, ferritin or haemosiderin (Guest 1976, Guest 1981, Bell et al. 1984). It is likely, however, that this doublet is due to the presence of a decomposition product of blood. In the Mössbauer spectra of the welding fumes, the two spinel phases are observed as a doublet and a magnetically split sextet respectively. The one spinel phase in MMA/SS(basic) exhibits superparamagnetic relaxation indicating a very small particle size (Mørup et al, 1980).

CONCLUSIONS

Magnetic, x-ray diffraction, and Mössbauer measurements indicate that the magnetic component of welding fumes is primarily substituted magnetite. It appears that the inorganic dust from ashed welders lungs contains a more perfect magnetite as compared to welding fumes produced by common welding processes. Similarly, in vivo measurements of SS welders suggest that the lung burdens do not always magnetically resemble welding fumes, but are occasionally closer to the magnetic constituent, magnetite. Thus, welding fumes as retained in the lungs appear to be enriched in refractory magnetite content, implying that other non-magnetic components have been depleted. This depletion is presumably due to the solubilization of material deposited on the lung air space surfaces by water and circulating metal-binding proteins, and the action of intracellular lysosomal enzymes, consistent with the incorporation of fume particulates in lung macrophages (Aitio and Järvisalo 1986).
For typical welding fumes from alloyed and stainless

steels, the initial content of Ni and Cr is approximately 50% of the effective content of magnetite. This suggests that to within this factor the thorax magnetic signal is a good measure of the initial dose of Cr and Ni, under the assumption that the magnetite is refractory to in situ corrosion. On the other hand, stainless steel welding fumes exhibit significant variability in the serum and water solubility of their Ni and Cr content and in the bioavailability and toxicity of the various species formed. However, if the nature of the exposure for a given individual is known, it is possible to reconstruct the initial composition of the lung deposit detected by an MPG measurement.

It is tempting to suggest that the presence of prompt water and serum soluble Ni and Cr in welding fumes are to be associated with the short term active deposit responsible for cross shift changes in urinary Ni and Cr, while the remaining Cr and Ni partially found in the substituted and degradable spinel are to be associated with the postulated slow active compartment responsible for elevated background levels. The demonstration of this effect would then strongly argue for a reexamination of previous environmental and biological monitoring data, and for the introduction of protocols for measuring biological exposure indicies which could differentiate between slow and fast compartments: e.g. between background levels and cross shift changes. Furthermore, the possible differentiation in species formation from these two compartments, one of which represents an excess in chemical reagent, the activity of which does not depend on quantity, but which might depend on age, could be associated with significant variation in early and delayed toxicities to organs containing the initial burden, and to those sensitive to circulating metabolites. A systematic assay for species toxicity remains to be performed, however, although preliminary in vitro studies show that solubilization of Ni and Cr from these matricies is accompanied by large changes in bioavailability and toxicity (Hansen and Stern 1984, Hansen and Stern 1985). Although it would appear that retention in the lungs of environmental Cr and Ni is a general phenomenon (Kollmeier et al 1985), the observation that only a limited fraction of occupationally exposed cohorts collect significant lung burdens of metallic aerosols could have important consequences for determination of individual risk, since for these individuals potentially toxic material is always present irrespective of further exposure.

ACKNOWLEDGEMENTS

This work was supported in part by the Danish Technical Science Research Council (STVF), the Health Foundation (Helsefonden), and the 4th ECSC Medical Research Program (Luxemburg). J.L. was a recipient of a study grant from the Danish Ministry of Education. Ashed lungs were kindly provided by Prof. H.J. Einbrodt, RWTH Aachen.

BIBLIOGRAPHY

A. Aitio and J. Järvisalo: Levels in welding fume components in tissues and body fluids. In: R.M. Stern et al. EDS, Health hazards and biological effects of welding fumes and gases. Excerpta Medica, Elsevier Amsterdam 1986 pp 169-180.

S. Bell, M.P. Weir, D.P.E. Dickson, J.F. Gibson, G.A. Sharp. T.J. Peters: Mössbauer spectoscopic studies of human haemosiderin and ferritin. Biochemica and Biophysica Acta 787 (1984) 227-236.

D. Cohen, T. Growther, G. Gibbs, M. Becklake: Magnetic lung measurements in relationship to occupational exposure in asbestos miners and millers in Quebec. Env. Res. 26 (1981) 535-550.

K. Drenck, R.M. Stern: Alternating current susceptibility bridge magnetopneumography. In: R.M. Stern et al. EDS, Health hazards and biological effects of welding fumes and gases, Excerpta Medica, Elsevier, Amsterdam 1986, pp 219-222.

P. Granjean, F.W. Sunderman Jr., S.K. Shen, I.J. Selikoff: Measurement of nickel in plasma and urine of shipyard workers. In: S.S. Brown Ed. Nickel Toxicology, Academic Press, NY 1980 pp 107-109.

L. Guest: Investigation into the endogeneous iron content of human lungs by Mössbauer spectroscopy. Ann. Occup. Hyg. 19 (1976) 49-62.

L. Guest: Quantitative determination of the exogenous and endogenous storage iron content of haematite workers'lungs. Analyst 106 (1981) 663-675.

K. Hansen, R.M. Stern: Toxicity and transformation potency of nickel compounds in BHK cells in vitro. In: F.W. Sunderman Jr. et al. EDS, Nickel in the human environment, IARC Lyon 1984, pp 193-200.

K. Hansen, R.M. Stern: Welding fumes an chromium compounds in cell transformation assays. J. Appl. Tox. 59 (1985) 306-314.

P.L. Kalliomäki, O. Kohonen, V. Vaaranen, K. Kalliomäki, M. Kohonen: Lung retention and clearence of shipyard arc welders. Int. Arc. Occup. Environ. Health 42 (1978) 83-90.

K. Koshi, T. Yagami: Chromosomes of cultured periphemal lymphocytes from stainless steel welders 2. Mutation Res. 130 (1984) 369.

K. Koshi, T. Yagami, Y. Nakanishi: Cytogenic analysis of perphirmal blood lymphocites from stainless steel welders. Industrial health 22 (1984) 305-318.

H. Kollmeier, C.W. Itting, J. Seemad, P. Wittig, R. Rothe: Increased chromium and nickel in lung tissue. J. Cancer Res. Clin. Oncol. 111 (1985) 173-176.

M. Lippman: Magnetopneumography as a tool for measuring lung burden of industrial aerosols, in: R.M. Stern et al. EDS, Health Hazards and Biological Effects of Welding Fumes and Gases, Excerpta Medica, Elsevier, Amsterdam 1986 pp 199-214.

A. Mutti, A. Cavatorta, C. Pedroni, A. Borgi, C. Giaroli, I. Franchini: The role of chromium accumulation in the relationship between airborne and urinary chromium. Int. Arc. Occup. Environ Health 43 (1974) 123-133.

S. Mørup, J.A. Dumesic, H. Topsøe: Magnetic Microcrystals. In Applications of Mössbauer Spectroscopy, Vol.II, Ed. R.L. Cohen (Academic Press, N.Y.)(1980) pp 1-53.

B. Sjögren: Respiratory disorders and biological monitoring among electric arc-welders and braziers. Arbete och Hälse 9 (1985) 1-48.

B. Sjögren, L. Hedström, U. Ulfvarson: Urine chromium as an estimator of air exposure to stainless steel welding fumes. Int. Arch. Occup. Environ. Health 51 (1983) 347-354.

R.M. Stern, E. Thomsen, E. Furst: Cr(VI) and other metallic mutagens in fly ash and welding fumes. Tox. Environ. Chem. 8 (1984) 95-108.

R.M. Stern, K. Drenck: Magnetic characteristics of metallic industrial aerosols including welding fumes, grinding dusts and asbestos materials, SVC report 86.23, The Danish Welding Institute, 1986.

E. Thomsen, R.M. Stern, B. Petersen: Exposure monitoring and chemical analysis of welding fume. In: R.M. Stern et al. EDS, Biological effects and health hazards of welding fumes and gases, Excerpta Medica, Elsevier Amsterdam 1986, pp 47-50.

S. Tola, J. Kilpig, M. Virtamo, K. Haapa: Urinary chromium as an indicator of the exposure of welders to chromium. Scand. J. Work. Environ Health 3 (1971) 192-202.

H. Welinder, M. Litorin, B. Gullberg, S. Skerfving: Elimination of chromium in urine after stainless steel welding. Scand. J. Work. Environ. Health 9 (1983) 397-403.

B. Åkesson, S. Skerfving: Exposure in welding of high nickel alloy. Int. Arch. Occup. Environ. Health 5 (1985) 111-118.

54

Chromosome aberrations in workers with exposure to organic solvents : a case referent study

F. Traversa, M. Girbino, F. Ottenga, A.D. Bonsignore – Institute of Occupational Medicine, University of Genova, Italy

ABSTRACT

A cytogenetic study was carried out among shipyard workers with exposure to styrene, and controls. Air concentration of styrene and biological indicators in urine were lower than the TLVs and BLVs. Chromosome aberrations were analyzed for each subject from cultures of peripheral blood lymphocytes. The mean frequencies of chromosome changes in styrene workers were not significantly different from controls. A significant difference was found between smoking styrene workers and nonsmoking referents. The possible additive genotoxic activity of styrene exposure and smoking is discussed.

INTRODUCTION

Analysis of acquired structural chromosome aberrations in cultured lymphocytes can be regarded as a method for the detection of early adverse effects of genotox icants at the group level, even though the biological significance of these changes is not yet clear (Evans et al., 1980).

The aim of our project is to point out the potentially hazardous job categories among the complex procedures of some factories in our country where genotoxic chemicals are being used. This could carry out a cytogenetic surveillance immediately useful for practical purposes, and suggest new plans for laboratory investigation.

In this paper we report the preliminary data of a study which we are performing among the workers of a small shipyard where boats made of glass-fiber reinforced plastics are built. Work procedures frequently need workers' direct attendance during polymerization of plastics, as for joining surfaces or forming small parts.

Styrene is known to be the most important chemical used in this industry. It has been observed to be mutagenic in some short-term tests and to be associated with an increase of chromosome aberrations and sister chromatid exchanges in workers exposed in the reinforced plastics industry (Meretoja et al., 1978).

SUBJECTS AND METHODS

The subjects of the study were 18 shipyard workers employed in manufacturing polyester plastic products and 20 referents working in various job categories in a mechanical building plant without any evident exposure to chemicals. The exposed workers, all men aged 21 to 51 years, had been employed for 2 to 12 years; the referents were all men aged 20 to 54 years. Each person was interviewed as to his health status and possible confounding factors (diagnostic radiographs, drug intake, smoking habits, alcohol consumption, recent viral infections, and previous occupational history); clinical examination and laboratory data (blood count, platelets, reticolocytes, and liver enzymes) were recorded. According to the obtained information, the subjects were healthy.

Styrene concentration in the workplace was estimated. TWA determinations, performed with the gaschromatographic NIOSH method, were 15 to 52 ppm (mean 35 $\pm$ 13). Spot urine samples were collected at the beginning of the work shift on Friday, and mandelic and phenylglyoxilic acids concentrations evaluated with a conventional gaschromatographic method

(Apostoli et al., 1983). Mean data from environmental and biological monitoring did not exceed the threshold limit values as recommended by ACGIH (1985-86).
The chromosome studies were performed according to conventional micromethods. The cells were harvested after a 48 h culture period and stained with a 10% solution of Giemsa. A total of 100 metaphases for each subject was analyzed on coded slides and chromosome changes were classifyed according to the ISCN (1978).

RESULTS

The results of the analysis of chromosome aberrations are given in table 1. The mean frequencies of aberrations, including gaps, were 5.7% for exposed workers and 4.6% for referents. The results were slightly higher in smokers than in nonsmokers both for exposed workers and referents, but no statistically significant difference was observed between the groups. However, the highest aberration frequencies were observed in smoking exposed workers, who had a significantly ($p < 0.01$) increased frequency of aberrations in comparison to the group of nonsmoking referents.

Table 1: Mean frequencies of structural chromosome aberrations among styrene workers (E) and referents (R), both divided into smokers (S) and nonsmokers (NS).

	number of subjects	age (years) (1)	years of exposure (1)	years of smoking (1)	MA + PGA (2)	total aberrations (3)
E NS	7	33.3±10.5	9.7±8.2	-	505.0±227.4	5.3±4.7
E S	11	37.9±10.3	8.3±5.6	19.6+9.5	556.4±252.4	6.2±5.4 (4)
R NS	6	35.8+10.8	-	-	-	3.3±3.9
R S	14	39.4±10.5	-	19.4±10.5	-	5.8±4.1

(1) mean ± SD.

(2) mandelic acid + phenylglyoxilic acid in urine, mg/g creatinine.

(3) including gaps, 100 cells per subject.

(4) p 0.01, compared with nonsmoking referents (chi-square test).

The most frequent types of aberrations were chromatid and chromosome breaks, fragments or acentric for all the groups; dicentric were extremely rare. Aneuploid cells were 2.8% for exposed workers and 3.4% for referents.

DISCUSSION

In the shipyard which we are studying the exposure situations are variable according to the work procedures, the size of the boats, the job categories etc. An accurate description of the risk can be made only after several determinations of atmospheric styrene and metabolites in urine, carried out in different periods of the year. At the same time, there is still a quite limited experience about cytogenetic assays in worker groups. Some more suggestive data will be available for our study when we can perform a continuous monitoring of biological parameters and exposure situations for several years, beginning from the engagement time.

Furthermore, the number of the subjects in the present study is too limited for any conclusions about the frequency of chromosome aberrations in relation to the degree and duration of exposure. However, our preliminary findings show that the mean frequencies of chromosome aberrations in styrene workers are slightly higher than in referents and, separately, in smokers without relation to the job, compared to nonsmokers, even though no statistically significant difference between the groups was observed. Emphasis must be placed on the fact that the exposure data, both from airborne determinations and from biological monitoring, were lower than the TLVs and BLVs.

On the contrary, we found a significant difference between the smoking exposed workers and nonsmoking referents. This seems to suggest that, in our conditions, styrene exposure and smoking may have had an additive genotoxic activity. According to our results, smoking workers exposed to styrene must be regarded as a population at cancer risk.

REFERENCES

Apostoli P., Brugnone F., Perbellini L., Cocheo V., Bellomo M.L.: Occupational styrene exposure: environmental and biological monitoring. Am. J. Ind. Med. 4, 741-754, 1983.

Buckton K.E., Evans H.J.: Methods for the analysis of human chromosome aberrations. World Health Organization, Geneva, 1973.

Evans H.J., Ishidate M., Leng M., Miller C.T., Mitclman F., Vogel E.: Cytogenetic damage as an endpoint in short-term assay systems for detecting environmental carcinogens. In: International Agcncy for Research on Cancer. Long-term and short-term screening assays for carcinogens: a critical appraisal. Lyon, 1980, 227-244 (IARC Monographs, suppl. 2).

Meretoja T., Järventaus H., Sorsa M., Vainio H.: Chromosome aberrations in lymphocytes of workers exposed to styrene. Scand. J. Work Environ. Health 4: suppl.2, 259-264, 1978.

— An International System for Human Cytogenetic Nomenclature (1978). Cytogenet. Cell Genet. 21, 1978.

55

Chromosomal abberations among workers engaged in the explosives industry

S.S Sawsan, M.M. El-Ghazali, M.M. El-Batanouni, M.M. Amr, A.A. Massoud – Department of Community, Environmental and Occupational Medicine, Ain Shams University, and Department of Industrial and Occupational Medicine, Cairo University, Egypt

INTRODUCTION

The concept of environmental chemicals resulting in damage to the genetic material has been a great stimulus to further investigation of this relationship. The cytogenetic surveillance of persons engaged in chemical production is considered to be a point of great importance in occupational health. Combined exposure to different metals and other chemicals are common; therefore, reliable risk assessment following occupational exposure to a single metal cannot be made at present (Karantzis and Lilly, 1979). The aim of the present study was to evaluate the cytogenetic effect of simultaneous exposure to lead, mercury and trinitrotoluene (TNT) among workers engaged in the explosives industry.

SUBJECTS AND METHODS

Subjects consisted of two main groups matched for age and sex (Table 1). Group 1: 14 exposed workers (10 males and 4 females). Group 2: 13 controls (9 male and 4 female office clerks), not occupationally exposed to chemicals. The workers under study were mainly exposed to lead and mercury and, to a

lesser extent, to TNT. Environmental studies were performed (El-Batanouni et al., 1985, in the press), but are not included in this paper.
All subjects had normal children, had not been exposed to X-rays, had not taken drugs, nor had experienced viral diseases in the previous six months; none suffered from malignancies. Smoking habits were taken into consideration.
Whole blood cultures were performed using Gibco Chromosome kits. Culture time was 72 hours, including 2 hours treatment with Colcemid. Slides were prepared by the air-drying method, stained by conventional Giemsa, coded and analyzed blindly for chromosomal aberrations. One hundred well-spread metaphases were examined for each subject at 1500 X. Evaluation of aberrations was performed according to the Edinburgh classification (Buckton and Evans, 1973). The results were analyzed statistically.

RESULTS

As shown in Table 1, there was no difference between the mean age of the exposed and control groups, matched for sex. The mean duration of exposure was higher among males than females ($p < 0.05$).
The percentage of metaphases with chromosome aberrations (Table 2) was significantly higher in the exposed workers than in the controls ($p < 0.001$). Both male and female workers had significantly higher percentages of aberrant metaphases than controls of the same sex (7.1 versus 3.1% for males; 4.4 versus 1.75% for females: $p < 0.05$ for both groups).

Table 1· Age, sex and duration of exposure among exposed and control groups.

Group	Age (years)		Duration of Exposure (years)		
	Males	Females	Males		Females
Exposed (n=14)	(30–46) 38±4.6	(25–34) 30.3±3.9	(16–30) 23.4±4.5	*	(11–20) 15.3±3.8
Control (n=13)	(30–45) 37.8±4.7	(27–33) 30.3±2.5	--		--
	$p > 0.05$	$p > 0.05$		* $p < 0.05$	

Table 2: Metaphases with chromosome aberrations among exposed and controls.

Group	Total examined metaphases	Aberrant metaphases No.	%
Exposed (n=14)	1400	89	6.4
Control (n=13)	1300	35	2.7

$p < 0.001$

The effect of smoking as a contributing cytogenetic risk factor is shown in Table 3. Exposed smokers showed higher aberration frequencies, but not at a significant level ($p > 0.05$).

Table 3: Number and percentage of metaphases with chromosome aberrations among exposed workers, according to smoking habits.

Exposed	Total examined metaphases	Aberrant metaphases No.	%
Smokers (n=6)	600	45	7.5
Non smokers (n=8)	800	44	5.5

$p > 0.05$

No correlation was found between percentage of aberrant metaphases and duration of exposure, as shown in Table 4. Table 5 shows the percentage of total aberrant metaphases and of the different types of aberrations recorded among the exposed. It is worth noting that chromatid-type aberrations (gaps, breaks and exchanges), and especially gaps, were more prevalent.

Table 4: Correlation between percentage of aberrant metaphases and duration of chemical exposure.

Case No.	Age (years)	Sex/smoking habit	% Aberrant metaphases	Duration of exposure (yrs)
1	42	MN	7	26
2	40	MS	6	25
3	37	MS	8	21
4	42	MS	10	26
5	39	MN	8	24
6	41	MN	7	26
7	33	MS	8	16
8	30	MS	6	16
9	46	MN	4	30
10	39	MS	7	24
11	34	FN	6	20
12	25	FN	4	11
13	30	FN	5	14
14	32	FN	3	16

$r = 0.3761$; $p > 0.05$

MN = Male non-smoker, MS = Male smoker, FN = Female non-smoker

DISCUSSION

Workers occupationally exposed to mixtures of chemicals are considered to be the most suitable groups for cytogenetic monitoring, to detect a possible genetic damage. The workers under study were simultaneously exposed to lead, mercury and TNT. Studies on cytogenetic effects of lead and mercury exist in the literature, but for TNT none were available. An increase of chromosomal aberrations among workers exposed to lead is reported in some studies (e.g., Schwanitz et al., 1970; Forni and Secchi, 1973; Deknudt et al., 1973; Bauchinger et al., 1976; Verschaeve et al., 1979). Deknudt and Léonard (1975) investigated the effect of lead in association with zinc and cadmium on lymphocytes <u>in vitro</u>, and concluded that lead could be responsible for chromosome aberrations. On the other hand, few reported on the effect of mercury in inducting chromosomal damage (Skerfving et al., 1974; Kato and Nakamura, 1976). In our cases, our assumption is that results may be due to the combined exposure to lead and mercury, but for TNT we cannot draw any conclusion, because information on agents that

Table 5: Number of examined metaphases, percentage of aberrant metaphases and of different types of aberrations recorded among the exposed.

Case No.	Sex/ smoking	No. exam. metaphases	% Aberrant metaphases	Chromatid gap	Chromatid br	Chromatid ex	isogap	isobr	C_u dic	C_u ace	C_s
1	MN	100	7	3	1	2	–	–	–	1	–
2	MS	100	6	3	–	1	–	1	–	–	1
3	MS	100	8	1	1	2	–	–	2	1	1
4	MS	100	10	4	3	–	1	–	1	1	–
5	MN	100	8	2	2	1	–	2	–	1	–
6	MN	100	7	2	–	2	1	–	1	1	–
7	MS	100	8	3	2	1	1	–	1	–	–
8	MS	100	6	2	2	–	1	–	1	–	1
9	MN	100	4	2	–	–	–	–	2	–	–
10	MS	100	7	3	2	1	1	–	–	–	–
11	FN	100	6	2	2	–	1	–	–	1	–
12	FN	100	4	2	1	–	1	–	1	–	–
13	FN	100	5	1	1	1	1	1	–	–	–
14	FN	100	3	3	–	–	–	–	–	–	–
Total		1400	89	33	17	11	8	4	9	6	3

MS = Male smoker, MN = Male non-smoker, FN = Female non-smoker.

br = break; isobr = isobreak; ex = exchange; dic = dicentric; ace = acentric fragment; C_u = chromosome-type unstable aberrations; C_s = chromosome-type stable aberrations.

exert an additive, potentiating or synergistic effect is lacking (Fishbein, 1976). O'Riordan et al. (1978) were of the opinion that heavy metals might act synergistically to enhance the mutagenicity of other compounds.

Smoking as a factor of cytogenetic damage was considered. Exposed subjects showed somewhat higher frequencies of aberrations than non-smokers, which is in accordance with Mäki-Paakkanen et al. (1981), while Nordenson et al. (1978) reported negative findings. A potential mutagenic effect of smoking was demonstrated by Obe and Herha (1978). Also El-Ghazali (1982) indicated smoking as a clastogenic agent among those exposed to pesticides.

The chromosome aberrations recorded were mainly of the chromatid-type, which suggests that the damage occurred after the DNA synthesis phase of the cell cycle.

In conclusion, from our data the simultaneous occupational exposure to lead, mercury and TNS seems to induce a cytogenetic effect on lymphocytes, and smoking also seems to be a contributing factor to chromosome damage. However the problem whether simultaneous exposure to more that one chemical has additive or synergistic effects on the genetic material, requires further investigation.

REFERENCES

Bauchinger M., Schmidt E., Inbrodt H.J. and Dresp J. (1976) Chromosome aberrations in lymphocytes after occupational exposure to lead and cadmium. Mutat. Res. 40, 57-62.

Buckton K.E. and Evans H.J. eds. (1973) Methods for the analysis of human chromosome aberrations. WHO, Geneva.

Deknudt G.H. and Léonard A. (1975) Cytogenetic investigations on leucocytes of workers from a cadmium plant. Environ. Physiol. Biochem. 5, 319-327.

Deknudt G.H., Léonard A. and Ivanov B. (1973) Chromosome aberrations observed in male workers occupationally exposed to lead. Environ. Physiol. Biochem. 3, 132-138.

El Ghazali S. (1982) Study of chromosomal aberrations in workers exposed to suspected mutagenic chemicals. M.D. Thesis, Ain Shams University, Egypt.

Fishbein L. (1976) Atmospheric mutagens. In: Chemical mutagens. Principles and methods for their detection, vol. 4. Hollander A. ed. Plenum Press, New York and London, p. 219.

Forni A. and Secchi G.C. (1973) Chromosome changes in preclinical and clinical lead poisoning and correlation with biochemical findings. In: Proc. Int. Symp. Environmental

Health Aspects of Lead, Amsterdam, October 1972. CEC, CID, EUR 5004, Luxembourg, pp. 473-485.

Karantzis G. and Lilly L.J. (1979) Mutagenic and carcinogenic effects of metals. In: Handbook on the toxicology of metals. Friberg L. ed. Elsevier-North Holland Biomedical Press, pp. 237-272.

Kato R. and Nakamura A. (1976) Jap. J. Hum. Genet. 20, 256-257.

Mäki-Paakkanen J., Sorsa M. and Vainio H. (1981) Chromosome aberrations and sister chromatid exchanges in lead-exposed workers. Hereditas 74, 269-275.

Nordenson I., Beckman G., Beckman L. and Nordstrom S. (1978) Occupational and environmental risk in and around a smelter in Northern Sweden. IV. Chromosomal aberrations in workers exposed to lead. Hereditas 88, 263-268.

Obe G. and Hehra J. (1978) Chromosomal aberrations in heavy smokers. Hum. Genet. 41, 257-263.

O'Riordan M.L., Hughes E.G. and Evans H.J. (1978) Chromosome studies on blood lymphocytes of men occupationally exposed to cadmium. Mutat. Res. 58, 305-311.

Schwanitz G., Lehnert G. and Gebhart E. (1970) Chromosome damage after occupational exposure to lead. Dtsch. Med. Wschr. 95, 1936-1941.

Skerfving S., Hansson K., Mangs C., Lindsten J. and Ryman N. (1974) Methylmercury-induced chromosome damage in man. Environ. Res. 7, 83-98.

Verschaeve L., Driesen U., Kirsch-Volders M. et al. (1979) Chromosome distribution studies after inorganic lead exposure. Hum. Genet. 49, 147-158.

56

Urinary screening of genotoxic substances with the use of the SOS-chromotest – first results on workers exposed to cutting fluids

E. Pospischil, P. Harmuth, Ch. Wolf, V. Meisinger, O. Jahn – Universität Klinik für Arbeitsmedizin, Vienna, Austria

Exposure to genotoxic substances in the environment today is one of the reasons for malignant disorders in man. During the last decade several methods and so-called In Vitro Short-Term Test systems have been developed for testing different chemicals for their mutagenic and genotoxic potentials. These assays have also been used for the detection of mutagenic agents in the urine of specifically exposed workers (1, 2). The group of employees so far examined includes workers exposed to volatile anesthetics, chemotherapeutical drugs and workers in foundries, steel plants and the carbon electrode industry.

Our study aimed at examining an occupational group that had been exposed to Cutting-fluids (CF), and that had so far hardly been considered in this respect.

For tribologic reasons, CF's are widley used to reduce the temperature of the metal pieces and to improve the metal interface. Operations such as cutting, screwing and grinding generate on oil mist which may be inhaled, ingested or absorbed via the skin by the workers. The majority of CF's used in industrial processes are synthetic or semi-synthetic. Apart from mineral oils, these substances contain emulsifiers, biocides, antifoamers, solvents, antioxidant agents and high-pressure additives (3).

Due to processing, mineral oils may be contaminated with polycylic aromatic hydrocarbons. It was demonstrated that CF's containing amines may form nitrosamines on heating or acidification (4). PAH's and nitrosamines are known to be carcinogenic. Epidemiologic studies, moreover, established a generally increased cancer risk for workers exposed to CF's (5, 6). Recent studies proved current CF's used in the metal industry and for glass manufacturing to be mutagenic in the AMES-test (7, 8).

METHODS

Our survey was carried out as a pilot study in a motor-construction plant. We examined 8 male employees working as toolsetters on semi-automatic drilling and grinding machines. The level of oil mist in the ambient air ranged from 0.7 to 1.2 mg/m^3, a value which is clearly below the TWA-level of 5 mg/m^3. The manufacturer of the CF used, specified the components of the brand as follows:
Mineral oils, fatty acid ester, fatty acid alkalol-amines, nonylphenolpolyglycolether and butyldiglycol.
8 non-exposed persons of the same plant were selected for the control group. Both groups were non-smokers and were organized to be homogenous in their dietary habits and anthropometric data.

Sampling protocol:
3 samples of urine were collected on Monday and Thursday, the first one at 6 a.m., the second one at 9 a.m. and the third one at 2 p.m. (end of shift). The control group was examined only on Thursday. The samples were stored at -20°C within one hour until determination.

Genotoxicity assay:
The genotoxic activity of the urine was measured with a colorimetric bacterial assay, which was described by Quillardet et al. (10). This in vitro short-term test is recently available as a commercial test-kit (Orgenics Ltd., Israel). Unlike the AMES test, detection of genotoxicity is performed by activation of the SOS-repair-system in E. coli PQ 37. In brief, the operon unit for the production of beta-galactosidase was submitted to the control of a repressor gene, so the beta-galactosidase activity depends strictly on a genotoxic event in the DNA of E.coli PQ 37 (further description in detail see 9). The determination of the beta-galactosidase can easily be performed photometrically. Lyophilized S-9 Mix was used as an exogenic metabolising system.
For preparation, the urine was neutralized to pH 7.5 and

subsequently centrifuged. One native portion of each sample was reserved for determination of urinary creatinine. Only samples up to 100 mg% urinary creatinine were used.
According to the instruction manual of the kit, unconcentrated urine was diluted in serial two-fold dilutions with DMSO. Each sample (10 µl) was checked with and without S-9 metabolisation. Upon termination of the different steps the micro-titer-plate was read at the absorption of 420 nm and 630 nm wavelength with a micro-ELISA-reader. The corresponding OD-values were plotted against the urine concentrations and the linear portion of the "dose-response" curve was determined. The slope of this curve was expressed as an "inducing-potency". For better evaluation of each kit, we corrected the results with the standard values (4-NQO, 2AA). The new values were called "corrected inducing potencies".

RESULTS

The following Table shows the average "corrected inducing potencies" of both groups. The filled bars indicate the S-9 metabolized samples (Table 1).

Table 1.

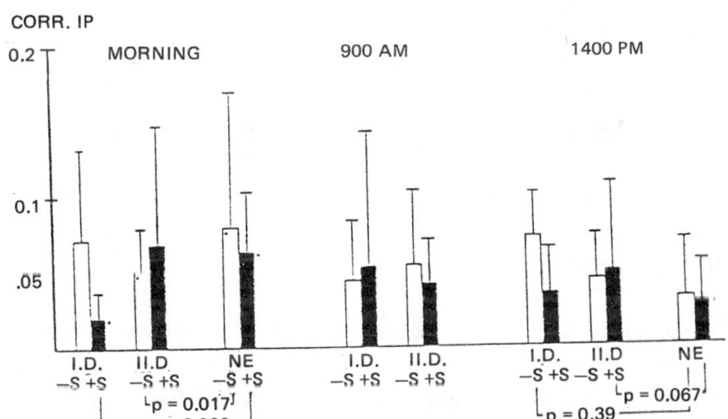

I.D., II.D. = Monday, Thursday, exposed group
NE = non-exposed group

The interindividual range is remarkably wide, especially in the morning. There is also a notable increasing activity of the exposed workers after S-9 metabolisation in vitro between the samples taken at 6 a.m. and those taken at 2 p.m., but there is only a statistical trend and no significance ($p = 0.08$).
If we assigned a positive and negative reaction to the

individual results with a threshold value of 0.05 - as postulated by Quillardet, the result would be the following table (it must be noted that these values were compared only hypothetically, since a direct comparison was only possible on the second day, when a protocol of the two groups was taken simultaneously):

Table 2.

		neg. -S9	neg. +S9	pos. -S9	pos. +S9	
I.Day:	6 a.m.					
	Exp.	3	7*	5	1*	*p = 0.04
	Non-exp.	3	3*	5	5*	
	2 p.m.					
	Exp.	1#	6	7#	2	#p = 0.01
	Non-exp.	6#	6	2#	2	
II.Day:	6 a.m.					
	Exp.	5	4	3	4	
	Non-exp.	3	3	5	5	
	2 p.m.					
	Exp.	4	4	4	4	
	Non-exp.	6	6	2	2	

DISCUSSION

In summary up it can be said that the present results only show a tendency but no significant increase of genotoxic activity in urine in workers exposed to CF's compared with the non-exposed control group at the end of shift by calculating the absolute values. These results were obtained only after metabolic exogenic activation. If we consider Table 2, an increasing genotoxicity was found in the group exposed to CF's by comparing the results hypothetically with the non-exposed group on the first day. It must be emphasized that both groups showed a high morning activity, and that the exposed workers showed especially high values on Monday. The average activity was even higher in the non-exposed group than in those exposed. Further the variation of the inter-individual levels was higher in the morning than at the end of shift. This increased genotoxicity of urine in the morning must certainly be ascribed to non-occupational exposures and cannot be ascribed to smoking habits, since both groups were non-smokers. This fact can hardly be evaluated retrospectively and generally hamper the use of genotoxic urinary screening surveys. In this study the relatively low number of persons examined is another limitation to the statistical calculation.

The SOS-CHROMOTEST is a method for monitoring the exposure of genotoxic agents like similar assays used in this field before. An advantage of this test system is certainly its simplified execution handling without microbiological facilities and the fact that unconcentrated and unpretreated urine is used, which in other bacterial test systems has not been possible so far. Further investigations, however, are needed in order to standardize the value of this assay for other bacterial test systems. Furthermore, it should be observed that data from urinary genotoxicity assays are to be considered as estimated indicators of recent exposure and must not be seen as indicators for biological effects.

REFERENCES

1. Ahlborg J.R. et al. (1985): Brit. J. Ind. Med. 42:691-699.
2. Vainio H. et al. (1984). In: Berlin A., Draper M., Henninki K., Vainio H. (eds.): Monitoring of human exposure to carcinogenic and mutagenic agents, IARC-Scientific Publ. No 59, p. 247 (Lyon).
3. Hausser H. et al. (1985): Zbl. Arbeitsmed., 35 (6):176-181.
4. Loeppky R.M. et al. (1981): Fd. Chem. Tox., 21(5):607-613.
5. Jarvholm B. et al. (1981): J. Occup. Med., 23 (5): 333-337.
6. Roush G.C. et al. (1982): Am. J. Epidemiol. 116 (1): 76-85.
7. Heger M. et al. (1986): Verh.d. Dt. Ges.f. Arbeitsmed., 26. Jahrestagung (Genter-Verlag, Stuttgart) in press.
8. Myslak Z. et al. (1986): Verh.d. Dt. Ges.f. Arbeitsmed., 26. Jahrestagung (Genter-Verlag, Stuttgart) in press.
9. Quillardet P. et al. (1982): Proc. Natl. Acad.Sci. 79:5971-5975.
10. Quillardet P. et al. (1985): Mutat. Res. 147: 65-78.

57

Quantitative studies on cytotoxicity and neoplastic transformation of BALB/3T3 cells by trace metals

F. Bertolero*, E. Sabbioni, J. Edel, R. Pietra – Radiochemistry Div., CEC, J.R.C.–ISPRA (Va) and *Istituto di Ricerche Farmacologiche M. Negri, Milano, Italy

INTRODUCTION

The assessment of risk factors to predict possible health effects induced by chemicals requires accurate experimental studies on the toxicological dose-effect relationship in target systems.
Our work continues to be addressed to the development of optimal culture conditions for transformation of mouse epidermal keratinocytes (MK cells) in vitro /1/ which at the time are not as well defined as they are for other systems of mouse embryo fibroblast lines.
The BALB/3T3 system was reported to be susceptible to transformation by a range of different organic chemical carcinogens /2, 3/, but data on metal induced transformation were lacking. We have therefore conducted some studies with the BALB/3T3 Cl A31-1-1 cell line in order to gain an appreciation of the potential of this system to design studies on the effects induced by different metal salts in vitro.

EXPERIMENTAL PROCEDURES

The BALB/3T3 Cl A31-1-1 cell line was kindly provided by U. Saffiotti (LEP, NCI, Frederick, MD, USA). Cells were grown in minimal Eagle's medium (MEM) supplemented with 10% fetal bovine serum (FBS). Batches of media and sera were prescreened. Standard procedures for cytotoxicity and neoplastic transformation assays were adopted from earlier studies /3/. Chromium uptake was investigated using 51-Cr-labelled Na_2CrO_4 (Amersham, England) of high specific activity applying radiochemical and nuclear techniques developed at the JRC-Ispra /4/. Cr^{+6} was estimated in the medium by solvent extraction with liquid anion exchanger according to Minoia et al. /5/.

RESULTS AND DISCUSSION

Dose-response studies of cytotoxicity induced by trace metals.

An example of the relative effectiveness of different trace metals expressed as micromolar concentrations needed to induce a given level of cytotoxicity under the same test conditions is reported in Figure 1.

Quantitative studies revealed a broad sensitivity range of this cell line to the induction of dose-dependent cytotoxicity by the metals tested.

CELLULAR UPTAKE AND METABOLISM OF CHROMATE

The cytotoxicity of CrO_4^{2-} in BALB/3T3, as reported in Figure 1 showed an estimated LD_{50} of $3x10^{-6}$ M Cr. This dose was selected to investigate the chemical stability in the medium and the cellular uptake using 51-Cr-labelled chromate.

After 1 h of incubation of BALB/3T3 cells with chromate approximately 70% of the radiolabelled chromium present in the medium was extractable as anionic Cr^{+6}.

The amount of extractable hexavelent chromium decreased with time so that at 72 h it reached 10% of the initial dose (Tab. 1). When chromate was incubated for 72 h in a cell-free system, no change occurred in the chemical form during the incubation period (data not shown), suggesting an active role of the BALB/3T3 cells in the chemical transformation process of chromate.

Chromate uptake and subcellular partition were studied in BALB/3T3 cells using radiotracers. The cellular incorporaton of 51-Cr expressed as percent of the dose/10^6 cells reached a maximum at 24 h and decreased thereafter (Tab. 2). The relative subcellular partition of absorbed chromium showed a major retention of 51-Cr in the total pellet as compared to the amount present in the soluble cellular compartment.

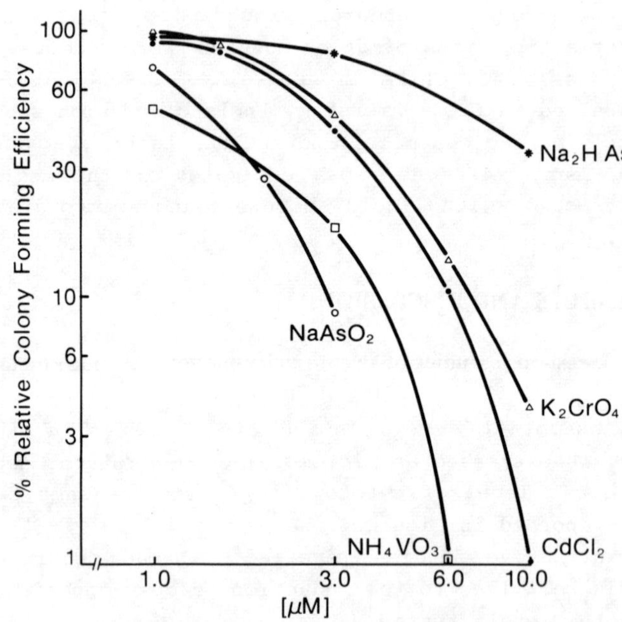

Fig. 1. – BALB/3T3 cells were seeded (2×10^2/60 mm dish) in MEM + 10% FBS, 24 h after seeding, medium was removed and the listed concentrations of metal salts were added with fresh medium for 72 h of incubation after which experimental media were replaced with regular medium. Cultures were fixed and stained on day 8; colonies/dish were scored.

Table 1: Hexavalent chromium extractable from the incubation medium.

Incubation time [h]	^{51}Cr extracted/dish [% of the dose][a]
1	68.4 ± 6.1
24	54.8 ± 1.8
48	24.4 ± 4.1
72	10.5 ± 0.4

BALB/3T3 cells were incubated with 51-Cr-labelled sodium chromate of known specific activity. At the listed timepoints medium was collected and the radioactivity was determined by gammacounting. Hexavalent chromium was estimated after solvent extraction with liquid anion exchanger.

a Mean ± SD, n =6.

The use of radiotracers with high specific activity permitted an accurate quantitative evaluation of the dose incorporated in the cells and it will be valuable for future investigations on dose-effect relationships at the cellular and subcellular level.

Table 2: Cellular uptake and subcellular partition of ^{51}Cr in BALB/3T3 cells.

	^{51}Cr incorporated [% of the dose/10^6 cells]		
Incubation time [h]	Total Homogenate	Pellet	Cytosol
1	0.23 ± 0.003[a]	0.08 (34.2)[b]	0.15 (65.8)[b]
5	1.70 ± 0.34	0.58 (34.4)	1.12 (65.6)
24	2.84 ± 0.32	1.19 (48.1)	1.29 (51.9)
48	2.05 ± 0.31	0.77 (59.8)	0.52 (40.2)
72	1.80 ± 0.79	1.50 (83.5)	0.30 (16.5)

24 h after seeding, BALB/3T3 cells (4×10^6 cells/100 mm dish) were exposed to 3×10^{-6} M 51-Cr-labelled CrO_4^{2-} for up to 72 h. At the listed timepoints medium was removed, cells were washed and collected for radioactivity counting and homogenization. The homogenate was centrifuged at 105000xgx90' and ^{51}Cr was determined in pellet and soluble fractions.

a Mean ± SD n = 4

b Data in parenthesis are the relative percentual distribution of the radioactivity recovered in the total homogenate.

Concurrent cytotoxicity and morphological transformation

BALB/3T3 cells were exposed to three different trace metals according to the standard assay procedures /3/. As^{3+}, Cd^{2+} and Cr^{6+}, three metals of major environmental concern, were found positive in the transformation assay (Tab. 3) showing a clear dose-related response. The relative effectiveness measured as the molar concentration required to induce a given level of biological response was similar for Cd^{2+} and Cr^{6+} and it was higher as compared to As^{3+}.

CONCLUSIONS

In recent years the use of isolated cells of animal and human origin was recognized as a valid approach for the quantification of the effects caused by xenobiotics.
The BALB/3T3 system is currently applied at the JRC-Ispra in

Table 3: Metal salts induced cytotoxicity and transformation in BALB/3T3 C1 A31–1–1 cell line (8° – 10° p).

Exposure [M]	Colony forming efficiency [%]	No.of type III foci/no.of dishes	Type III foci positive dishes/no. of dishes	Transformation frequency$\times 10^{-4}$ cells
$NaAsO_2$ 2×10^{-6}	64 ± 3	0/18	0/18	0.0
$NaAsO_2$ 3×10^{-6}	57 ± 4	2/18	2/18	0.2
$NaAsO_2$ 6×10^{-6}	27 ± 2	3/18	3/18	0.6
$CdCl_2$ 1.5×10^{-6}	84 ± 7	7/24	6/24	0.3
$CdCl_2$ 3.0×10^{-6}	48 ± 3	15/22	9/22	1.5
$CdCl_2$ 6.0×10^{-6}	11 ± 2	24/24	16/24	9.2
K_2CrO_4 1.5×10^{-6}	83 ± 5	3/18	3/18	0.2
K_2CrO_4 3.0×10^{-6}	63 ± 7	33/23	17/23	2.2
K_2CrO_4 6.0×10^{-6}	14 ± 2	49/21	19/21	16.0
Bidistilled H_2O 0.1%	100 ± 6	0/21	0/21	0.0

BALB/3T3 cells were seeded (2×10^{2}/60 mm dish for cytotoxicity or 1×10^{4}/60 mm dish for transformation) and treated 24 h later at the listed concentrations of AS^{3+} or Cd^{2+} or Cr^{6+} for 72. Dishes were fixed and stained on day 8 for cytoyoxicity and after 4 wks for transformation.

the frame of the Environmental Protection Porgram for research actions on Trace Metal Pollution. The research is carried out combining nuclear and radiochemical techniques and cell biology methodology.

The present work shows that the cellular system studied provides a valuable model for investigating the correlations between the cellular uptake, the metabolism, the cytotoxicity and the transformation induced by trace metals in vitro.

REFERENCES

/1/ Bertolero F., Kaighn M.E., Gonda M.A. and Saffiotti U.: Mouse epidermal keratinocytes: Clonal proliferation and response to homones and growth factors in serum-free medium. Exp. Cell Res. 155: 64-80, 1984.

/2/ Saffiotti U., Bignami M., Bertolero F., Cortesi E., Ficorella C. and Kaighn M.E.: Studies on chemically induced neoplastic transformation and mutation in the BALB/3T3 C1 A31-1-1 cell line in relation to the quantitative evaluation of carcinogens. Toxicol. Pathol. 12: 383-390, 1984.

/3/ Cortesi E., Saffiotti U., Donovan P.J., Rice J.M. and Kakunaga T.: Dose-response studies on neoplastic transformation of BALB/3T3 clone A31-1-1 cells by aflatoxin B_1, benzidine, benzo (a) pyrene, 3-methylcholanthrene and N-methyl-N -nitro-N-nitrosoguanidine. Teratogenesis Carcinog. Mutagen. 3: 101-110, 1983.

/4/ Sabbioni E., Radiotracer and nuclear techniques: application in metallobiochemical investigations of environmental levels of trace elements Trace Elements in New Zealand, Proceedings of the International Workshop on Trace Elements in New Zealand, Dunedin, New Zealand (1981), J.V Dunckley ed., University of Otago, Dunedin, New Zealand, pp. 322-351., 1981.

/5/ Minoia C., Mazzucotelli A., Cavalleri A. and Minganti V.: Electrothermal Atomization Atomic-absorption Spectrophotometric Determination of Chromium (VI) in Urine by Solvent Extraction Separation with Liquid Anion Exchangers. Analyst. 108: 481-484, 1983.

Session V:

Biological indices for assessment of human genotoxicity induced by exogenous chemicals

D. Henschler – Institute of Toxicology, University of Würzburg, F.R.G.

Monitoring of genotoxic effects differs from non-genotoxic events in two respects: (a) the effects are in principle irreversible, (b) a long latency period elapses between (start of) exposure and the final manifestation: malignant tumors or inheritable disease through mutations in germ cells. The sequence of events between exposure and final effect, together with possibilities of monitoring at the different stages of this cascade, are outlined in Figure 1. Because the quantitative interrelationships between the last two steps are unknown, none of the presently available methods allow for quantitative risk estimation; they only demonstrate and quantify exposure. The possibility of determining effective doses at certain targets within an individual is a decisive step away from conventional monitoring methods, which quantify chemicals or their metabolites in biological materials. Provided a quantitative relationship between the change at the target and tumor formation exists, such methods supply suitable indicators of higher risk individuals or groups.

Protein binding (Table 1), mostly measured with hemoglobin, represents the first target. A variety of nucleophilic sites in hemoglobin may form adducts with electrophilic ultimate carcinogens. A number

Dose ⟶ Effect(s)	Methods of monitoring
External dose ↓	Ambient air monitoring Skin exposure
Internal dose ↓	Agent(s) or metabolites in blood, tissue, exhal. air, urine, feces Mutagenicity in urine
Biol. effective dose ↓	Protein binding DNA (RNA) binding DNA excision products
Early biol. effects ↓	Chromosomal aberrations SCE Micronuclei UDS
Final effects ↓	Cancer Mutations in germ cells

Fig. 1. – Sequence of events from exposure to effects, and levels of monitoring.

of methods to determine these adducts, mostly after derivatisation, have been described but not dealt with in detail in this conference. Reference substances have to be prepared for comparison, but recently modifications have been proposed which allow for the determination and identification of "unknowns" at terminal valine amino groups (Törnquist and Ehrenberg). The fact that hemoglobin has a half-life of approximately one hundred days provides the opportunity of determining the accumulated adducts over this period of time. The stability of the adducts has to be predetermined. Major advantages of protein binding are: easy availability of the target, linear dose-response-relationships, proportionality of protein binding to RNA-binding in other organs over a wide dose-range. Shortcomings become apparent with methylating and ethylating agents, due to high and widely varying background levels. Monitoring with this method has been practised to date with ethylene oxide, propylene

Table 1: Methods and criteria of monitoring exposure to genotoxic chemicals (mo=month, y=year, d=day, hr=hour; ly=lympocytes).

Methods	Material	Persistence	Sensitivity	Specificity	Confounding
Protein-binding	blood	mo			
chemical			+++	++	(+)
immunologic			++++	++	(+)
DNA-binding	ly.	hr			
chemical	tissue		+	++	(+)
immunologic		y	+++	++	(+)
^{32}P-postlabelling			++++	-	?
DNA-excis.prod.	urine	hr			
chemical			++	++	(+)
immunologic			+++	++	(+)
Cytogenetics	ly.				
chromos.aberr.		y	+	-	++
SCE		d-mo	++	-	++
micronuclei		y	+	-	++
Mutagenicity in urine		hr d	+++	-	+++

oxide, vinyl chloride, ethene and propene and some aromatic amines.

DNA-adduct determination may be performed in lymphocytes and other available cells. Chemical determination lacks sensitivity; sensitive immunological methods sometimes show cross reactivity within related groups of chemicals, such as PAHs. ^{32}P-postlabelling techniques are the most sensitive methods described so far; recent developments have produced methods with a minimum detection limit of 1 adduct in 10^{10}. The method lacks specificity but may be used to monitor populations for exposure to "unknowns". Little is known about the persistence of DNA-adducts in lymphocytes and other relevant tissues except for some model animal experiments but extensive biomonitoring of human populations with exposure to known genotoxic agents may soon fill this gap.

Determination of DNA-excision products in the urine constitutes an elegant method of determining the repair efficiency of the organism. This method is now widely used in large scale monitoring of populations exposed to aflatoxins; unfortunately, it does not produce quantitative data on the DNA-damage and repair in the target organ(s) since the origin of the excision products cannot be determined.

Cytogenetic studies were introduced some twenty years ago and have been most frequently employed up till now. Some 120 human studies have been published, a few with results providing dose response relationships such as in cigarette smokers. The persistence of the lesions is shorter with sister chromatid exchanges (days to months) than with chromosomal aberrations (CA) and micronucleus (MN) formation (years). There are many confounding factors such as smoking, infectious diseases and age. All three endpoints lack any compound specificity. The interrelationships between CA, SCE and MN are little understood. The sensitivity is rather low, as compared to the other methods.

Mutagenicity studies with human urine are highly sensitive but completely non-specific. Many confounding factors have been identified (drug intake, physical stress, smoking, natural diet constituents etc.). Despite of all these shortcomings which render the results rather difficult to evaluate, the method may be used with advantage to uncover unexpected exposures. If strong mutagenic activity is found, analytical treatment of the urine may reveal the chemical identity of the mutagenic principles, and thus lead to the detection of exposure to "unknowns" with their final identification.

The question whether and to what purpose these monitoring methods should be used has been discussed controversially in this meeting and elsewhere. There is general agreement that all these methods need to be better validated before they are used on a large scale. Interlaboratory calibration should be organized on an international level, and more information on control data and their significance must be collected. As to the present state of knowledge, they can only quantify internal exposure but not necessarily quantify cancer risk. But that is what is urgently needed in occupational toxicology since the presently practised methods of testing for carcinogenicity and classifying substances as potentially bearing cancer risk for humans reveal an ever-increasing number of proven carcinogens which will forseeably exceed the proven non-carcinogens - a situation which discredits today's strategies of protection and prevention. As long as the quantitative relationship between early biochemical and biological lesions (DNA-binding, chromosomal damage) and final cancer risk is unknown or at least very uncertain, these methods can only be used as quantitative tools together with prospective epidemiology, in that populations at exposure are monitored and the data obtained are correlated with cancer incidence. For this application, high specificity and high sensitivity are prerequisites for any correlation of data since mixed exposures are the rule in occupational reality. Under these auspices, the non-specific methods, such as cytogenetic changes in lymphocytes, ^{32}P-postlabelling, and mutagenicity in urine are of little or no value, except when they are combined with other more specific methods. Therefore, one may predict that, for prospective epidemiology monitoring methods for protein binding and DNA binding will be given preference.

The meeting also discussed briefly and with a controversial outcome ethical problems connected with biomonitoring of exposure to carcinogens. It has been postulated that since the problem of how to handle the data obtained is unresolved further studies should not be initiated before a final and internationally accepted solution has been achieved. On the other hand, this restrictive policy will not hinder scientific progress in this area. There will always be scientists who will perform the studies they want to do in their own scientific interest. It has been recommended that this type of biomonitoring should be encouraged, rather than restricted, with two provisions: complete transparency in the planning, conduct and evaluation of the study, and information of people being tested about the benefit of these studies for the prevention of damage to health.

Part VI

Biochemical indices of nervous tissue toxicity and exposure to neurotoxicants

Coordination:
W.N. Aldridge (Carshalton, UK)
F. Clementi (Milano, Italy)

Chemical and biochemical indices of changes in the nervous system

W.N. Aldridge – Toxicology Unit, Medical Research Council Laboratories, Carshalton, Surrey, SM5 4EF, United Kingdom

ABSTRACT

The principles behind the biomonitoring methods used to indicate neurotoxic potential of exposure to chemicals are described. For exposure to anti-cholinesterases and organophosphorus compounds causing delayed neuropathy rational biomonitoring methods are available. At the moment other less direct methods have to be used and these require prediction from studies on experimental animals.

INTRODUCTION

The nervous system is only 2-3% of the body weight. Of course there are other organs of similar or lesser size but the nervous system is unique in the complexity of its specialised structures and connections to other parts of the body. Thus functional or morphological changes in a few cells in localised areas of the central nervous system can produce profound and/or localised signs and symptoms. I shall consider only those exposures which are toxic by specific effects in the nervous system and exclude tumours. Substances which produce symptoms arising from the nervous system, e.g. the mitochondrial poison cyanide or the haemoglobin combining carbon monoxide, are excluded. Practical toxicology has changed much over the last fifty or so years. On the industrial scene workpeople were exposed and when undesirable signs of poisoning were seen, methods for prevention were instigated. Thus the primary experiment was often done in man. This is no longer acceptable and now society demands that evidence (from experimental studies using animals or *in vitro* systems) is produced that under proposed conditions of use, the risk is acceptably small.

The main aim of biochemical or chemical monitoring as an index of tissue toxicity is not only to identify the cause of illness in exposed people but more importantly to identify in the workplace those potentially at risk - i.e. to measure exposure or biochemical effects due to exposure below the threshold of toxicity (Aldridge, 1986).

The remarkable specificity of many toxicants to cells in selected areas of the nervous system is the reason why it is unlikely that circulating enzymes may be used to detect cellular damage. For example lactic dehydrogenase or the transaminases which are used to detect changes in the liver or cardiac infarcts cannot be used to identify damage in the nervous system. As the macromolecular structure of the differentiated cells of specialised areas of the nervous system are established using the methods of molecular biology, it may become possible to monitor in blood release of macromolecules specific for certain cells (O'Callaghan and Miller, 1983).

At present discriminating histopathology is the only method we have, even in animals, to identify cell specific damage in the nervous system. For example in the rat trimethyltin compounds produce cell necrosis only in the hippocampus, amygdaloid nucleus and pyriform cortex (Brown *et al*, 1979, 1984). The type of response is not species specific for similar pathology is seen in other animals. Symptoms are seen in rats as cell necrosis is developing (particularly aggression) and in poisoning incidents in man the same clinical course is seen (Ross *et al*, 1981; Fortemps *et al*, 1979; Ray *et al*, 1984). There is no method at the present of biochemical or clinical monitoring to indicate whether the exposure has led to depletion of neurones from specific areas of the nervous system either in animals or man. Even with severe neuronal loss, sophisticated tests are required to detect behavioural changes.

Whatever the mode of exposure to chemicals (inhalation, oral, skin) the course of developing toxicity may be described as the four stages in Figure 1, going from left to right. These four stages are (1) all those processes which influence the delivery to the toxic chemical to its site of action, (2) the reaction of the toxic chemical with targets (usually macromolecules), one or more of which lead to stage (3) the biochemical, physiological and morphological consequences of reaction(s) in (2) and (4) the clinical consequences as shown by the development of signs, symptoms or syndrome of the disease state.

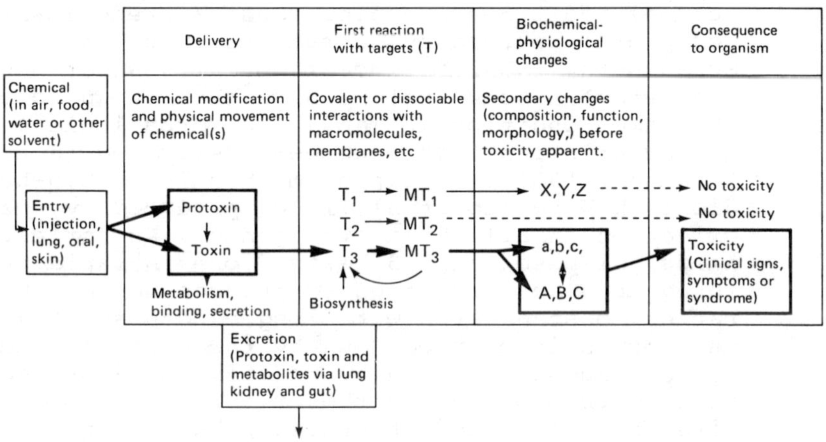

Fig. 1 – Scheme for developing toxicity due to chemicals. Heavy lines or letters = toxicity, light lines = less or no toxicity. T_1, T_2, T_3 = targets; MT_1, MT_2, MT_3 = modified targets; A,B,C,X,Y,Z,a,b,c = independent or linked changes. Acute or chronic toxicity differ by the time in one or more phases (Aldridge, 1986)

RECEIVED DOSE OF TOXIC CHEMICAL (Figure 1, stage 3)

Acrylamide causes in animals and man dying back lesions in long nerves particularly, though not exclusively, in the peripheral nerves (LeQuesne, 1980). The morphological characteristics have been well described (Cavanagh and Gysbers, 1983). Although there have been many cases of acrylamide neuropathy in man the dose-response relationship is not known (WHO, 1985). A measure of absorbed dose is essential since acrylamide is a powder and presents many problems of determination in the external environment. Acrylamide is also a neutral and water soluble compound which is readily excreted; assessment of the extent of chronic exposure would require frequent measurements of the parent compound in body fluids. However, acrylamide reacts with haemoglobin (Hashimoto and Aldridge, 1970) and these adducts can now be determined (Bailey et al, 1986). Thus, because of the long life of haemoglobin in erythrocytes in man (120 days), chronic dosimetry is possible yielding an integrated dose over a rather long period.

Hexane, 2-hexanol, 2-hexanone (methyl n-butyl ketone) and 2,5-hexanediol cause a predominantly peripheral neuropathy in rats (Spencer and Schaumberg, 1975; DiVincenzo et al, 1980). Hexane and methyl n-butyl ketone have been shown to produce the same condition in man. The morphological characteristics have been well described (Jones and Cavanagh, 1983). The proximal toxin is 2,5-hexanedione (DiVincenzo et al, 1980) produced by enzymic oxidation of all the above agents. As for acrylamide, disposal of 2,5-hexanedione is rapid but the integrated dose required to produce neuropathy is rather constant even with long exposures. Assessment of an integrated dose by the determination of pyrrole adducts formed by reaction of 2,5-hexanedione with proteins in the erythrocyte - presumably with the most abundant protein, haemoglobin - may be possible (Anthony et al, 1983; Mattocks and White, 1970). Much development work will be required.

Neuropathies produced by acrylamide and 2,5-hexanedione have been used as examples of dose measurement involving stage 1 in Figure 1, i.e. delivery of the proximal toxin. The method of measurement of dose utilises a reaction with a target (haemoglobin) which leads to no toxicity (e.g. T_2 and MT_2 in Figure 1).

REACTION WITH PRIMARY TARGETS (Figure 1, stage 2)

Other methods of biomonitoring rely on knowledge of mechanisms of toxicity and will be the most reliable. Two examples are poisoning by organophosphorus compounds and carbamates (Vandekar, 1980) and the neuropathy caused by some organophosphorus compounds (Johnson, 1982). For the former initiation of toxicity is brought about by inhibition of acetylcholinesterase and for the latter by reaction with neuropathy target esterase (NTE). For biomonitoring purposes these enzymes in the nervous system are inaccessible but they are also present in blood - acetylcholinesterase in erythrocytes and NTE in leucocytes (Lotti and Johnson, 1978; Lotti, 1983; cf. Lotti, M., in this volume).

Much work has been done to validate the biomonitoring method using acetylcholinesterase in erythrocytes and the principles are well understood. For chronic exposure new issues arise. For example the dimethylphosphorylated acetylcholinesterase arising from exposure to many pesticides spontaneously reactivates (half-life 0.85 hr) and ages (loses a methyl group to become monomethylphosphorylated enzyme: half-life 3.84 hr). The aged enzyme is irreversibly inhibited and requires resynthesis of new enzyme for a return of enzyme activity. If we may judge from the results of experiments in animals the turnover of acetylcholinesterase in brain will be faster than that of erythrocytes. After chronic exposure aged enzyme will accumulate faster in erythrocytes than in brain and measurement of the activity of erythrocyte cholinesterase will overestimate risk. The validation of the method using NTE in lymphocytes will require attention to similar aspects. These two examples emphasize the possibilities for biomonitoring when the primary target (e.g. T_3 and MT_3; stage 2) is known.

BIOCHEMICAL CONSEQUENCES (Figure 1, stage 1)

Quantitative assessment of the degree of damage in the peripheral nervous system is difficult using morphological and histological methods. Damage to nerves is followed by increases in β-glucuronidase and β-galactosidase (Dewar and Moffett, 1979). Since these enzymes are associated with repair processes, in experimental animals the degree of previous damage can be inferred. The method can be used for both peripheral and central regions of the nervous system (Rose et al, 1980, 1983), to distinguish distal, proximal and evenly distributed damage in the sciatic/posterial tibial nerve (Rose and Dewar, 1983) and has been tested on neuropathies produced by acrylamide, methyl mercury and misomidazole and the minimal changes produced by pyrethroids (Rose and Dewar, 1983). Increases in β-glucuronidase and β-galactosidase are consequences of reaction with the primary target (Figure 1, stage 3). When in experimental animals the relationship is known between delivered dose and β-glucuronidase and β-galactosidase a firmer base for prediction to man is available.

CONCLUSIONS

Few opportunities are available for biochemical monitoring of humans for neurotoxicity. Monitoring exposure to anticholinesterases and the potential method using neuropathy target esterase in leucocytes for delayed neuropathy caused by organophosphorus compounds are based on knowledge of primary targets (Figure 1, stage 2).

For sometime to come monitoring of man will rely on measurement in body fluids of received dose of chemical or bioactivated metabolite (Figure 1, stage 1). Such measurements may be correlated in experimental animals with signs and symptoms of poisoning or, better, to other biochemical, physiological or morphological change (Figure 1, stages 3 or 4). Such relationships provide a base for prediction of risk of exposure to man.

REFERENCES

Aldridge WN (1986). The Biological Basis and Measurement of Thresholds. Ann. Rev. Pharmacol. Toxicol., 26:39-58.

Anthony DC, Boekelheide K, Anderson CW & Graham DG (1983). The effect of 3,4-dimethyl substitution on the neurotoxicity of 2,5-hexanedione. 2. Dimethyl substitution accelerates pyrrole formation and protein cross linking. Toxicol. Appl. Pharmacol., 71:371-382.

Bailey E, Farmer PB, Bird I, Lamb JH & Peal JA (1986). Monitoring exposure to acrylamide by the determination of S-(2-carboxyethyl)cysteine in hydrolysed haemoglobin by gas chromatography-mass spectrometry. Anal. Biochem. In the press.

Brown AW, Aldridge WN, Street BW & Verschoyle RD (1979). The behavioural and neuropathologic sequelae of intoxication by trimethyltin compounds in the rat. Amer. J. Pathol., 97: 59-82.

Brown AW, Verschoyle RD, Street BW, Aldridge WN & Grindley H (1984). The neurotoxicity of trimethyltin chloride in hamsters, gerbils and marmosets. J. Appl. Toxicol., 4:12-21.

Cavanagh JB & Gysbers MF (1983). Ultrastructural features of the Purkinje cell damage caused by acrylamide in the rat: a new phenomenon in cellular neuropathology. J. Neurocytol., 12:413-437.

Dewar AJ & Moffatt BJ (1979). A biochemical approach to neurotoxicity - short review. Pharmacol. Ther., 5:545-562.

DiVincenzo GD, Krasavage WJ, O'Donoghue JL (1980). Role of metabolism in hexacarbon neuropathy. In: The scientific basis of toxicity assessment. Ed. H. Witschi, pp.183-200, Amsterdam: Elsevier/North Holland Biomedical, 329pp.

Fortemps E, Amand G, Bourbon A, Lauwerys R & Laterre EC (1979). Trimethyltin poisoning: report of two cases. Int. Arch. Occup. Environ. Health., 41:1-6.

Hashimoto K & Aldridge WN (1970). Biochemical studies on acrylamide, a neutoxic agent. Biochem. Pharmacol., 19:2591-2604.

Johnson MK (1982). The target for initiation of delayed neurotoxicity by organophosphorus esters: biochemical studies and toxicological applications. Rev. Biochem. Toxciol., 4:141-212.

Jones HB & Cavanagh JB (1983). Distortion of the nodes of Ranvier from axonal distension by filamentous masses in hexacarbon intoxication. J. Neurocytol. 12:439-458.

LeQuesne PM (1980). Acrylamide. In: Experimental and Clinical Neurotoxicology. Ed. P.S. Spencer & H.H. Schaumberg, pp.309-325, Baltimore/London, Williams & Wilkins, 929pp.

Lotti M (1983). Lymphocyte neurotoxic esterase. A biochemical monitor of organophosphorus induced delayed neurotoxicity in man. Adv. Biosci., 46:101-108.

Lotti M & Johnson MK (1978). Neurotoxicity of organophosphorus pesticides: predictions can be based on in vitro studies with hen and human tissues. Arch. Toxicol. 41:215-221.

Mattocks AR & White INH (1970). Estimation of metabolites of pyrrolizidine alkaloids in animal tissues. Anal. Biochem., 38:529-535.

O'Callaghan JP & Miller DB (1983). Nervous-system specific proteins as biochemical indicators of neurotoxicity. Trends Pharmacol. Sci., 4:388-390.

Ray C, Reinecke HJ & Besser R (1984). Methyltin intoxication in six men: toxicologic and clinical aspects. Vet. Hum. Toxicol., 26:121-122.

Rose GP & Dewar AJ (1983). Intoxication with four synthetic pyrethroids fails to show any correlation between neuromuscular disfunction and neurobiochemical abnormalities in the rat. Arch. Toxciol., 53:297-316.

Rose GP, Dewar AJ & Stratford IJ (1980). A biochemical method for assessing the neurotoxic effects of misomidazole in the rat. Brit. J. Cancer, 42:890-899.

Rose GP, Dewar AJ & Stratford IJ (1983). Protection against misomidazole-induced neuropathy in rats: a biochemical assessment. Toxicol. Lett., 17:181-185.

Ross WD, Emmett EA, Steiner J & Tureen R (1981). Neurotoxic effects of occupational exposure to organotins. Am. J. Psychiatry, 138:1092-1095.

Spencer PS & Schaumberg HH (1975). Experimental neuropathy produced by 2,5-hexanedione - a major metabolite of the neurotoxic industrial solvent methyl n-butyl ketone. J. Neurol. Neurosurg Psychiatry, 38:771-775.

Vandekar M (1980). Minimising occupational exposure to pesticides: cholinesterase determination and organophosphorus poisoning. Residue Rev., 75:67-79.

WHO (1985). Acrylamide. In: Environmental Health Criteria. WHO, Geneva, 49:1-121.

Biochemical and clinical toxicology of organophosphates in the development of new biomonitoring tests

Marcello Lotti – Istituto di Medicina del Lavoro dell'Università di Padova, Via Facciolati 71, 35127 Padova, Italy

The aim of this Symposium, according to the title, is to bring biochemists and physicians together in order to develop new biochemical tests to evaluate human exposures to chemicals. In this paper, the relationship between experimental and clinical toxicology and how these two aspects of toxicology might influence each other, will be discussed; the toxicology to the nervous system of organophosphorus (OP) esters will be taken as an example.

OP esters were developed at the end of the second world war as chemical warfare agents and now are widely used as pesticides, plasticisers and hydraulic fluids. The biochemistry and the toxicology of OP esters on the nervous system have been extensively studied (Holmstedt 1959, O'Brien 1960, Koelle 1963, Aldridge & Reiner 1972, Taylor 1985) and several cases of poisoning in man have been reported (Hayes 1982).

A number of analytical procedures have been developed to measure the "internal" dose of OPs; a well known test to monitor the acute effects of OPs is the measurement of red blood cell acetylcholinesterase activity (RBC AChE).

I will discuss in this paper the development of tests to investigate OP detoxification, and the validation of a test to assess another toxicity due to some OPs, the organophosphate-induced delayed polyneuropathy (OPIDP).

Furthermore the paper is divided into two parts to exemplify how the understanding of the mechanism of action might lead to new biomonitoring tests, and how the use of biomonitoring tests might offer clues to mechanistic studies.

UNDERSTANDING THE MECHANISM OF ACTION OF TOXIC CHEMICALS AND THE DEVELOPMENT OF BIOMONITORING TESTS

Biochemistry and pharmacology of esterases

The biochemistry of OPs and esterases is summarized in FIGURE 1.

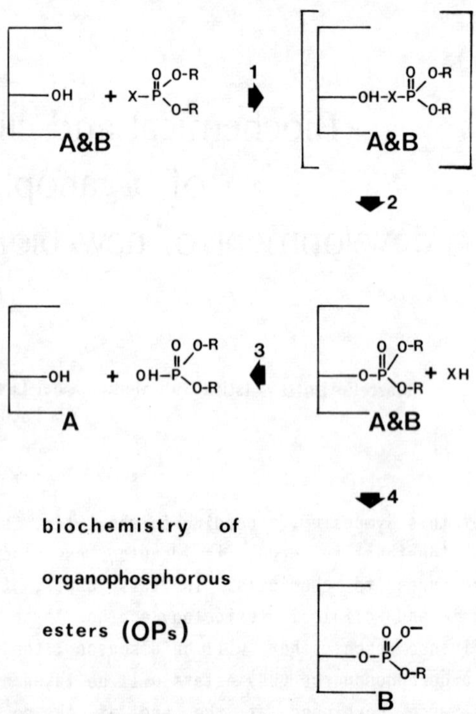

Fig. 1. – Interactions of A and B esterases with esters of organophosphorus acids. –OH is the serine residue at the catalytic site of the enzymes. –R are alkyl groups of various types and –X is the leaving group of the organophosphate.

The interaction of OPs with esterases is analogous to part I of the reaction of substrate hydrolysis. Reactions (1) & (2) lead to the phosphorylated esterase through the formation of the Michaelis complex. The speed of reaction (3) differentiates OPs which are substrates of esterases from those which are inhibitors of esterases; high for the former and slow for the latter. A further reaction (4) called "aging" occurs when certain inhibitors phosphorylate esterases; it is a non-enzymatic, time-dependent de-alkylation of the phosphoryl residue attached to the active centre of the enzyme and is strictly dependent on the chemical nature of the phosphorylating agent.
On the basis of their interaction with OPs, esterases can therefore be

subdivided in two main groups with different pharmacological significance: A-esterases are those enzymes which hydrolyze OPs and are therefore involved in the detoxification mechanisms. B-Esterases are the targets of OPs toxicities (Aldridge & Reiner 1972).

Several A and B- esterases are present in human blood. Will the measurement of A-Estersaes activities in plasma reflect the hydrolyzing capability of an individual for a given OP? Will the measurement of B-Esterases activities in blood mirror the activities of these enzymes at the target organ?

A–ESTERASES IN PLASMA

Several OPs are hydrolyzed by human plasma including paraoxon (Aldridge 1953), DFP (Augustinsson & Casida 1959), tabun (Augustinsson & Heimbuerger 1954), ethyl 4-nitrophenyl alkylphosphonates and ethyl 4 nitrophenyl phenylalkylphosphonates (Becker and Barbaro 1964), soman (de Bisschop et al 1985), dichlorvos (Reiner et al 1980, Pierini et al 1985). Carbamates and carboxylic acid esters are also hydrolyzed by human serum (Augustinsson & Casida 1959, Reiner & Skrinjaric-Spoljar 1968, Erdos et al. 1960). The hydrolysis is due to a group of enzymes called A-esterases (arylesterases, EC 3.1.1.2) which can be differentiated from B-esterases not only quantitatively but also because of certain biochemical differences. This group of enzymes is widely distributed in many animal tissues (Aldridge 1953) and the activities vary with age (Augustinsson & Barr 1963). A-esterases have no pH optimum, and some are inhibited by EDTA; for certain enzymes the hydrolyzing activity is restored by calcium, suggesting that calcium is a cofactor. Other ions might be activators or inhibitors of A-esterases (Aldridge & Reiner 1972). Compounds which react with SH groups inhibit A-esterases, suggesting that they have an -SH group in the active site (Aldridge 1953). The substrate specificity of A-esterases is very complex and not well understood.

The biochemical characteristics of A-esterases in human plasma have to be studied in detail before the use in human biomonitoring. Several criteria for characterization have been suggested (Aldridge & Reiner 1972) and an example is given in FIGURE 2 where the pH and temperature sensitivities of two A-estersaes in human plasma are compared (Moretto, Traverso and Lotti unpublished results).

A-esterases vary with age and might be influenced by environmental factors. They show a remarkable polymorphism in human populations (Playfer et al. 1976). The question arises whether subjects with low plasma A-esterase activity are more susceptible to the toxic effects of OPs (Chemnitius et al 1983). In order to answer this question however, certain basic information is still lacking.

The organ distribution of paraoxonase is different among species; in man, that distribution is unknown (Aldridge 1953). In particular, the ratio between plasma and other organ A-esterase activities should be known.

The relative importance of A-esterase activities versus other metabolic pathways is also unknown; it will certainly be characteristic for a given OP, but other factors, like the dose, might be critical (Plapp & Casida 1958).

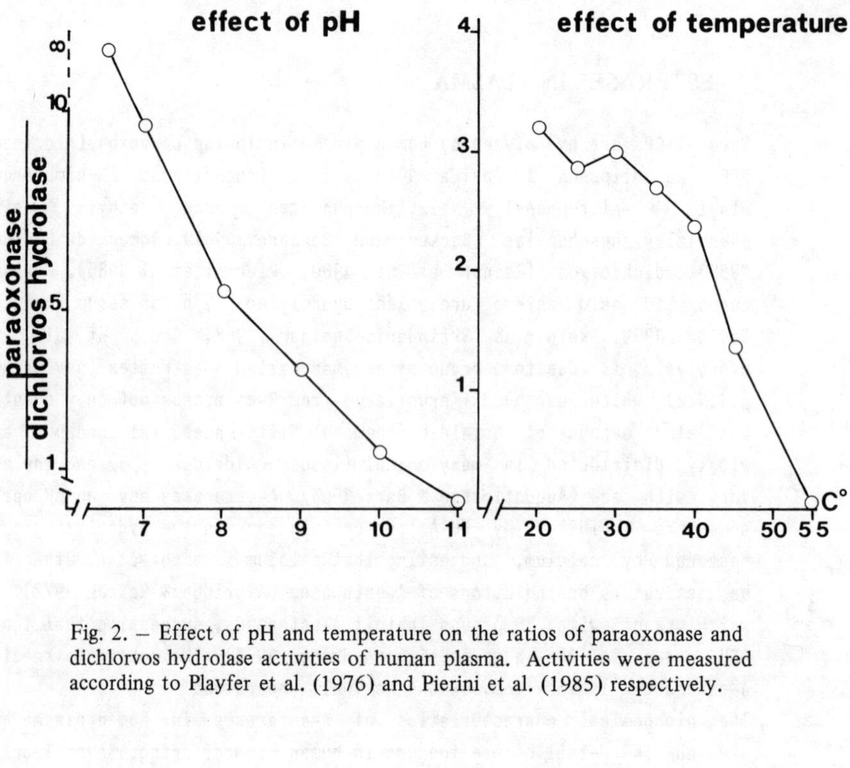

Fig. 2. – Effect of pH and temperature on the ratios of paraoxonase and dichlorvos hydrolase activities of human plasma. Activities were measured according to Playfer et al. (1976) and Pierini et al. (1985) respectively.

B–ESTERASES IN BLOOD

Three B-esterases in human bloood are linked to OP toxicities: acetylcholinesterase, the target for acute toxicity is present in red blood cells and plasma. The measurement of RBC AChE is widely used for monitoring the acute effects of OPs. Plasma butyrrylcholinesterase is another B-esterase but its physiological significance is unknown. Measurement of its inhibition after OP exposure, might indicate the extent but certainly not the effects of the exposure. The third B-esterase is Neuropathy Target Esterase (NTE) in lymphocytes, the enzyme thought to be the target in the nervous system for the initiation of OPIDP (Johnson 1975). The following discussion deals with the validation of this test in clinical practice for the biomonitoring of OPIDP.

The mechanism of initiation of OPIDP has been extensively reviewed (Davis & Richardson 1980, Johnson 1982, Lotti et al. 1984). In brief OPIDP is initiated by the phosphorylation of a protein in the nervous system called Neuropathy Target Esterase (NTE). A second step is then required to produce the toxic effect, the "aging" of the phosphoryl-enzyme complex (see Figure 1). Compounds that age cause OPIDP if the threshold of inhibition of NTE in vivo is reached, whereas compounds that do not age do not cause OPIDP. A high level of inhibition 70-80% of NTE in peripheral nerve of the experimental animal, usually soon after dosing with an OP capable of causing OPIDP, predicts the onset of clinical symptoms approximately 2 weeks later. The most convincing experimental evidence available for NTE as molecular target for OPIDP is represented by protection experiments. When high inhibition of NTE is caused in animals by phosphinates, carbamates or sulphonates, no neuropathy occurs because the NTE-inhibitor complex cannot age. If similarly treated animls are further dosed with high doses of an OP capable to cause OPIDP, neuropathy still fails to develop, because the catalytic site of NTE is already phosphorylated and therefore aging cannot occur.

It is therefore clear that the catalytic activity of NTE is not vital in maintaining the health of the neuron, because it can be blocked by protective agents without overtly harming experimental animals. Furthermore, clinical symptoms develop despite restoration of NTE activity in the nervous system. Aging of sufficiently inhibited NTE is therefore the critical trigger reaction.

This mechanism contrasts with that of acute toxicity; inhibition of acethylcholinesterase by OP leads to accumulation of acetylcholine at nerve endings and thus to cholinergic symptoms.

In a study of the occurrence of NTE in tissues of the adult hen by Dudek and Richardson (1982), NTE had limited distribution among tissues and was present in lymphatic tissue as well as neural tissue. A subsequent analogous study in man showed a wider distribution, and also a measurable activity in lymphatic tissues (Moretto et al. 1983). Dudek and Richardson (1982) suggested that if the inhibition of NTE in blood lymphocytes correlates with that in nervous tissue, it might be possible to monitor OPIDP hazard by measuring NTE activity in blood lymphocytes. In their study, inhibition of NTE in hen lymphocytes correlated with that in the brain 4 hours after a single exposure to certain OPs. However this correlation was not evident at any other time after an acute exposure or during a chronic exposure in an early study (Lotti & Johnson 1980). The lack of correlation in chronic studies might be due to a less accurate method of expressing the activity or to a different dynamics of the circulating lymphocyte pool.

Further basic information is required to validate the use of this test in man (Lotti 1983). Some have been collected so far.

There is indirect evidence that NTE represents the target of OPIDP in man as in the experimental animal. Data have been collected for an OP with negligible neurotoxic potential in both hen and humans, as derived from in vitro experiments (Lotti et al. 1981). In an autopsy sample of human brain after omethoate (phosphorodithioic acid, 0,0-dimethyl S- 2- methilamino -2- oxoethyl ester) poisoning, NTE activity was 100% and AChE was highly inhibited, as expected from data obtained in hens; omethoate preferentially inhibits acetylcholinesterase rather than NTE and does not cause OPIDP.

Inhibition of peripheral blood lymphocyte NTE correlated with that in peripheral nerve in a case of suicidal poisoning with chlorpyrifos (0,0-diethyl- 0- 3,5,6-trichloro - 2- pyridyl phosphorothioate) (Osterloh et al 1983). Furthermore in a case of attempted suicide with the same organophosphate, inhibition of lymphocytic NTE correlated with the development of OPIDP (Lotti et al 1986b).

The threshold of NTE inhibition which is required to initiate OPIDP in man is unknown. This threshold is known to be 70-80% in the hen. It might be that a smaller proportion of NTE inhibition is necessary in man to initiate OPIDP. This possibility is raised from observation of OPIDP in man due to methamidophos (Senanayake & Johnson 1982). In fact, a dose about twice the unprotected median lethal dose causes 50% inhibition of NTE and no neuropathy in hens.

Methods to measure NTE activity in human peripheral lymphocytes (Bertoncin et al. 1985) and also platelets (Maroni & Bleecker 1986) are now available. Studies of non-exposed populations show that there is a rather large interindividual variation for lymphocytic NTE activity, and therefore in occupational biomonitoring studies, baseline pre-exposure values are required to assess the effects.

Biomonitoring studies of occupational exposures by measuring lymphocytic NTE have beeen attempted on two occasions (Lotti et al. 1983, Emmett et al. 1985).

CLINICAL BIOMONITORING AND NEW INSIGHT INTO THE MECHANISMS OF TOXICITY

How might the use of biomonitoring tests generate information useful for mechanistic studies and improve the use of the test itself?

Some unexpected observations were made in a case of attempted suicide with chlorpyrifos, where the patient developed OPIDP (Lotti et al 1986b).

The first was the development of the peripheral neuropathy itself. In fact, chlorpyrifos was reported as not causing delayed polyneuropathy in the hen (Hayes 1982), even though some limited data suggested the contrary (Osterloh et al. 1983, Johnson 1981).

The second was the observation, on day 30, of a similar inhibition of RBC AChE and lymphocytic NTE (about 50-60%). In fact, by comparing the I50 for the two enzymes obtained in vitro, chlorpyrifos is much more potent as an AChE inhibitor rather than as NTE inhibitor; therefore a higher inhibition of AChE was expected. These observations led to the following studies.

The first experiment was aimed to verify whether man was more susceptible to OPIDP than the hen. Chlorpyrifos (150 mg/kg/p.o.) to the hen causes OPIDP, and the reason for differences with previous reports might be due to a much higher dose used in our experiment (Lotti et al. 1986a). This toxic effect did not correlate however with the inhibition of NTE 24 hours after dosing, as it usually occurs. In fact, we observed at that time an inhibition of nervous system NTE below the threshold (about 50%), which accords well with previous results (Johnson 1981). However observing the birds, we noticed that the severity of acute symptoms increased during the early days after dosing. We thought therefore that the effects on NTE were also increasing with the time. This was in fact the case, as shown in FIGURE 3 (Bertoncin, Capodicasa and Lotti, unpublished results).

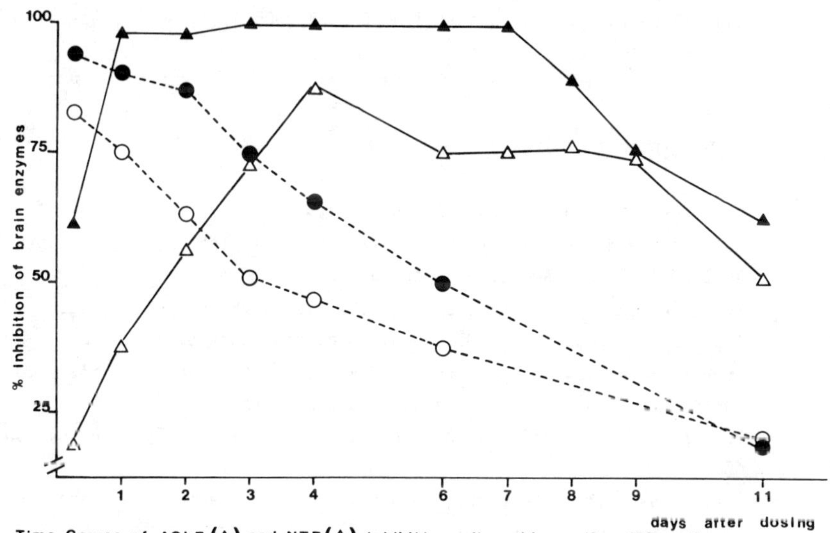

Time-Course of AChE (▲) and NTE (△) inhibition after chlorpyrifos, 150 mg/kg p.o.
For comparison AChE (●) and NTE (○) inhibition after DFP, 1.5 mg/kg s.c.

Fig. 3. – Effects of chlorpyrifos on target enzymes of hen brains.

This figure shows also that the recovery of inhibited AChE is faster than that of inhibited NTE and therefore the level of AChE inhibition approaches that of NTE from day 8 and thereafter. This observation was confirmed in experiments with lower doses of chlorpyrifos. These results might be interpreted in several ways including the presence of impurities, a different access of chlorpyrifos to the two targets, spontaneous reactivation of inhibited

acetylcholinesterases etc. Some of these hypothesis have been already tested, but so far we do not have an explanation for this observation.
It is evident however that the clinical use of a biomonitoring test triggered experiments in the pharmacokinetics and the mechanism of toxicity of chlorpyrifos.
These studies however, already have practical relevance. It is possible by tuning the dose and the time of dosing of chlorpyrifos, that OPIDP might occur without cholinergic symptoms; this will be very important in the toxicological evaluation of occupational exposures to certain OPs.
The timing of NTE measurement has been further defined by these observations in man and subsequent studies in animals. In fact, for certain OPs the usual 24-48 hours interval after exposure might not be appropriate to observe the maximal effect.

ACKNOWLEDGEMENTS

I wish to thank Mrs. Christina Drace-Valentini for the preparation of this manuscript.
Supported in part by grants from CNR (85.00581.56, SP5) , Italian Ministry of Education and Regione Veneto.

REFERENCES

Aldridge W.N. (1953) Biochem.J. 53, 117.
Aldridge W.N. & Reiner E. (1972) Enzymes inhibitors as substrates. North Holland Publ. Co. Amsterdam, London.
Augustinsson K.B. & Casida J.E. (1959) Biochem.Pharmacol. 3, 60.
Augustinsson K.B. & Barr M. (1963) Clin.Chim. Acta 8, 568.
Augustinsson K.B. & Heimbuerger (1954) Acta Chem.Scand. 8, 753.
Becker E.L. & Barbaro J.F. (1964) Biochem.Pharmacol. 13, 1219.
Bertoncin D., Russolo A., Caroldi S., & Lotti M. (1985) Arch.Environ.Hlth. 40, 139.
Chemnitius J.M., Losch H., Losh K., & Zech R.(1983) Comp.Biochem.Physiol. 76, 85.
Davis C.S., & Richardson R.J. (1980) Organophosphorus Compounds in P.S.Spencer and H.H.Schaumburg (Eds), Experimental and Clinical Neurotoxicology, Williams & Wilkins, Baltimore.
de Bisschop H.C., Mainil J.G., & Willems J.L. (1985) Biochem.Pharmacol. 34, 1895.
Dudek B.R., & Richardson R.J. (1982) Biochem.Pharmacol. 31, 1117.
Emmett A., Lewis P.G., Tanaka F., Bleecker M., Fox R., Darlington A.C., Synkowski D.R., Dannenberg A.M., Taylor W.J., & Levine M.S. (1985) J.Occup.Med. 27, 905.
Erdos E.G., Debay C.R., & Westerman M.P. (1960) Biochem.Pharmacol. 5, 173.

Hayes W.J. Jr. (1982) Pesticides studied in man. Williams & Wilkins, Baltimore.
Holmstedt B. (1959) Pharmacol.Rev. 11, 567.
Johnson M.K. (1975) CRC Crit.Rev.Toxicol. 3, 289.
Johnson M.K. (1981) Proc.Meeting European Society of Toxicology, Dublin, Ireland.
Johnson M.K. (1982) Rev.Biochem.Toxicol. 4, 141.
Koelle G.B. (1963) Cholinesterases and anticholinesterase. Handbuch der experimentellen Pharmakologie. Springer Verlag, Berlin.
Lotti M., & Johnson M.K. (1980) Arch.Toxicol. 45, 263.
Lotti M., Ferrara S.D., Caroldi S., & Sinigaglia F.(1981) Arch.Toxicol.48, 265.
Lotti M. (1983) Lymphocytic Neurotoxic Esterase-A biochemical monitor of organophosphorus-induced delayed neurotoxicity in man. Neurobehavioural Methods in Occupational Health, Advances in the Biosciences vol. 46, ed. Renato Gilioli et al., Pergamon Press, Oxford and New York.
Lotti M., Becker C.E., Aminoff M.J., Woodrow J.E., Seiber J.N., Talcott R.E., & Richardson R.J. (1983) J.Occup.Med. 25, 517.
Lotti M., Becker C.E., & Aminoff M.J. (1984) Neurology 34, 658.
Lotti M., Bertoncin D., & Moretto A. (1986a) Toxicologist 6, 22.
Lotti M., Moretto A., Zoppellari R., Dainese R., Rizzuto N., & Barusco G. (1986b) Arch. Toxicol. in press.
Maroni M., & Bleecker M.L. (1986) J.Appl.Toxicol. 6, 1.
Moretto a., Fassina A., & Lotti M. (1983) Proceedings 2nd Internatl. Meeting on Cholinesterases, Bled Yugoslavia.
O'Brien R.D. (1960) Toxic Phosphorus Esters. Chemistry, metabolism and toxic effects. Academic Press, New York.
Osterloh J., Lotti M., & Pond S.M. (1983) J.Anal.Toxicol. 7, 125.
Plapp F.W. & Casida J.E. (1958) J. Econ. Entomol. 51, 800.
Playfer J.R., Eze L.C., Bullen M.F. & Evans D.A.P. (1976) J.Med.Genet. 13, 337.
Pierini M., Moretto A., Lotti M (1985) Toxicologist 5, 142.
Reiner E., Simeon V., & Skrinjaric-Spoljar M. (1980) Comp.Biochem.Physiol. 66, 149.
Reiner E., & Skrinjaric-Spoljar M. (1968) Croat. Chem. Acta 40, 87.
Senanayake N., & Johnson M.K. (1982) N.Engl.J.Med. 306, 155.
Taylor P. (1985) Anticholinesterase agents. In Goodman and Gilman's The Pharmacological basis of Therapeutics. Seventh Edition. Macmillan Publ. Co. New York.

60

The assessment of the dose-response relationship for low-level exposure to neurotoxicants in man

Margit L. Bleecker*†, Jacqueline Agnew† – *Johns Hopkins Medical Institutions, Department of Neurology. †Environmental Health Science, Division of Occupational Medicine.

The main goal of biological monitoring of exposure is either to insure that the current or past exposure of the worker is "safe" or to detect potential excessive exposure before the occurrence of detectable adverse health effects. However, since an internal exposure considered safe according to the present state of knowledge may still cause some harmful effects in susceptible individuals, a program for the detection of adverse biological effects in exposed workers must be applied concommitantly (7).

Limitations of biological monitoring do exist since a biological measurement evaluating the total uptake of a foreign chemical or its degree of retention in the individual requires that the relationship between these parameters (uptake and body burden) and the rate of variation of the biological measurement be known. Unfortunately, for many industrial chemicals knowledge of the relationship between uptake and quantitative changes in the proposed biological parameter is still lacking or insufficient. In addition, various factors may influence the fate of an industrial chemical "in vivo". Exposure to industrial chemicals that modify the activity of the biotransformation enzymes may also influence the fate of another compound; interference may occur with alcohol, food additives, pesticide residues, drugs or tobacco; biological conditions (sex, weight, fatty mass, etc.) may modify the metabolism of an industrial chemical. These various environmental and biological factors must be taken into consideration when interpreting the results of biological exposure tests (1).

The best indicator of risk is the amount of the chemical or its active metabolite at the site of action. A direct measurement of the latter is not usually feasible and therefore the concentration of the pollutant or its metabolite in another body compartment (blood, urine) or the amount bound to another molecule may be used for this purpose if the latter parameter is in equilibrium with the amount at the site of action (2).

Neurotoxins are heterogeneous, acting on multiple cell types and complex organ systems, each of which differs in dose-response relationships. Markers of internal dose in the nervous system are difficult to achieve. In the case of metals, even the use of provocative mobilization by chelation is unlikely to yield information on nervous system concentration. Since most chelants do not cross the blood-brain barrier or blood-nerve barrier, they tend to draw only from larger peripheral body stores.

Given the limitations of biological monitoring, markers of neurotoxic effect are usually obtained simultaneously. It is assumed that neurotoxic injury proceeds from primary biochemical alteration, through reversible cellular injury, to irreversible cytotoxicity and permanent organic lesion. Markers of neurotoxic effect hopefully provide reliable indicators of early-stage effects, related to these biochemical and morphological alterations, that precede actual cell damage.

Neurotoxicant-induced alterations in functional measures, be they neurobehavioral, neurophysiological or neurochemical are believed to precede morphological indexes of toxicity and therefore are more sensitive. However, functional indicators of neurotoxicity can be compromised by the adaptive capacity of the individual especially with moderate to low levels of exposure.

If neurological functioning is viewed as an adaptive process operating within some upper and lower limits, exposure to a neurotoxicant might alter the dynamic equilibrium of the individual's functioning, which up to some point may be brought back to a normal range. It is logical to predict that at some point during exposure the functional reserve of the individual will be depleted and function will deteriorate (8).

If the variance in measures of biological effect is so great due to influences by genetics, age, sex, prior experience, overall health and adaptive capacities, only an overall group effect from exposure may be observed and the dose-response effect may be lost in the noise. This is important to appreciate since the absence of a dose-response effect is frequently used as an argument against the association between an exposure and a biological effect.

The following is an example of a group effect secondary to lead exposure with no dose-response. The performance of thirty male lead workers on simple visual reaction time was compared to that of 60 age, sex and education matched control workers. The exposed men had worked with lead for one year or less. Subjects ranged in age from 18 to 60 with a median age of 34. Blood lead concentrations (PbB) of exposed subjects at the time of testing ranged from 42 to 152 ug/dl with a median of 60 ug/dl. Erythrocyte protoporphyrin (EP) levels ranged from 44 to 310 ug/dl with a median of 162 ug/dl. Blood lead levels of controls did not exceed 21 ug/dl and their maximum erythrocyte protoporphyrin level was 36 ug/dl.

Simple visual reaction time was measured with the Simple Reaction Timer, type RT3. The RT3 contains a counter mechanism, lighted time display, and response button. To initiate the visual stimulus, the counter displayed lights at 000 milliseconds (msec) and began to count at 1 msec increments. The subject pressed the response button as quickly as possible with the index finger of his dominant hand. This immediately froze the display and the reaction time was recorded by the examiner. A practice set of at least three responses preceded the actual test period. If necessary, further practice was allowed until the subject clearly understood the procedure. The test period consisted of eleven responses. Stimuli were presented at random intervals of one to ten seconds.

Simple Visual Reaction Time was scored in this study by three different measures. Two measures of central tendency, mean and median, were used because they have commonly been used in previous investigations of potential neurotoxic effects. The median score is probably more realistic as it is not strongly influenced by extreme outliers. The minimum score, however, was of particular interest. It signified the individual's best performance and therefore thought to be most representative of the physiologic events involved between stimulus presentation and response. Reaction time tasks which are performed continuously over a prolonged time period can also give a measure of visual attention or vigilance based on the variance of intraindividual scores (3). Unfortunately, the trial of only eleven responses used in this study was likely to be too short to be affected by attention problems.

The reaction time measures as indicated in Table I were significantly prolonged in the lead-exposed group ($p < .05$) on all measures.

Table 1.

Reaction Time	Range (msec)	
	Lead-Exposed	Controls
Mean	197-360	182-434 *
Median	199-345	168-414 *
Minimum	161-291	135-290 *

*$p < .05$

Data for the lead-exposed workers were examined to determine whether exposure indices were associated with test score outcomes.

Five lead exposure measures were: total time exposed; blood lead concentrations - initial and maximum; and erythrocyte protoporphyrin levels - initial and maximum. Four new measures were created by multiplying the first one, time, by each of the remaining variables. This procedure weighted the single PbB and EP values according to the duration of the subject's lead exposure.

Age was positively correlated with all of the measures except PbB initial and PbB maximum. Because age was not considered a possible confounder for the PbB measures, Spearman rank correlations were computed for the entire group for PbB Initial and PbB Maximum versus Median Simple Visual Reaction Time (Table 2). None of the correlations were statistically significant.

Table 2: Spearman rank correlation coefficients for lead exposure indices on simple visual reaction time.

Reaction Time	Lead Exposure Index: Pb B (initial)	Lead Exposure Index: Pb B (maximum)
Median	-.06 (.781)	-.03 (.878)
Mean	-.13 (.521)	-.12 (.562)
Minimum	-.01 (.950)	-.02 (.917)

(r followed by p value in parentheses)

Since age was associated with EP values, and, more particularly, with all of the variables which incorporated time, the data were stratified by age into those under 35 years old and those aged 35 and older. Correlations between exposure measures and reaction time are shown in Table 3 with none demonstrating statistical significance.

Table 3: Spearman rank correlation for lead exposure indices vs median simple reaction time for lead exposed subjects.

FOR LEAD EXPOSED SUBJECTS

Lead Exposure Index	< Age 35	> Age 35
Exposure Time	.13 (.648)	.44 (.174)
EP (initial)	.01 (.980)	-.14 (.689)
EP (maximum)	-.05 (.864)	-.15 (.670)
Time x PbB (initial)	.17 (.554)	.47 (.142)
Time x PbB (maximum)	.12 (.675)	.47 (.142)
Time x EP (initial)	.15 (.593)	.25 (.450)
Time x EP (maximum)	.14 (.611)	.18 (.593)

(r followed by p value in parentheses)

This study is an example of one which identifies statistically significant differences between an exposed group and a group well matched on major known confounding factors yet fails to demonstrate a dose response relationship within the exposed group.

Reaction times for the lead-exposed group were consistently prolonged over those of the control group, for every measure observed. Two studies of lead workers which have reported measurement of reaction time did not find statistically significant differences in performance when lead workers were compared to controls on simple reaction time (5, 6). Reaction time techniques varied in every study, however.

A cross-sectional study does not allow inferences to be made due to cause and effect. Without baseline studies of an individual's performance during the period which preceded lead exposure, it is difficult to determine whether a change has truly taken place. Interindividual differences are particularly difficult to measure when the observed outcomes are continuous scores for neurobehavioral measures rather than simply the presence or absence of disease.

Because the temporal sequence of events is unknown for cross-sectional results, the best approach is to use a comparison group which is expected to be as similar as possible with respect to baseline ability. This was accomplished in this study by the use of age and education level matched controls. Other individual factors which may affect neurobehavioral test performance, however, cannot be identified and thus may obscure dose-response relationships within the exposed group.

Reaction time in the present study was not correlated with blood lead, erythrocyte protoporphyrin, duration of exposure or combinations thereof. Within the relatively small group of exposed subjects, high values on these measures were not well represented. The size of the group and limited distribution of scores over the range of lead index values may have influenced the lack of statistical associations.

The differences found between study groups on Reaction Time remained even after individuals with the highest blood lead and erythrocyte protoporphyrin values were removed. This suggests that the measures of exposure used may not have been equivilant for each exposed individual with respect to representation of intensity of exposure or absorption.

The studies by Hanninen et al. (5) Grandjean et al (4), and Valciukas et al. (9) did identify associations within lead-exposed groups between several other neurobehavioral tests and various indices of lead exposure, particulary on tests of visual perception. In the present study, which dealth with relatively low levels and brief lead exposure, reaction time measurements did not successfully demonstrate a similar association with indexes of exposure within the lead worker group.

This study was supported by the National Institute for Occupational Safety and Health, Grant Award Number 1 RO3 OHO1893-01.

REFERENCES

1. Aitio, A., Riihimaki, V., Vainio, H. 1984. Biological Monitoring and Surveillance of Workers Exposed to Chemicals. Hemisphere Publishing Corporation.

2. Baselt, R.C. 1980. Biological Monitoring Methods for Industrial Chemicals. Davis, California. Biomedical Publications. A handbook of laboratory techniques for monitoring personal exposure to toxic substances.

3. Cherry, N., Venables and Waldron, H.A. 1983. The use of reaction time in solvent exposure. In Neurobehavioral Methods in Occupational Health. R. Gilioli, M.G. Cassitto and V. Foa (eds.), Pergamon Press, New York, pp. 191-195.

4. Grandjean, P., Arnvig, E., Beckmann, J. 1978. Psychological dysfunctions in lead-exposed workers: relation to biological parameters of exposure. Scand. J. Work Environ. Health 4:295-303.

5. Hanninen, H., Hernberg, S., mantere, P., Vesanto, R. and Jalkanen, M. 1978A. Psychological performance of subjects with low exposure to lead. J. Occup. Med. 20:683-689.

6. Johnson, B.L., burg, J.R., Xintaras, C., Handke, J.L. 1980. A neurobehavioral examination of workers from a primary nonferrous smelter. Neurotoxicology 1:561-581.

7. Lauwerys, R.R. 1983. Industrial Chemical Exposure: Guidelines for Biological Monitoring. Biomedical Publications, Davis, California.

8. Tilson, H.A. and Mitchell, C.L. 1983. Neurotoxicants and adaptive responses of the nervous system. Federation Proceedings 42:3189-3190.

9. Valciukas, J.A., Lilis, R., Fischbein, A., Selikoff, I.J., Eisinger, J. and Blumberg, W.E. 1978B. Central nervous system dysfunction due to lead exposure. Science 201:465-467.

61

Occupational screening of olfactory function

Richard L. Doty, Ph.D, Carl Monroe, M.D, M.P.H. – Smell and Taste Center, Department of Otorhinolaryngology and Human Communication, School of Medicine, University of Pennsylvania, Philadelphia, PA 19104 and Safety, Health and Environmental Affairs Department, Rohm and Haas Company, Philadelpia, PA 19105

ABSTRACT

To assess the feasibility of mass screening of olfactory function in a corporate setting, we asked workers frequenting the cafeteria of the corporate offices of a major chemical manufacturing company to voluntarily self-administer the University of Pennsylvania Smell Identification Test. Although the majority agreed to participate in the project, only about half returned completed tests to the examiners, possibly reflecting the informal atmosphere in which the tests were distributed. On average, the corporate subjects performed slightly better than previously-tested controls matched on the basis of gender, age, smoking habits, and ethnic background. As in previous studies, women significantly outperformed men, and age was significantly related to the test scores. Approximately 1% of the group evidenced major olfactory dysfunction relative to published percentile norms. Thus, in addition to demonstrating that self-administered olfactory testing is practical and feasible within the corporate setting, these data suggest that the incidence of olfactory dysfunction is quite low in corporate groups not exposed to airborne chemicals used in manufacturing processes, providing a meaningful baseline for the assessment of test scores from chemically exposed occupational groups.

INTRODUCTION

Until recently, routine evaluation of olfactory nerve (CN I) function was rarely performed in occupational settings in spite of evidence that (a) employees who are unable to detect low concentrations of poisonous or explosive vapors are at risk in a number of occupational classifications and (b) losses or distortions in smell ability can occur as a result of job-related exposure to some chemicals (for reviews see Amoore, 1986; Doty, 1979; Schiffman, 1983). Aside from aiding in the detection of a variety of medical disorders, including major dementias, routine olfactory testing can alert the employer to problematic conditions within a plant, as well as minimize the impact of litigation claims from persons with olfactory problems unrelated to the work environment.

In the present chapter we briefly describe the application of a recently developed self-admininstered test of olfactory function to the workers of the corporate offices of a major chemical manufacturing company (for a complete description of the study, see Doty, Gregor & Monroe, 1986). This reliable and well-validated microencapsulated smell test, which was developed at our Center, is now used routinely in hundreds of otolaryngologic and neurologic clinics in North America and has been shown to be sensitive to numerous subject variables, including age, gender, and smoking habits. The fact that this test detects both gross and suble alterations in olfactory perception [including decrements associated with Alzheimer's disease (Doty, Reyes & Gregor, 1986), cystic fibrosis (Weiffenbach & McCarthy, 1984), Korsakoff's psychosis (Mair et al., 1986) and Parkinson's disease (Doty, Stellar & Gregory, 1986)] suggests it is likely sensitive to neurotoxic olfactory problems caused by chronic or acute exposure to some environmental chemicals.

The primary goals of this pilot project were to (a) establish the level of acceptance of a voluntary olfactory testing program within a corporate setting, (b) determine the proportion of the sample that reported or evidenced marked olfactory impairment, (c) develop guidelines for using the test instrument as a medical surveillance tool for future testing of persons occupationally exposed to potentially toxic volatiles, and (d) compare the results of such testing to scores obtained from control subjects from other non-exposed populations matched on the basis of age, gender, and general occupational level.

DESCRIPTION OF MEASURING INSTRUMENT

The University of Pennsylvania Smell Identification Test (UPSIT; commercially termed The Smell Identification Test™, Sensonics, Inc., Haddonfield, NJ) consists of 4 booklets containing a total of 40 "scratch and sniff" odorants, one odorant per page (Figure 1). The odorants are embedded in 10-50 um diameter microencapsulated crystals located on brown strips at the bottom of each page. Above each strip is a multiple choice question with four response alternatives. For example, one of the items reads, "This odor smells most like: (a) chocolate; (b) banana; (c) onion; or (d) fruit punch". The subject is required to answer one of the alternatives, even if no smell is perceived (i.e., the test is forced-choice). The reasons for the specific choice of odorants, response alternatives, and other elements of the test's format are described elsewhere (Doty, 1983; Doty, Shaman & Dann, 1984). The internal consistency and the test-retest reliability have been shown to be very high (e.g., test-retest r = 0.95, Doty, Newhouse & Azzalina, 1985). In general, scores on this test correlate strongly with those of traditional olfactory detection threshold tests (Doty, Shaman & Dann, 1984). Furthermore, as can be seen in Figure 2, administration of this test to large numbers of individuals reveals, on the average, (a) greater female than male performance (particularly in the later years) and (b) marked age-related alterations in the ability to smell.

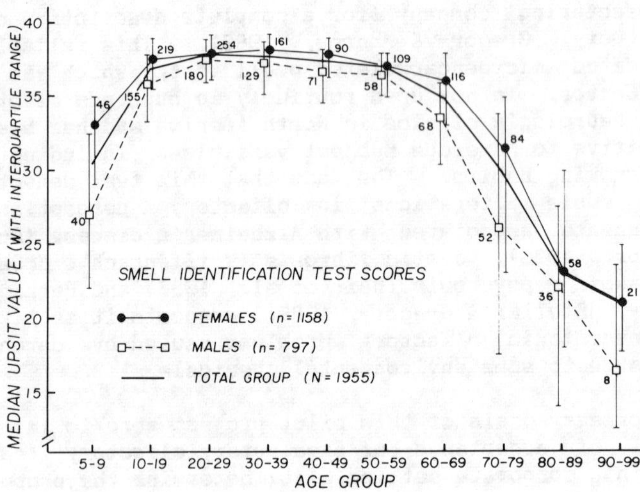

Fig. 1. – Picture of the microencapsulated forced-choice University of Pennsylvania Smell Identification Test (UPSIT). Reprinted with permission from Doty et al., Physiol. Behav., 1984, 32, 489–502.

STUDY POPULATION AND GENERAL PROCEDURES

The UPSITs were initially distributed to 640 volunteers frequenting the cafeteria of the corporate offices of a major chemical manufacturing company. The total workforce of the building was estimated to be approximately 1,000. Although we encouraged the participants to complete the test during the coffee break or lunch period in which they volunteered, we also allowed them to take the tests back to their offices with the instructions to complete and return them at their convenience. Unfortunately, only 381 of the 640 volunteers returned the tests to us. Of those which were returned, 47 had been given to family members and 11 were incomplete, resulting in a final subject sample size of 323 or 50% of the initial sample. Only 10 of these individuals were in occupations where at least some contact with chemicals might be expected (8 chemists and 2 chemical engineers), and all were reportedly in good health. Details of the administration procedure and demographics of the study population are presented elsewhere (Doty, Gregor & Monroe, 1986).

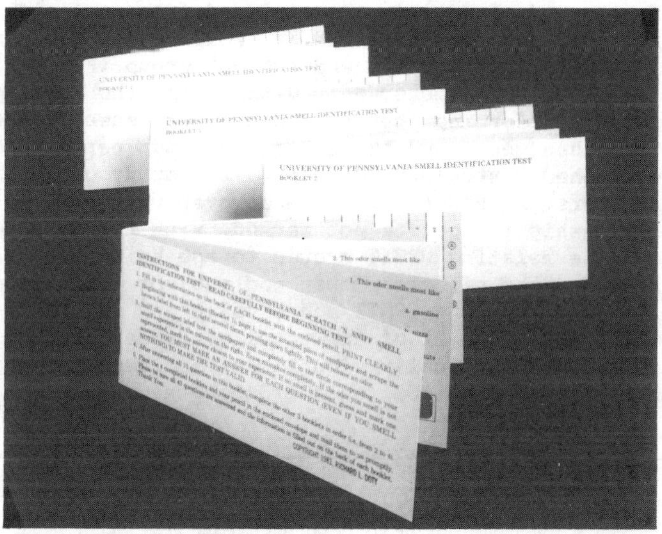

Fig. 2. – University of Pennsylvania Smell Identification Test scores as a function of subject age and gender. Numbers by date points indicate sample sizes. From Doty et al., Science, 1984, 226, 1441–1443. Copyright 1984 by the American Association for the Advancement of Science. Reprinted with permission.

The control subjects consisted of 323 persons selected from a computerized subject pool maintained by the Smell and Taste Center. These subjects were matched to each employee on the basis of gender, age, smoking behavior, and general occupational category, and had previously been tested at health fairs and other public events. In cases where more than one control subject met the criteria of a needed match, a random process was used to select the matched control.

RESULTS

Seven of the corporate subjects reported, at the time of the collection of informed consent, histories of allergies or sinus problems. However, the test scores of these individuals were within the normal range for persons of their sex and age, as indicated by published test norms (Doty, 1983). Three other subjects (aged 20, 24, and 61) scored considerably below the normal range (UPSIT scores of 28, 15, and 18, respectively), with only the 24 year old being aware of a smell problem.

To establish whether the corporate subjects differed from their matched controls, as well as whether subject age, gender, or smoking behavior influenced the test scores, an analysis of covariance was performed using age as a covariate and sex, current smoking behavior (yes, no), and subject group (corporate, control) as factors. The data from the three subjects with olfactory dysfunction were excluded from analysis. Significant effects of age ($F = 5.72$, $df = 1/315$, $p < 0.017$), sex ($F = 4.32$, $df = 1/315$, $p = 0.038$), and subject group ($F = 8.61$, $df = 1/316$, $p < 0.004$) were observed, with -- on the average -- younger persons performing better than older ones, women performing better than men, and the corporate group performing better than the matched controls. However, the study group did not contain many persons under 21 years or over 55 years of age, so the relationship to age was not marked, with the slope of the regression line relating UPSIT scores to age being very slight (-0.02). Likewise, the effects of sex and subject group were quite small, being less than one UPSIT score value and accounting individually for no more than 2% of the total variance (Table 1). In this study, no statistically significant influence of smoking behavior was apparent ($F = 1.66$, $df = 1/315$, $p = 0.20$).

Table 1: University of Pennsylvania Smell Identification Test Scores (+SD) of men and women from the study population and age–, gender–, race–, and smoking habit-matched controls.

	Men	Women
Corporate Subjects	37.80 (2.08)	37.98 (2.06)
Matched Controls	37.12 (2.00)	37.88 (1.74)

DISCUSSION

This pilot study taught us several important lessons regarding voluntary self-administered olfactory testing within a corporate population. First, while volunteering behavior for testing can be expected to be quite high, actual compliance in completing and returning tests may be rather low. Greater compliance might occur if such testing is performed either on a non-voluntary basis or on a voluntary basis within a more formal context, such as that of a medical examination. Second, UPSIT scores of corporate subjects with no exposure to industrial chemicals will be close, if not identical, to those obtained from samples of other populations of non-exposed persons, such as those frequenting health fairs and numerous public events (see Doty et al., 1984). Third, only a small percentage of working-age persons appear to have marked smell dysfunction (in accord with threshold studies indicating a value of approximately 2%; cf. Amoore, 1986). Such low incidence provides a sound baseline for assessing the likelihood of olfactory alterations within a given occupational environment. Finally, this study indicates that self-administered olfactory testing is practical and feasible within the corporate setting, making rapid, inexpensive, and widespread screening of the olfactory function of industrial workers possible.

Because the receptors of the olfactory system are rather directly exposed to the outside environment, they are susceptible to adverse effects of viruses and environmental toxins (Doty & Kimmelman, 1986). For example, many viruses enter into the central nervous system via the olfactory primary receptor cells (Monath, Croop & Harrison, 1983; Stroop, Rock & Fraser, 1984; Tomlinson & Esiri, 1983) and damage, in some cases, the olfactory pathways (Goto et al., 1977; Reinacher et al., 1983). Because, in part, of the highly active transport mechanisms within the olfactory nerve cells (see Shipley, 1985), environmental toxins can similarly alter or damage olfactory structures (Buckley et al., 1985; Jiang, Buckley & Morgan, 1983; Rehn et al., 1981). Such alterations are presumably the basis of smell dysfunction following acute or chronic exposure to a number of solvents, heavy metals, and dusts (cf. Doty, 1979; Emmett, 1976). Unfortunately, there have been few sound quantitative studies of such problems to date, despite the large anecdotal literature on this topic.

The present study demonstrates the feasibility of mass olfactory screening in the industrial environment, and provides an initial baseline from which to interpret dysfunction in occupationally-exposed groups. Since the UPSIT (a) readily detects olfactory dysfunction associated with numerous diseases and medical problems and (b) correlates strongly with olfactory detection thresholds, it provides a convenient and quantitative means of quantitatively determining the consequences of exposure to air-borne chemicals in the workplace.

REFERENCES

Amoore, J.E. Effects of chemical exposure on olfaction in humans. In C.S. Barrow (Ed.), Toxicology of the Nasal Passages. Washington: Hemisphere Publishing Corp., 1986, pp. 155-190.

Buckley, L.A., Morgan, K.T., Swenberg, J.A., James, R.A., Hamm, T.E. and Barrow, C.S. The toxicity of dimethylamine in F-344 rats and B6C3F1 mice following a 1-year exposure. Fund. Appl. Toxicol., 1985, 5, 341-352.

Doty, R.L. The Smell Identification Test™ Administration Manual. Philadelphia: Sensonics Inc., 1983.

Doty, R.L. A review of olfactory dysfunctions in man. Amer. J. Otolaryngol., 1979, 1, 57-79.

Doty, R.L., Gregor, T. & Monroe, C. Quantitative assessment of olfactory function in an industrial setting. J. Occupat. Med., 1986, 28, 457-460.

Doty, R.L. & Kimmelman, C.P. Smell and taste and their disorders. In A.K. Asbury, G.M. McKhann & W.I. McDonald (Eds.), Diseases of the Nervous System. Philadelphia: W.B. Saunders, in press.

Doty, R.L., Newhouse, M.G., and Azzalina, J.D. Internal consistency and short-term test reliability of the University of Pennsylvania Smell Identification Test. Chemical Senses, 1985, 10, 297-300.

Doty, R.L., Reyes, P. & Gregor, T. Olfactory dysfunction: Reliable presence in Alzheimer's disease. Submitted, 1986.

Doty, R.L., Shaman, P., and Dann, M. Development of the University of Pennsylvania Smell Identification Test: A standardized microencapsulated test of olfactory function. Physiology and Behavior (Monograph), 1984, 32, 489-502.

Doty, R.L., Shaman, P., Applebaum, S.L., Giberson, R., Sikorsky, L. & Rosenberg, L. Smell identification ability: Changes with age. Science, 1984, 226, 1441-1443.

Doty, R.L., Stellar, S. & Gregor, T. Smell dysfunction in Parkinson's disease. Submitted, 1986.

Emmett, E.A. Paroxmia and hyposmia induced by solvent exposure. Brit. J. Indust. Med., 1976, 33, 196-198.

Goto, N., Hirano, N., Aiuchi, M., Hayashi, T. and Fijiwara, K. Nasoencephalopathy of mice infected intranasally with a mouse hepatitis virus, JHM strain. Jap. J. Exp. Med., 1977, 47, 59-70.

Mair, R.G., Doty, R.L., Kelly, K.M., Wilson, C.S., Langlais, P.J., McEntee, W.J. & Vollmecke, T.A. Multimodal sensory discrimination deficits in Korsakoff's psychosis. Neuropsychologia, 1986, in press.

Monath, R.P., Croop, C.B. and Harrison, A.K. Mode of entry of a neurotropic arbovirus into the central nervous system. Reinvestigation of an old controversy. Lab. Invest., 1983, 48, 399-410.

Rake, G. The rapid invasion of the body through the olfactory mucosa. J. Exp. Med., 1937, 65, 303-315.

Rehn, B., Breipohl, W., Schmidt, C., Schmidt, U. & Effenberger, F. Chemical blockade of olfactory perception by N-methyl-formimino-methylester in albino mice. II. Light microscopical investigations. Chem. Senses, 1981, 6, 317-328.

Reinacher, M., Bonin, J., Narayan, O. and Scholtissek, C. Pathogenesis of neurovirulent influenza A virus infection in mice. Route of entry of virus into brain determines infection of different populations of cells. Lab. Invest., 1983, 49, 686-692.

Schiffman, S.S. Taste and smell in disease. New England Journal of Medicine, 1983, 308, 1275-1279, 1337-1343.

Shipley, M.T. Transport of molecules from nose to brain: Trans-neuronal anterograde and retrograde labeling in the rat olfactory system by wheat germ agglutinin-horseradish peroxidase applied to the nasal epithelium. Brain Res. Bull., 1985, 15, 129-142.

Stroop, W.G., Rock, D.L. and Fraser, N.W. Localization of herpes simplex virus in the trigeminal and olfactory systems of the mouse central nervous system during acute and latent infections by in situ hybridization. Lab. Invest., 1984, 51, 27-38.

Tomlinson, A.H. and Esiri, M.M. Herpes simplex encephalitis. Immunohistological demonstration of spread of virus via olfactory pathways in mice. J. Neurol. Sci., 1983, 60, 473-484.

Weiffenbach, J.M. & McCarthy, V.P. Olfactory deficits in cystic fibrosis: distribution and severity. Chem. Senses, 1984, 9, 193-199.

62

Peripheral models for the study of neurotransmitter receptors: their potential applications to occupational health

Lucio G. Costa – Department of Environmental Health, University of Washington, Seattle, WA

Studies on the toxicity of chemicals to humans are limited by the inaccessibility of most tissues, with the exception of certain biopsies and post-mortem samples. Usually, urinary or hematic levels of a chemical and/or its metabolites serve as indicators of exposure, while the activity of enzymes or other parameters present in the blood can sometimes offer indications of certain biological effects or physiological alterations.

Among the thousands of chemicals used in the industrial setting and contaminating the environment, many are known to affect the central and/or the peripheral nervous system (O'Donoghue, 1985). Transmission of messages between cells in the nervous system and from these to various organs, involves the release of a chemical neurotransmitter from one cell and its subsequent recognition by a second cell. Specific enzymes synthesize the neurotransmitter from one or more precursors. After their release and their interaction with specific receptors, neurotransmitters are rapidly inactivated by enzymatic degradation or by an uptake mechanism. Several of the parameters involved in chemical neurotransmission are the target for a variety of drugs used for the treatment of central nervous system diseases, as well as for heart, lung and GI tract disorders. Furthermore, several environmental and occupational toxins, particularly metals and pesticides, are also known to affect chemical neurotransmission. In order to investigate these parameters in humans, it is necessary to study them in peripheral tissues which can be sampled and studied repeatedly in exposed populations, to establish the effects of exposure and to determine the eventual rate of recovery. Therefore, new strategies are being developed to study the same enzymes and receptors in tissues which can be easily and ethically obtained from living patients. Cell types which have been used include platelets, red blood cells, lymphocytes and fibroblasts (Stahl, 1985). This new strategy has already found some preliminary applications in the field of psychiatric disorders, in particular in the study of depressive illnesses, and in the study of asthmatic patients.

A basic assumption of these studies is that the neurochemical parameters measured in these peripheral tissues are valid markers of the same parameters present in the brain or in other organs. This assumption needs to be validated by studies which show that the parameter studied has the same biochemical and functional characteristics of that present in other tissues, and that alterations of this parameter in various disease states and/or as a result of drug or chemical exposure, occur in a parallel fashion in the marker tissue and in the target organ. A few examples will illustrate the pitfalls and the successes of such validation studies.

Human skin fibroblasts, have been increasingly considered as models for studying neurotransmitter functions with particular regard to the cholinergic system. However, there is controversy on the presence of muscarinic cholinergic receptors on fibroblasts. In 1984 a paper described the presence of a binding site for the cholinergic muscarinic ligand ^{3}H-quinuclidinyl benzilate on cultured human skin fibroblast (Nadi et al. 1984). This binding site exhibited some pharmacological specificity and was functional, since muscarinic agonist inhibited the beta-adrenoceptor stimulated adenylate cyclase. Furthermore, since in 18 patients with major affective disorders there was a higher density of muscarinic binding sites than in controls, it was proposed that muscarinic cholinergic receptors on fibroblasts could be used clinically as a model for studies of affective illnesses. However, additional studies failed to reproduce these findings (Van Riper et al. 1985; Kelsoe et al. 1985). Some binding of the muscarinic ligand ^{3}H-N-methylscopolamine, but not of ^{3}H-QNB, was observed, and there was no evidence that these binding sites were functional receptors. Indeed, stimulation with muscarinic agonists did not cause inhibition of prostaglanding E_1-stimulated adenylate cyclase, stimulation of cyclic GMP accumulation, or increased metabolism of phosphoinositides, three cellular response systems coupled to muscarinic receptors in other tissues. Thus, the use of human skin fibroblasts for the study of muscarinic receptors in other tissues appears inappropriate, at least until any different evidence is obtained.

On the positive side, on the other hand, are the results obtained by a large number of studies on beta-adrenoceptors in lymphocytes. Binding of adrenergic ligands and the presence of a norepinephrine-stimulated adenylate cyclase have been consistently demonstrated, and the biochemical and pharmacological characteristics of beta-adrenoceptors in lymphocytes have been found to be the same as in the CNS and other tissues (Williams et al. 1976; Niaudet et al. 1976). A recent study established definitely the existence of a highly significant linear relation between the density of beta-adrenoceptors in atrial membrane and lymphocytes, as well as between the maximal contracting responses of the atria to isoprenaline and the corresponding atrial and lymphocyte beta-adrenoceptors (Brodde et al. 1986). Treatment of animals and humans with beta-adrenergic agonist and antagonists caused a decrease and increase, respectively, of beta-adrenoceptors in lymphocytes and solid tissues such as brain, heart and lung (Aarons et al. 1980; Aarons and Molinoff, 1982). Furthermore, alterations of the density and function of beta-adrenoceptors have been found in lymphocytes from asthmatic patients, as well as in lung of guinea pigs in an experimental model of asthma (Brooks et al. 1979; Barnes et al. 1980).

These two examples show the enormous potentials, as well as the pitfalls and false hopes, that the research or peripheral markers for neurotransmitter receptors encounters. However, since some of the positive findings are being successfully applied in various clinical studies, the same techniques and approaches could be used to investigate problems related to environmental and occupational health. There are three areas in which such approach would be useful.

First, mechanistic studies on the interaction of certain neurotoxicants with neurotransmitters could be done directly using human models such as the cell types illustrated before. With the exception of cholinesterases, there are not many examples of this kind of approach. However, recent experiments which investigated the interaction of various toxicants (e.g. organotins; Johnson and Knowles, 1983) with the serotonin transport system in platelets, indicate that certain peripheral models should have some useful applications.

A second potential application of such peripheral markers to environmental and occupational health, is their use in identifying particular susceptibility to certain chemicals. It has been shown that the activities of various enzymes involved in xenobiotic metabolism, which can be measured in blood cells or in serum, can also be genetically determined, and that identification of individuals with particularly high or low activity mey help in determining their possible higher sensitivity to certain pesticides or carcinogens. A similar approach could be utilized with regard to neurotransmitters. An example of a potential application of this concept to occupational health is in the area of occupational lung diseases. In particular, workers who have an exaggerated response to the bronchoconstriction caused by the muscarinic agonist methacholine, are also usually more susceptible to the effects of various bronchoconstrictors such as sulfur dioxide or diisocyanate (Brooks, 1983). The interesting correlations found with beta-adrenoceptors in lymphocytes and lung of asthmatic patients, are an incentive to investigate such correlation with regard to muscarinic receptors. Animal studies should first establish a good correlation between muscarinic receptors in, for example, lymphocytes (Adem et al. 1986), and in the respiratory tract. Then, one can hypothesize that workers with higher muscarinic receptors in peripheral cells, will have the same abnormality in the respiratory tract, and will be more sensitive to the action of methacholine. If that is true, it would be possible to detect workers which might be at higher risk for exposure to bronchoconstrictor substances encountered in the workplace and thus to recommend protective measures.

Finally, a third area in which the availability of peripheral markers of neurotransmitter function could find useful application is the detection of insults occurring in the CNS or other tissues following exposure to toxic chemicals. Several neurotoxicants, particularly metals and pesticides, are known to interact with various parameters of neurotransmission. An example is represented by organophorphorus insecticides. While their acute toxicity is well characterized, less is known on the effects of chronic exposure, particularly on the nervous system. Various investigations have shown that repeated administration of organophosphates to experimental animals, which results in the development of tolerance to their toxicity, causes decrease in the density of cholinergic muscarinic receptors in brain and in peripheral tissues (Costa et al. 1982). Tolerance to anticholinesterases could be considered as a protective phenomenon by which the organism normalizes functions despite challenge from the external environment. However, the alterations of cholinergic receptors present in tolerant animals, could lead to altered response to drugs and other environmental pollutants including other insecticides, and/or to an imbalance of delicate central nervous system functions, such as cognitive processes. Since tolerance to organophosphorus insecticides has been reported to occur also in human, it is possible that similar alterations of cholinergic receptors might have occurred. It is interesting that one of the recurrent complaints in workers chronically exposed to organophosphorus pesticides is memory impairment, which might be associated with an alteration of cholinergic receptors in brain, since the cholinergic system is known to have a major function in the processes of learning and memory (Costa, 1986). Recently, we have found that repeated exposure of rats to an organophosphate, which induced a decrease in muscarinic receptor density in various brain areas, also caused an impairment of memory (McDonald, Murphy, Costa, unpublished), thus confirming experimentally some of the findings of the epidemiological studies. We are now investigating whether a similar receptor alteration can be found in circulating peripheral lymphocytes. If these studies will show a correlation between muscarinic receptor alterations in the CNS and in lymphocytes, in animals which have a behavioral impairment (i.e., memory loss), this will open an exciting possibility for monitoring agricultural workers chronically exposed to organophosphorus insecticides.

In conclusion, while neurobehavioral and electrophysiological tests are being developed and successfully used in the identification of neurotoxic effects, the use of biochemical markers for detection of neurotoxicity in human population is at a very early stage. Studies in this area, using the approaches that I have proposed, or others, have good potentials for several applications in occupational health. There is obviously a need for the development of more of such tests and, in particular, for their validation in animals, before results can be applied to human.

REFERENCES

AARONS R.D. and MOLINOFF P.B., J. Pharmacol. Exp. Ther. 221:439, 1982.

AARONS R.D., NIES A.S., GERBER J.G. and MOLINOFF P.B., J. Pharmacol. Exp. Ther. 224:1, 1980.

ADEM A., NORDBERG A. and SLANINA P., Life Sci. 38:1359, 1986.

BARNES P.J., DOLLERY C.T. and MACDERMOT J., Nature 285:569, 1980

BRODDE O.E., KRETSCH R., IKEZONO K., SERKOWSKI H.R., and REIDMEISTER J.C., Science 231:1584, 1986.

BROOKS S.M., in Environmental and Occupational Medicine (W.N. Rom ed.), Little Brown and Company, Boston, 1983, p. 233.

BROOKS S.M., McGOWAN K., BERNSTEIN I.L., ALTENAU P. and PEAGLER J., J. Allergy Clin. Immunol. 63:401, 1979.

COSTA L.G., in Recent Advances in Nervous System Toxicology (C.L. Galli, L. Manzo and P.S. Spenser, eds.), Plenum Press, NY, 1986 (in press).

COSTA L.G., SCHWAB B.W. and MURPHY S.D., Toxicology 25:79, 1982.

JOHNSON T.L. and KNOWLES C.O., Toxicology 29:39, 1983.

KELSOE J.R., GILLIN J.C., JANOWSKI D.S., BROWN J.H., RISCH S.G. and LUMKIN B., New England J. Med. 312:86, 1985.

NADI N.S., NURNBERGER J.I. and GERSHON E.S., New England J. Med. 311:225, 1984.

NIAUDET P., BEAURAIN G. and BACH M., Eur.J. Immunol. 6:834, 1976.

O'DONOGHUE J.L. (Ed.), Neurotoxicity of Industrial and Commercial Chemicals, CRC Press, Inc., Boca Raton, FL, 2 vols., 1985.

STAHL S.S., Psychopharmacol Bull. 21:663, 1985.

VAN RIPER D.A., ABSHER M.P. and LENOX R.H., J. Clin. Invest. 76:882, 1985.

WILLIAMS L.T., SNYDERMAN R. and LEFKOWITZ R.J., J. Clin. Invest. 57: 149, 1976.

Biological monitoring of exposure to organophosphates – A field study using cholinesterase estimation of whole blood dried on filter paper

B. Kolmodin-Hedman – Professor, M.D., Medical Division, National Board Occupational Safety & Health, Box 6104, S-900 06 Umeå, Sweden

K. Eriksson – M.Sc., Occupational Medicine, National Board Occupational Safety & Health, Box 6104, S-900 06 Umeå, Sweden

BACKGROUND

At the Medical Division of the National Board of Occupational Safety and Health in Umeå, Sweden, monitoring of various groups occupationally exposed to pesticides have been studied in order to develop chemical/analytical methods of enough sensivity and stability to be used for monitoring of the workers. For the control of persons handling organophosphorus compounds (OP-compounds) the cholinesterase (ChE) estimation in capillary blood described by Augustinsson and Holmstedt (1965), and further developed by Eriksson and Faijersson (1978,1980), has been used. To be used adequately as a biological monitoring a pre-exposure value is compared with an exposure value in each individual. Comparing group means of enzyme activity is less apt to follow small changes, and the big inter-individual differences in pre-exposure values is another reason.
Organophosphates: Several OP-compounds have been used: dimethoate, formothion, isofenphos and occasionally chlorfenvinphos. They are formulated as granulates, WP (wettable powders) and as solutions.

SCREENING OF SYMPTOMS

A standardized questionnaire was filled in, in close connection to the spray period, asking for allergic disposition, previous illnesses and possible acute symptoms connected to mixing and spraying of the pesticides. Pre-exposure cholinesterase activity was estimated at the first contact in the spring.

MATERIAL

One group of 8 orchard workers, working with azinphosmethyl, were controlled and also a group of 9 gardeners using the above mentioned OP-compounds.

CHOLINESTERASE ESTIMATION

Cholinesterase activity is estimated according to the method by Eriksson and Faijersson (1978,1980), where both AcChE and BuChE are estimated on capillary blood, and 50 µl is sampled on filter paper (Munktell No. 4, Mölnlycke Ltd). The method uses a hydrolyzing step of propionylthiocholine to thiocholine, which is reacted with 4-thiopyridon at 30°C. 4-thiopyridon absorbance is read at 324 nm with a spectrophotometer (Shimadzu Comp UV 190). BuChE is selectively inhibited by the Astra synthesis compound 1397. By absorbance reading before and after inhibition both AChE and BuChE is estimated. The method is checked by inter-laboratory comparison, with a coefficient of variation of 5-8%. Storage of the paper should be prompt in a freezer. Checks at our lab showed a good reproducibility for one month of storage.

RESULTS

Pre-exposure values are shown in Figure 1, where inter-individual variance could be shown for both AcChE and BuChE varying between 21-76 nmol/s·ml and 21-95 nmol/s·ml at 30°C.
The orchard workers exposed to mean air concentration of $65 \cdot 10^{-6}$ g|m^3 of azinphosmethyl showed no decrease after work both of AcChE and BuChE. Irritative symptoms from skin and upper respiratory airways were judged to be due to mites, concomitant fungicide use or due to climatic conditions.
Table 1 shows the individual values for the gardeners before and after exposure and the calculated difference. The mean difference ChE was +0.56±3.57 for AChE and +2.67±6.24 for BuChE. This is not significant tested against zero in a Student's t-test. In the studied group of 9 gardeners, 5 did not report any symptoms at all, 2 had headaches and correlated that to the mixing procedure of the pesticides, 2 experienced eye irritations and 1 person cough. However, the last man correlated this and his eye symp-

toms to handling a manganese sulphate.

DISCUSSION

All Swedish workers handling insecticides of risk class I (toxic OP-compounds) must have a week-long certificate course of education. Most workers do their work properly by using respirators and rubber gloves, but mostly a penetrable nylon overall. The absence of any pronounced symptoms and a non-significant change in ChE activity supports the idea that protection is enough. Fungicides and herbicides used during the same cultivating period might have irritative properties, and might explain symptoms as eye irritation and cough. Also, early spring and late autumn might create weather conditions in Sweden with chilly temperatures and strong winds. Desinsectors regularly using a mixture of malathion, dichlorvos and bromophos in kerosene type solvents might also be exposed to solvent vapors, which must be taken into account when evaluating any health effects.

CONCLUSION

Cholinesterase activities are estimated simply by the filter paper method, estimating both AcChE and BuChE in capillary blood. Information and education mostly prevent an effect clinically or in ChE activity. Symptoms must be evaluated in the light of all occupational exposure.

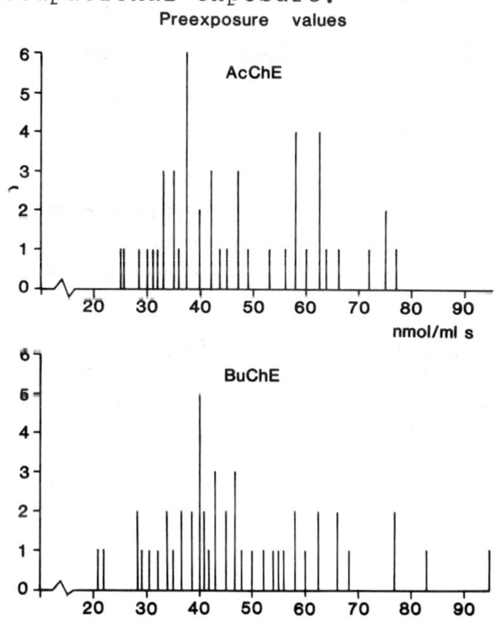

Fig. 1. – Pre-exposure ChE values, n=47, in Swedish male workers.

Table 1.

Preexp – exp

	AcChE nmol/ml s	AcChE △	BuChE nmol/ml s	BuChE △
OS	51 – 54	–3	64 – 74	–10
SOM	55 – 56	–1	56 – 51	+5
NT	54 – 56	–2	39 – 37	+2
IM	56 – 55	+1	69 – 58	+11
HM	28 – 26	+2	39 – 40	–1
MH	38 – 32	+6	35 – 25	+10
DW	36 – 41	–5	27 – 29	+3
BEW	31 – 28	+3	28 – 24	+4
AA	37 – 33	+4	39 – 39	–0

	AcChE	BuChE
Mean value	+ 0.56	+ 2.67
Standard variation	± 3.57	± 6.24
T–value, 95% level	0.47	1.28

REFERENCES

Augustinsson K-B, Eriksson H, Faijersson Y (1978) A new approach to determining cholinesterase activities in samples of whole blood. Clin Chim Acta 89:239-252.

Augustinsson K-B, Holmstedt B (1965) Determination of cholinesterase in blood samples dried on filter-paper and its practical application. Scand J Clin Lab Invest 17:573-583.

Eriksson H, Faijersson Y (1980) A reliable way of estimating cholinesterases from whole blood in the presence of anti-cholinesterases. Clin Chim Acta 100:165-171.

Eriksson K (1984) Exposition för bekämpningsmedel med en organisk fosforförening som aktiv substans.vid besprutning av fruktodlingar. Undersökningsrapport 1984:4 (In Swedish) NBOSH, Umeå, Sweden.

Stålberg E, Hilton-Brown P, Kolmodin-Hedman B, Holmstedt B, Augustinsson K-B (1978) Effect of occupational exposure to organophosphorus insecticides on neuromuscular function. Scand J Work Environ Health 4:255-261.

64

Lead decreases lymphocyte β-adrenergic receptors in humans

A. Padovani, S. Govoni, M.S. Magnoni, C. Fernicola*, L. Coniglio*, M. Trabucchi** –
Institute of Pharmacological Sciences, University of Milan, *Occupational Health Unit, Brescia and **Chair of Toxicology, 2nd University of Rome, Italy

SUMMARY

The characteristics of the lymphocyte β-adrenergic receptors (β-R) were measured in a group of male workers professionally exposed to inorganic lead.
According to the blood lead (PbB) and plasma zincoprotoporphirin (ZPP) concentration the whole group was divided in subjects having: PbB and ZPP $<$ 40 mcg/dl; PbB and ZPP $>$ 40 mcg/dl; ZPP $>$ 40 mcg/dl and normal PbB. The lymphocyte β-R Bmax was decreased either in subjects having both PbB and ZPP $>$ 40 mcg/dl or in those having only ZPP $>$ 40 mcg/dl, indicating a long-term effect of the metal on the receptor.

INTRODUCTION

Prolonged lead exposure has a documented association with toxic effects on multiple organs, including cardiovascular and nervous systems.
Experimental chronic lead exposure modifies different neurochemical parameters in the brain before any detectable damage to body organs (1,2). Many experimental reports indicate that lead induces pronounced alterations of the catecholaminergic (CA) system (3,4). Studies measuring urinary CA metabolites, i.e. vanylmandelic acid and homovanillic acid, indicate that chronic

lead intoxication alters CA activity also in man (5,6).
In our study, lead exposure effect on CA receptors was investigated utilizing an easily accessible tissue, i.e. lymphocytes. In fact, lymphocytes bear CA receptors which in selected experimental situations behave as neuronal neurotransmitter receptors (7). Lymphocytes may therefore be utilized as a predictive model of ß-R changes in less accessible tissues, as recently indicated by Brodde et al. (8).

METHODS

A group of 37 workers occupationally exposed to lead was selected. From each fasting volunteer 20 ml of blood was withdrawn by venipuncture. Lymphocytes were isolated according to the method of Boyum (9).
Fresh heparinized blood was diluted with an equal volume of phosphate buffered saline (PBS). Aliquots of the mixture were layered on Ficoll Paque (Pharmacia) and centrifuged at 400 g for 20 min. The lymphocyte band was washed with PBS by centrifugation. The washed lymphocytes were homogenized in ice cold distilled water using an Ultra-Turrax and centrifuged at 45,000 x g x 30 min. The resulting pellets were diluted with 10 ml distilled water and kept for 45 min. at 4°C. The preparation was then centrifuged at 45,000 x g x 30 min.; the final pellets were resuspended in ice cold incubation buffer. The ß-R characteristics were measured using ^{125}I-Iodocyanopindolol (ICYP) as radiolabelled ligand. Protein content was measured according to Lowry et al. (10).
On the same blood samples PbB and ZPP were measured by spectrophotometry and hematofluorimetry.

RESULTS

^{125}I-ICYP binding to lymphocyte membranes is inhibited by lead acetate. The maximal inhibition is 42% in vitro and the IC_{50} 8.7 µM (fig. 1).
The results obtained using the lymphocytes derived from a group of workers exposed to inorganic lead show that the ^{125}I-ICYP Bmax is decreased (-35%) in subjects having both PbB and ZPP > 40 mcg/dl and in those having ZPP > 40 mcg/dl and normal PbB (fig. 2). The ß-adrenoceptor affinity (Kd) is unmodified.

DISCUSSION

The data support the concept of an action of lead on the CA

transmission which involves also receptor changes. Whether the decrease in ß-R density may be ascribed to a direct effect of the metal at membrane level or represents an adaptive response to an impaired CA transmission has to be established.

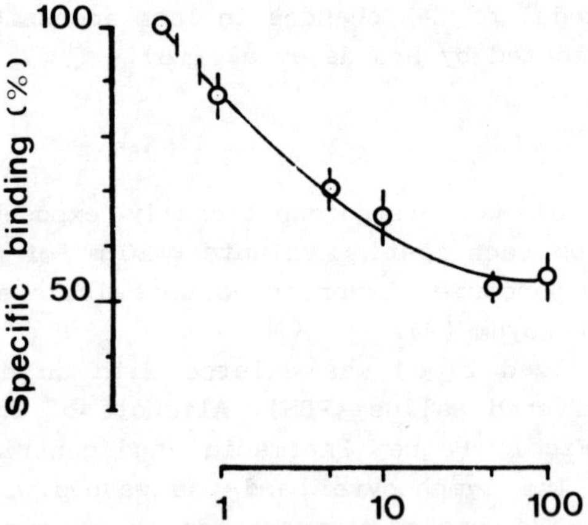

Fig. 1. – In vitro effect of lead on ICYP specific binding to human lymphocytes.

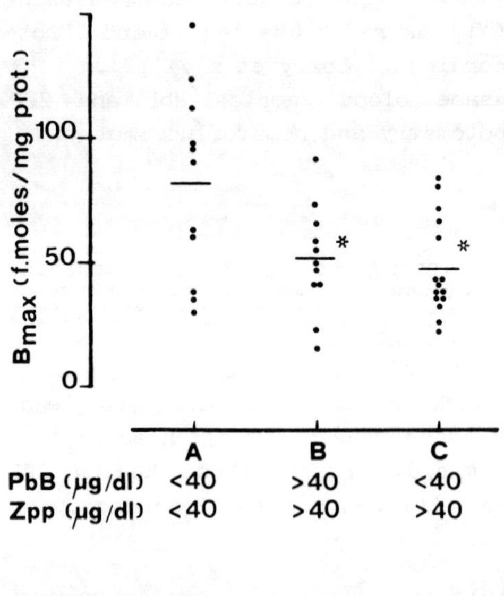

Fig. 2. – Distribution of Bmax values in the 3 groups of workers occupationally exposed to lead.

It has been shown that CA in vitro inhibit the mitogenic response of lymphocytes to phytohaemoagglutinin, suggesting that lead-induced changes in ß-R may have consequences on lymphocyte mediated immune response (1).
Furthermore, the study of ß-R at lymphocyte level may represent a peripheral model which mirrors similar events occurring in other tissues, such as the brain or the heart. Along this line, Brodde et al. (8) have shown parallel changes in cardiac and lymphocyte ß-R.
It is important to note that the changes in ß-R are observed also in subjects having normal plasma lead levels and high ZPP levels, indicating a past exposure. This fact is suggestive of a long-term effect of lead on biological systems, not directly linked to the persistence of the metal in biological fluids. A further analysis of the data reveales that there is no linear correlation between changes in the number of lymphocyte ß-R and PbB or ZPP blood levels. This finding confirms that the biological alterations induced by lead are not directly dependent on exposure indices but may be the consequence of more complex toxic processes.
Although the molecular mechanisms of lead toxicity remain to be elucidated, the data reported contribute to underline the risk of long-term derangements in biological omeostasis consequent to environmental or occupational exposure to lead.

REFERENCES

1. G.P. COOPER and R.S. MANALIS: Influence of heavy metals on synaptic transmission: a review. Neurotoxicol., 4:6984, 1983.
2. C. WINDER and I. KITCHEN: Lead neurotoxicity: a review of the biochemical, neurochemical and drug induced behavioral evidence. Progress in Neurobiol., 22:59-87, 1984.
3. C. WINDER: The interaction between lead and catecholaminergic function. Biochem. Pharmacol., 31:3717-3721, 1982.
4. S. GOVONI, L. LUCCHI, M.S. MAGNONI, P.F. SPANO and M. TRABUCCHI: Decreased density of beta-receptors in brain capillaries of lead exposed rats. Dev. in the Science and Practice Toxicol., 521-524, 1983.
5. E.K. SILBERGELD and J.J. CHISOLM: Lead poisoning: altered urinary catecholamine metabolites as indicators of intoxication in mice and children. Science, 192:153-158, 1976.
6. J.J. CHISOLM and E.K. SILBERGELD: The effects of lead exposure in young children on urinary excretion of HVA.

Proc. Int. Symp. on Heavy Metals, Amsterdam, pp. 546, 1981.

7. J.F. KRALL, M. CONNOLLY and M.L. TUCK: Acute regulation of beta-adrenergic catecholamine sensitivity in human lymphocytes. J. Pharm. Exp. Ther., 214:554-560, 1980.
8. O.E. BRODDE, R. KRETSCH, K. IKEZONO, H.R. ZERKOWSKI, J.C. REIDEMEISTER: Human beta-adrenoceptors: relation of myocardial and lymphocyte beta-adrenergic receptor density. Science, 231:1584-1585, 1986.
9. A. BOYUM: Isolation of mononuclear cells and granulocytes from human blood. Scand. J. Chr. Lab. Inves., 21 (97): 77-89, 1968.
10. O.H. LOWRY, N.J. ROSEBROUGH, A.L. FARR and R.J. RANDALL: Protein measurement with the Folin phenol reagent. J. Biol. Chem., 193:265-275, 1951.
11. C. ROCHETTE-EGLY and J. KEMPF: Cyclic nucleotides and calcium in human lymphocytes induce to divide. J. Physiol., 77:721-725, 1981.

Biochemical indices of nervous tissue toxicity and exposure to neurotoxicants Summary report

M. Maroni – Institute of Occupational Health, Clinica del Lavoro L. Devoto, University of Milan, via S. Barnaba 8, Milano, Italy

Neurotoxicity area seems to be progressing more slowly than other fields of toxicology and surely too slowly with respect to the needs of our society, where the great majority of occupational and environmental chemicals (pesticides, metals, organic pollutants, etc.) are primarily neurotoxic.

Nervous system presents exclusive structural and functional characteristics and neurotoxicity has distinctive peculiarities when compared with other toxic effects dealt with in this Symposium.

The nervous system is a complex and integrated system, scarcely accessible for biochemical investigations in man. Most neurotoxicants show selective toxicity for cells or areas of the central or peripheral nervous system, but the mechanisms underlying such a selectivity are mostly unknown and often not related to unequal distribution of the toxin within the nervous system. The relationships intercurring among biochemistry, anatomy and physiology of the nervous system are not fully elucidated, either in normal or in abnormal conditions. Thus, it is difficult to assess the importance and specificity of certain biochemical modifications observed in animals or man after intake of neurotoxins.

In neurotoxicology, animal experiments are used to predict neurotoxicity for man, but we know the animal model may be unsuitable for the study of some nervous functions which are present only in the human species. The same functional and pathological nervous tissue changes induced by exogenous chemicals can also be induced by several other unrelated factors (stress, life style, social conditions, etc.) which can act independently or as co-agents together with the exogenous substances. Therefore human neurobehavioral changes are to be considered effects with potential multifactorial ethiology the nature of which may be difficult to ascertain in many cases.

The events leading to neurotoxic effects by exogenous chemicals can be summarized in subsequent phases as follows:

1. penetration of the neurotoxin into the organism and distribution to reach the target cells;
2. interaction of the neurotoxin with the critical targets responsible for toxicity (receptors) and with other non-critical targets;
3. induction of biochemical and functional modifications of the nervous cells;
4. appearance of structural and/or functional changes of the nervous system.

STATE OF THE ART ON BIOCHEMICAL INDICES FOR MONITORING NEUROTOXICITY IN MAN

The tests exploring the phase-1 are the most conventional and those more frequently used so far for human biological monitoring.

They consist of measuring the toxin concentration in body fluids, typically blood or urine, to predict the possibility of occurrence of adverse effects in the nervous system. This approach suffers many limitations and uncertainties. First of all, the dose-effect relationships, expecially at low doses, between serum or urine concentration and nervous system effects are not known for many substances. Then, there is a great interindividual variability in metabolism of exogenous substances and this may be cause of wide dose-response curves. Moreover, what is measured, that is the "total concentration", is not the active fraction responsible for toxicity at the target cells. Finally, interpretation of some of these tests may change according to modality of exposure, the same results having different meaning according to whether exposure was current or past and, respectively, short or prolonged.

The phase-2 tests are very interesting because they assess the interaction of toxins with their specific targets (receptors) or with other non-critical targets that are similar to the true receptors or can adequately inform on the dose delivered to them. These non-critical targets when present in peripheral tissues can be easily and ethically obtained from living patients.
The measurement of erythrocyte acetylcholinesterase activity is the most known effect test and the only one so far entered into the practice of human monitoring in neurotoxicology. Biological monitoring of exposure to organophosphorus esters or carbamates can also be performed by the corresponding plasma enzyme, butirrylcholinesterase, whose activity however is not related to the toxic effects occurring in the nervous tissue but only to the presence of the toxin in the organism. In this respect, butirrylcholinesterase is a good example of a non-critical target usable for internal dose estimation in biological monitoring.
Among the new bioindicators candidates to enter into use in a near future, remarkable hopes are put on protein adducts (expecially hemoglobin adducts) and on target enzymes relevant to the nervous system which are present in platelets, red blood cells or lymphocytes (f.i., neuropathy target esterase (NTE), dopaminergic system enzymes, muscarinic receptors, etc.). While significant experimental results has already been achieved with adducts of some neurotoxins such as acrylamide and n-hexane metabolites, the use of new target enzymes appears to be still far-off, NTE being the only target for which substantial experimental data are available.

The phase-3 tests should reveal early biochemical changes occurring in the nervous system before the underlying toxic mechanisms can produce nervous system illnesses of clinical significance. The approaches so far pursued have mainly consisted of measuring substances supposedly released from the nervous system when injured (f.i., acetylcholinesterase in plasma) or evaluating body fluid neurotransmitters (f.i. aminergic mediators in urine), the concentration of which should precociously be altered by chemicals that specifically interfere with their metabolism. However, the results of application of these tests have been so far rather poor and this kind of studies, despite good technical faesibility, seems not to be as promising as other investigations in providing an early and specific evaluation of human neurotoxicity.

In this broad category of tests may also be included the great number of neurophysiological or behavioral techniques which can be used to study selective central or peripheral nervous system functions (electroneurography, brain evoked potential study, olfactory function measurement, etc.). These tests, which are receiving more and more applications in the field because of their remarkable sensitivity, show no specificity for neurotoxic agents and resent very much of external interfering factors as well as of methodological variations. Moreover, reversibility is a critical pre-requisite for bioindicators to be used in human monitoring and not much is known on significance and reversibility of some of the altered functions which are revealed by these tests.

FUTURE NEEDS

The difficulties in human neurotoxicity monitoring reside in the delay with which basic neurosciences are providing information on biochemistry, physiology and physiopathology of the nervous system. Therefore great progresses can hardly be expected to occurr in biological monitoring without a concurrent remarkable expansion of basic knowledge in neurotoxicology and other related neurosciences.

In the specific field we are examining, future needs for research can be derived from a priority-list of objectives:

1. **to understand the biochemical mechanisms of action of neurotoxicants.**
   This is necessary to identify critical and non-critical targets and to elaborate dose-effect relationships, thus providing a basis to develop specific bioindicators. This objective requires animal and in vitro biochemical toxicology studies;
2. **to develop and validate biochemical indices usable in man.**
   Protein adducts and other indices of the phase 2 seem to be the most reliable candidates for a short-term practical utilization. These tests should receive priority in human testing and validation. Development and validation of new indices require comparative animal-man investigations and human field studies in carefully controlled conditions;
3. **to understand the relative importance of toxic and non-toxic factors in determining human nervous system modifications.**
   This objective can be pursued with multifactorial field

investigations in which potential social or individual confounders are carefully taken into consideration while studying groups of subjects with occupational or environmental exposure to neurotoxins. Analytical assessment of internal dose of the neurotoxins is a key point to render these studies meaningful and demonstrative;

4. **to address the "individual susceptibility" issue.**
This topic is very important to identify and predict high risk groups in the population. Genetically-determined and acquired variations in metabolism and biodisposition appear the mechanisms more likely to explain interindividual variability in neurotoxic responses. Investigations on the biochemical basis of these mechanisms seem to be the most probable source of appropriate biological indices to be used in man in the future.

List of participants

ABBRITTI G.	Istituto Medicina Lavoro	Perugia (I)
AGNELLO B.	Via Laurentina 472	Roma (I)
AL-NAKIB T.M.	Dept. Pharm. Toxicol. P.O.B. 24923 Safat	Kuwait
ALDRIDGE W.N.	MRC Toxicol. Unit Woodmansterne Road	Carshalton (UK)
ALESSIO L.	Cattedra di Medicina Lavoro Piazza Spedali Civili	Brescia (I)
ANNONI G.	Clinica Medica III Università di Milano	Milano (I)
ASHBY J.	Central Toxicol. Lab. ICI-PLC Alderley Park	Macclesfield Cheshire(UK)
AUDISIO R.	U.S.S.L. 65 Via Matteotti 83	Sesto S. Giovanni (I)
BALDUZZI M.	Via Spegazzini E/2	Roma (I)
BALSAT A.	34 Rue F. Huet	Embourg (B)
BARBIERI P.G.	USSL 36	Iseo (I)
BARCOVIC D.	Rotvlatkebacic 2	Rieka (YU)
BARUFFINI A.	Via Belvedere 35	Lecco (I)
BAULANDE G.	IBM France, 224 BD Kennedy	Corbeil Essones (F)
BENETTI C.	Via Bissolati 7	Legnano (I)
BERGAMASCHI E.	Via Marzabotto 12	Parma (I)
BERLIN A.	Commission of European Comm. Batiment J. Monnet	Luxembourg (L)
BERTAZZI P.A.	Istituto Medicina Lavoro Via S. Barnaba 8	Milano (I)
BERTOLERO F.	Centro Comunitario di Ricerca	Ispra (I)
BERTOLETTI C.	Via E. Natali 5	Concesio (I)
BIGGIO A.	Enichem	S. Donato Milanese (I)
BINI A.	Istituto Medicina del Lavoro Via Campi 287	Modena (I)
BLOCH P.	21 Oatlands Rd	Bothey Oxford (UK)
BLOEMAN L.	P.O. Box 48	Terneuzen(NL)
BOLOGNESI C.	Via XXV Aprile 287	Pieve Lig. (I)
BONETTI P.	Via Piave 6	Rosignano Solvay (I)
BORRELLI A.	Ciba Geigy	Origgio (I)

BOURDEAU P.	Commission of European Comm. Rue de la Loi200	Brussels (B)
BRACCO A.	Manufacture Michelin	Clermond Ferrand (F)
BRAGT P.	Medical Biological Lab. TNO	Rijswijk(NL)
BRESSA G.	Centro Studi Uomo e Ambiente	Padova (I)
BRIGNARDELLO R.	Via Colle di Mezzo 19	Roma (I)
BULLA G.	Ind. Vernici Italiane Via G. La Masa 20	Milano (I)
CACCIABUE E.	Via Imperia 4	Genova (I)
CANDELA S.	Via Antica 26	Reggio Em.(I)
CANNAVO' R.	Viale Italia 69	Lodi (I)
CAPODAGLIO E.	Clinica del Lavoro	Pavia (I)
CASSINA T.	Centro Diagnostico Italiano	Milano (I)
CATENACCI G.	Clinica del Lavoro	Pavia (I)
CAZZOLI F.	Via F.lli Bandiera 35	Mantova (I)
CHECOWETH M.	3066 E. Gordon Ville Rd	Midland Michigan(USA)
CHEVALIER B.	Roussel Uclaf 35 Bld des Invalides	Paris (F)
CHIAPPINO G.	Istituto Medicina del Lavoro Via S. Barnaba, 8	Milano (I)
CLEMENTI F.	Cattedra di Farmacologia Via Vanvitelli 32	Milano (I)
CLERICI L.	Centro Comunitario di Ricerca	Ispra (I)
CLONFERO E.	Istituto Medicina del Lavoro Via Facciolati 71	Padova (I)
COLOMBI A.	Istituto Medicina del Lavoro Via S. Barnaba 8	Milano (I)
COLOMBI A.M.	Industrie Vernici Italiane Via G. La Masa 20	Milano (I)
COSTA G.	Dep. Envir. Health Univers. of Washington	Seattle(USA)
CURATOLO R.	USSL 29 via Suzzani 239	Milano (I)
DAVID A.	28 Chemin F. lehmann	Grand Saconnex(CH)
DE KORT W.I.	Dept. Directorate General Inspector of Labour	Holland
DE MATTEIS F.	MRC Toxicology Unit Woodmansterne Road	Carshalton (UK)
DELABARRE P.	Passagiersstraat 100	Gent (B)
DELLA PORTA G.	Istituto dei Tumori Via Venezian 1	Milano (I)

DELLA ROSA H.V.	Univ. de Sau Paulo Toxicologia POB 30786	San Paulo (BR)
DELLA TORRE L.	Via Galilei 13	Imbersago(I)
DEVECCHI M.	Via Mezzano 23	Trecate (I)
DI CREDICO N.	Sisas, L.go Corsia dei Servi	Milano (I)
DI STEFANO G.	Via O. Corbino 1	Ragusa (I)
DIOGUARDI N.	Clin. Med. e Terapia Med. III, via della Pace 9	Milano (I)
DISCALZI G.	Via Guidobono 21	Torino (I)
DORIGATTI F.	Ospedale S. Raffaele Via Olgettina 46	Milano (I)
DOSS M.O.	Abt. Klin. Bioch. F.H.P. Univ. Deutchhaustrasse 17	Marburg (FRG)
DUSCIO D.	Via dei Vespri 242	Misterbianco (I)
EINISTO P.	Inst. of Occup. Health Haartamanikatu 1	Helsinky (SF)
EL GHAZALI S.	Fac. of Medicine Ain Shams University POB 38	Il Cairo (ET)
EL HADRAMY D.	Serv. Medical du Travail	Nunakihott (Mauritania)
ELDER G.H.	Dept. Med. Biochem. Univ. of Waleshealth Park	Cardiff (UK)
EMMETT E.A.	The Johns Hopkins Univ. 3100 Wyman Park Drive Bldg 6	Baltimore (USA)
FABRI G.	Istituto Medicina del Lavoro Univ. Cattolica S. Cuore	Roma (I)
FARINA G.F.	Gruppo Lepetit Via R. Lepetit 8	Milano (I)
FAUSTINI A.	Via Busalla 6	Roma (I)
FAVA G.	Via Dondero 2/19	Sampierdarena (I)
FERIOLI A.	Istituto Medicina Lavoro Via S. Barnaba 8	Milano (I)
FERNICOLA C.	Via Ugoni 6/A Via Vittor Pisani 26	Milano (I)
FITZPATRICK J.M.	Room and Haas Italia S.p.A. Via Vittor Pisani 26	Milano (I)
FOA' V.	Istituto Medicina Lavoro Via S. Barnaba 8	Milano (I)
FORMENTINI V.	ENEL, Via Carducci 1-3	Milano (I)
FORNI A.	Istituto Medicina Lavoro Via S. Barnaba 8	Milano (I)

FRANCHINI I.	Istituto Clinica Medica Nefrologia dell'Università Via Gramsci 14	Parma (I)
FRANCO G.	Via Vignazza 11 C	Pavia (I)
FREGOSO M.	Alfa Romeo, V. Regina Elena 299	Roma (I)
FRONTALI N.	Istituto Superiore Sanità Viale Regina Elena 299	Roma (I)
FROSTLING H.	Svenska Arbeitsgivareforenin	Stockholm (SF)
GARASTO G.	P.M.P. USSL 31 Corso Giovecca 169	Ferrara (I)
GARATTINI S.	Via Sette Camini 1 USSL 37	Cogno (I)
GENNARO V.	Viale Benedetto XV 10	Genova (I)
GERIN M.	Dept. Occup. Med. Fac. Med. Univ. of Montreal POB 6128	Montreal Quebec (CDN)
GIACHINO G.	Via Quintino Sella 92	Torino (I)
GIANI G.	Via B. Gosio 7	Roma (I)
GITTER S.	Tel Aviv Univ. P.O. Box 3360	Tel Aviv(IL)
GOGGI E.	Ospedale Civile	Carate Brianza (I)
GOLF S.	Inst. Klin. Chem. Pathobiochemie, Friedrichstr. 24	Giessen (FRG)
GOYER R.A.	N.I.E.S.H. P.O. Box 12233	Research Triangle Park (USA)
GREEN G.M.	The Johns Hopkins Univers. 2943 N. Charles St.	Baltimore (USA)
GRESCH E.	3200 Travis Court	Midland(USA)
GRIECO A.	Istituto Medicina Lavoro Via S. Barnaba 8	Milano (I)
HARDT F.	Centralsyge Hudet	Hillerad(DK)
HART R.W.	140 William Street	Melbourne Victoria (AUS)
HEERING H.	Utrechtseweg 310	Arnhem (NL)
HENSCHLER D.	Inst. Pharmakol. Toxikol. Univ., Versbacher Str. 9	Wurzburg(FRG)
HERNBERG S.	Inst. Occup. Health Haartmaninkatu 1	Helsinki (SF)
HESSL S.M.	Div. Occup. Medicine Cook Country Hosp.	Chicago(USA)
HOGBERG J.	Natl. Board of Occup. Safety Health	Solna (S)

HOGBERG M.	Nat. Board of Occup. Safety Health	Solna (S)
HOMANN J.	Lilienweg 19	Giesseb(FRG)
HOUCHINS H.R.	16602 E Laxford Rd.	Azusa Calif. (USA)
IACOVONE M.T.	Pirelli - Viale Sarca 222	Milano (I)
IARIARTE M.J.	Arizmendiarrieta Lagun Aro	Mondragon Guipuzcoa(E)
INSERRA A.	Istituto Medicina del Lavoro	Catania (I)
JACKSON J.R.	Albright Wilson House P.O. Box 3 Hagley Road	Oldbury Warley Mids (UK)
JARUP L.	Svantensgatan 6	Stockholm(SF)
JOVY D.	Rhoendorferstrasse 12	Bonn (FRG)
JULITTA O.	USSL 67 Serv. Igiene Via Forlanini 121	Garbagnate Milan.(I)
KARACIC V.	M. Pijade 158	Zagreb (YU)
KAROL M.	283 Jefferson Drive	Pittsburg (USA)
KEMKES M.B.	Klinikstrasse 36	Giessen(FRG)
KOLMODIN H. B.	Vintergatan 7	Umea (S)
KUSTERS E.	Med. Dienst BASF Antewerpen NN Scheldelaan	Antewerpen (B)
LAGUZZI B.	Via Livorno 57	Torino (I)
LAUWERYS R.	Univ. de Louvain Clos-Chapelle aux Champs 30	Brussels (B)
LEWALTER J.	Aerztilche Abt. Geb. L9	Leverkusen Bayerwwerks (FRG)
LINDAHL M.	Dept. Occup. Medicine University Hospital	Linkoping(S)
LOMBARDI A.	Soc. It. Poliestere C. da Pagliarone	Acerra (I)
LOTTI M.	Istituto Medicina Lavoro Via Facciolati 71	Padova (I)
LUOTAMO M.	Arinatie 3	Helsinki(SF)
MACCHI P.	Enichem, P. Boldrini 1	S. Donato Milanese (I)
MALAGUTI L.	Istituto Superiore Sanità Via Regina Elena 266	Roma (I)
MALORNI W.	Via S. Ciprano 20	Roma (I)
MALTONI C.	Instituto di Oncologia V.le Ercolani 4/2	Bologna (I)
MANNO M.	Istituto Medicina Lavoro Via Facciolati 71	Padova (I)

MANZO L.	Dip. Med. Interna Terap. Medica, Piazza Botta 10	Pavia (I)
MARANELLI G.	Via dell'Abetone 21/B	Rovereto (I)
MARONI M.	Istituto Medicina Lavoro Via S. Barnaba 8	Milano (I)
MARTINEZ L.	Monte 1103 Cerro	Ciudad Habana(Cuba)
MATTIUSSI R.	Fondaz. Carlo Erba Via Cino del Duca 8	Milano (I)
MAZZELLA DI BOSCO M.	Via B. Stringer 27	Roma (I)
MECAN A.	Bul. Brat. I. Jed. 288	Zenica (YU)
MILIGI L.	Via di Novoli 5	Firenze (I)
MINOIA C.	Via I maggio 15	La Rotta di Travaco (I)
MOCELLIN G.	625 Swanston Street	Carlton Victoria (AUS)
MONROE C.B.	Rohom and Haas Co. P.O. Box 584	Bristol PA (USA)
MONTALBETTI N.	Lab. di Biochimica Ospedale Niguarda	Milano (I)
MONTRASI A.	Grace Italiana S.p.A. V. Trento 7	Passirana di Rho (I)
MOSCONI G.	Via Merula 2	Bergamo (I)
MULLER G.	Schamale Str.	Munster(FRG)
MURRAY R.	London School of Hygiene Keppel Street	London (UK)
MUTTI A.	Istituto Clin. Med. Nefrol. Via Gramsci 14	Parma (I)
MYSLAK Z.	Ardeystraße 67	Dortmund(FRG)
NASLUND P.E.	Solbanksgatan 40	Goteburg (S)
OKAZAKI I.	583 Nakamachi Setagayaku	Tokyo (J)
OLIVERI A.	Villa Glori 42	Vicenza (I)
ONG C.M.	Dept. of Public Health Dutram Hill	Singapore
ORBAEK P.	Yrkesmed Klin.	Malmo (S)
PALOTIE A.	Arinatie 3	Helsinki(SF)
PANI P.	Ist. Farmacologia Biochimica	Cagliari (I)
PARDINI M.C.	V.le Gorizia 24/E	Roma (I)
PARLATO G.	Enichem Augusta	Augusta (I)
PARMEGGIANI L.	I.C.O.H.	Geneva (CH)
PERBELLINI L.	Istituto Medicina Lavoro Policlinico Borgo Roma	Verona (I)
PETAZZI A.	Via De Amicis 18	Rho (I)

PICCIONI P.	Via Mombarcaro 69	Torino (I)
PIERRE F.	I.N.R.S. Av. de Bourgogne	Vandoeuvre Les Nancy(F)
POGNA R.	Bozzetto Ind. Chimiche S.p.A. Via Mazzini 11	Pedrengo (I)
POSPICHIL E.	Spitalgasse 23	Vienna (A)
PRICE R.G.	Biochemistry Dept. King's College Campden Hill	Londond (UK)
RONGIER E.	CEN/Valrho Serv. Hyg. Ind.	Bagnols/Ceze Cedex (F)
RUBIN R.J.	The Johns Hopkins Univers. 615 North Wolfe Str.	Baltimore (USA)
RUBINO G.	Istituto Medicina Lavoro	Torino (I)
SABBIONI E.	Centro Comunitario di Ricerca	Ispra (I)
SACCO P.	Via del Monte Palmo 6	Roma (I)
SALE P.	Enichem - Porto Torres	Porto Torres (I)
SALVEMINI A.	Immont Italiana via Devizzi 51/A	Cinisello Balsamo (I)
SANGUINETTI F.	Via T. Gulli 34	Ravenna (I)
SANTELLA R.M.	College of Physicians Columbia University	New York (USA)
SANTERNI N.	Ciba Geygy	Origgio (I)
SARTO F.	Istituto Medicina Lavoro Via Facciolati 71	Padova (I)
SASSI C.	Pirelli, viale Sarca 222	Milano (I)
SCHIEBELER H.	BASF Aktiengesellschaft	Ludwigshafen/ Rhein (FRG)
SCHNEIDER S.	Liegistrasse 5	Frankfurt (FRG)
SCHULTE P.A.	N.I.O.S.H. Mail Stop R-13 4676 Columbia Parkway	Cincinnati (USA)
SECCHI C.C.	Istituti Clinici Perfezionam. Div. Med. Gen Via S. Barnaba 8	Milano (I)
SESANA G.	Servizio Medicina Lavoro Ospedale Civile	Desio (I)
SEVIERI G.	Centro Med. Regionale Lomb. Via Longhena 7	Milano (I)
SKENDER L.	M. Pijade 158	Zagreb (YU)
SODERKVIST P.	Dept. Occup. Med. University Hospital	Linkoping (S)
STERN R.	Danish Welding Institute Park Alle 345	Brondby (DK)
STONARD M.	ICI PLC Cent Tox. Lab. Alderley Park	Macclesfield Cheshire(UK)

STREICHER M.	Sanofi Recherches 095 Route d'Espane	Toulouse (F)
STRIK J.J.T.W.A.	Rijksint Voor Volksgezon- dheid A.V. leeuwenhoeklaan	Bilthoven(NL)
SUMA'MUR P.K.	Jl. A. Yani 69 70	Jakarta (RI)
SUNDERMAN JR. W.	Dept. Lab. Med. Pharmacol. Univ. 263 Farmington Av.	Farmington (USA)
TIROLESE M.	USSL 29 Via Suzzani 239	Milano (I)
TOFFOLETTO F.	Serv. Med. Lavoro Ospedale	Desio (I)
TRAVERSA F.	Via F. Filzi 2A/8	Genova (I)
TRESCA G.	P. Agricoltura 24 Servizio Sanitario	Roma (I)
TRUSENDI M.	Via Moneta 13/B	Carrara (I)
TURNER B.	46 Main South Road	Chust Church (N. Zealand)
TURRI P.V.	Via delle Pietre 21	Verona (I)
VAINIO H.	I.A.R.C. - 150 Cours Albert Thomas	Lyon (F)
VAN SITTERT N.	P.O. Box 162	The Hague (NL)
VESTERBERG O.	Arbertarskyddsstyrelsen Ekelundsvagen 16	Solna (S)
VIGLIANI E.C.	Fondazione Carlo Erba Via Cino del Duca 8	Milano (I)
VILLA L.	USSL 23	Sondrio (I)
VILLICH O.	PMP Via Alberoni 17	Ravenna (I)
VIVEK CHANDRA R.	Hyderbad Und. Ltd	Sanatnagar Hyderabad (India)
WOOD S.	Dept. of Labour, Mespil R.	Dublin (IRL)
WOOTEN KNAPP L.	6808 24th Street	Fort Hood (USA)
WRIGHT D.S.	BP Research Centre Chertsey Road	Sunsbury On Thames (UK)
WRONSKA-NOFER T.	Dep. Biochem. Dept. of Pathomorf., Inst. Occup. Hlth	Lodz (PL)
ZANGIROLAMI A.	Via 10 luglio 44	Rovigo (I)
ZIGLIO G.	Istituto di Igiene Via F. Sforza 35	Milano (I)
ZILOCCHI R.	Viale Abbadia 2	Piacenza (I)
ZOOK B.	Holzmarkt 4	Hannover 1 (FRG)
ZSCHIESCHE W.S.	Schillerstrasse 29	Erlangen(FRG)

Index